GEL CHROMATOGRAPHY

GEL CHROMATOGRAPHY

THEORY, METHODOLOGY, APPLICATIONS

BY

TIBOR KREMMER, C.Sc.Ph.D.
Department of Biochemistry
National Oncological Institute
Research Institute of Oncopathology, Budapest

AND

LÁSZLÓ BOROSS, D. Sc.
Department of Biochemistry
József Attila University, Szeged

A WILEY-INTERSCIENCE PUBLICATION

JOHN WILEY & SONS
CHICHESTER · NEW YORK · BRISBANE · TORONTO

THIS BOOK IS THE REVISED VERSION OF THE ORIGINAL HUNGARIAN
"GÉLKROMATOGRÁFIA", MŰSZAKI KÖNYVKIADÓ, BUDAPEST

TRANSLATED BY
MRS M. GÁBOR

© AKADÉMIAI KIADÓ, BUDAPEST 1979

JOINT EDITION WITH AKADÉMIAI KIADÓ, BUDAPEST

ALL RIGHTS RESERVED.

NO PART OF THIS BOOK MAY BE REPRODUCED BY ANY MEANS, NOR TRANSMITTED, NOR TRANSLATED INTO A MACHINE LANGUAGE WITHOUT THE WRITTEN PERMISSION OF THE PUBLISHER.

Library of Congress Cataloging in Publication Data
Kremmer, T.
   Gel chromatography.

  'A Wiley-Interscience Publication.'
  Bibliography: p. 299
  Includes index.
  1. Gel permeation chromatography. I. Boross, L.
  Joint author. II. Title.
QD272.C444K73   543'. 08   77-24994

ISBN 0-471-99548-7

PRINTED IN HUNGARY

CONTENTS

PREFACE ... 9

HISTORICAL SURVEY ... 11

ACKNOWLEDGEMENTS ... 13

PART I. THEORY (By T. Kremmer)

1. THE FUNDAMENTALS OF GEL CHROMATOGRAPHY ... 17
1.1. Gel formation; the structure of gels ... 17
1.2. Gel-forming substances; their production ... 20
1.2.1. Natural gel-forming substances ... 21
1.2.1.1. Agarose gels ... 21
1.2.2. Semi-synthetic gel formers ... 24
1.2.2.1. Sephadex gels ... 26
1.2.2.2. Molselect gels ... 30
1.2.2.3. Organophilic dextran gels ... 30
1.2.2.4. The Sephadex LH-20 and LH-60 ... 31
1.2.2.5. Other organophilic dextran gels ... 32
1.2.3. Synthetic gel-forming substances ... 32
1.2.3.1. Acrylamide-acrylate gels ... 33
1.2.3.2. Bio-Gel P polyacrylamide gels ... 34
1.2.3.3. Acrilex polyacrylamide gels ... 38
1.2.3.4. Organophilic and hydrophilic acrylate gels ... 38
1.2.3.5. Polystyrene gels ... 39
1.2.3.6. Other column packings in gel chromatography ... 42

2. THEORY ... 48
2.1. The volumetric distribution of the gel bed ... 49
2.1.1. The geometric model of the volumetric distribution of the gel bed ... 50
2.2. The steric-volumetric theory of gel chromatography ... 55
2.3. The terminology of gel chromatography ... 56
2.4. The volumetric equations of gel chromatography ... 60
2.5. The geometric model of molecular exclusion ... 63
2.5.1. The Porath and Squire equations ... 63
2.5.2. The steric model of Laurent and Killander ... 66

2.5.3.	The Laurent and Laurent electric analogue model	68
2.5.4.	Molecular sieving theories	68
2.6.	The kinetic theories of gel chromatography	69
2.6.1.	The hydrodynamic parameters of gel chromatography	69
2.6.1.1.	Resolution in gel chromatography	69
2.6.1.2.	The selectivity of gel columns	71
2.6.1.3.	The capacity factor of resolution	71
2.6.1.4.	Theoretical and effective HETP	72
2.6.1.5.	The influence of the flow rate on HETP	75
2.6.1.6.	The HETP in gel chromatography	76
2.6.1.7.	Calculation of the column dimensions on the basis of the HETP	78
2.6.2.	The theory of restricted diffusion: Ackers' investigations	79
2.6.3.	The correlation between molecular dimensions and the size of the gel pores	82
2.6.3.1.	Molecular weight determination by gel chromatography: Andrews' investigations	82
2.6.3.2.	The role of the molecular size and shape in gel chromatography: Siegel and Monty's investigations	86
2.6.3.3.	The correlation between the pore size and the chromatographic properties of gels: the Determann equations	89
2.6.3.4.	Determination of the molecular weight by thin layer gel chromatography	92
2.6.3.5.	The examination of polydisperse systems	93
2.6.3.6.	The empirical equations and general correlations of the molecular weight determination	95
2.7.	The thermodynamic aspects of gel chromatography	97
2.7.1.	Similarity between the distribution equilibrium of gel chromatography and immiscible-phase polymer solutions	98
2.7.2.	Examination of the osmotic properties of gels	100
2.7.3.	The influence of adsorption and temperature in gel chromatography	101
2.7.4.	The influence of the concentration of solutes in gel chromatography	103

PART II. METHODS AND TECHNIQUES (By T. Kremmer)

1.	**GEL CHROMATOGRAPHIC METHODS**	107
1.1.	Selection and preparation of gels	108
1.2.	Particle size	109
1.3.	Swelling	112
1.4.	Preservation and sterilization	114
1.5.	Purification and drying	115
1.6.	The moving phase of gel chromatography	116
2.	**THE METHODS USED IN COLUMN GEL CHROMATOGRAPHY**	119
2.1.	The selection of the chromatographic column	120
2.1.1.	Column types and their applications	120
2.1.2.	Column sizes and proportions	122
2.1.3.	The wall effect and its elimination	124
2.1.4.	The mixing volume and its reduction	124
2.2.	Instruments and auxiliaries used in column chromatography	125

2.2.1.	Tubing and joints: the safety loop	125
2.2.2.	Solvent reservoirs and Mariotte flasks	128
2.2.3.	Pumps	128
2.2.4.	Thermostating the column: heating and cooling mantles	131
2.3.	Packing and checking the column	131
2.3.1.	The method and conditions of packing	132
2.3.2.	Regulation of the flow rate	135
2.3.3.	Checking the homogeneity of the gel bed	142
2.3.4.	Preparation and application of the sample	143
2.3.5.	Determination of the volumetric parameters of the gel bed	145
2.4.	General processes of gel chromatography	147
2.4.1.	Fraction collecting	148
2.4.2.	Detection and determination of the separated substances	149
2.4.3.	Documentation and appraisal of the results	152
3.	**SPECIAL TECHNIQUES**	155
3.1.	Recycling gel chromatography	155
3.2.	Gradient elution in gel chromatography	158
3.3.	Gel chromatographic methods based on solubility	161
3.3.1.	Solubility chromatography	161
3.3.2.	Zone precipitation	162
3.4.	Complex formation in gel chromatography	164
3.5.	Application of gel particles for adsorption and liquid-liquid partition chromatography	167
3.6.	The preparative (batch) procedures of gel chromatography	169
3.6.1.	Concentration of solutions of macromolecules using xerogels	169
3.6.2.	Desalting and buffer change	171
3.7.	Thin layer gel chromatography	172
3.7.1.	The materials and techniques used in thin layer gel chromatography	173
3.7.2.	Sample application and inspection of the gel layer	176
3.7.3.	Determination of the flow rate	176
3.7.4.	Development and evaluation of the chromatograms	178

PART III. APPLICATIONS OF GEL CHROMATOGRAPHY (By L. Boross)

1.	**APPLICATIONS IN PROTEIN CHEMISTRY**	183
1.1.	Separation of proteins from salts and small-molecule substances	183
1.2.	Protein fractionation on gel columns	184
1.3.	Concentration of protein solutions	187
1.4.	Characterization of isolated proteins	188
1.5.	Gel chromatography in the examination of protein complexes	195
1.6.	Gel chromatography of peptides and amino acids	200
2.	**GEL CHROMATOGRAPHY IN THE CHEMISTRY OF NUCLEIC ACIDS**	203
2.1.	Purification and characterization of nucleic acids	203
2.2.	Fractionation of transfer ribonucleic acids on gel columns	207
2.3.	Gel chromatography of oligonucleotides, mononucleotides, nucleosides and nucleic acid bases	208

3.	**GEL CHROMATOGRAPHY OF CARBOHYDRATES**	213
3.1.	Gel chromatography in the chemistry of polysaccharides	213
3.2.	Gel chromatography of oligosaccharides	219
4.	**GEL CHROMATOGRAPHY OF SYNTHETIC POLYMERS**	224
5.	**GEL CHROMATOGRAPHY OF SMALL-MOLECULE ORGANIC COMPOUNDS**	232
5.1.	Gel chromatography of aliphatic and aromatic hydrocarbons	232
5.2.	Gel chromatography of organic acids, their derivatives and lipids	239
5.3.	Fractionation of alcohols and ethers on gel columns	245
5.4.	Gel chromatography of phenols	248
5.5.	Gel chromatography of organic bases	253
6.	**GEL CHROMATOGRAPHY OF INORGANIC IONS**	257
6.1.	Gel chromatography of inorganic salts from a distilled water solution	257
6.2.	The influence of background electrolytes on the behaviour of inorganic ions during gel chromatography	264
6.3.	Gel chromatography of inorganic compounds using organic solvent	271

BIBLIOGRAPHY 273

SUBJECT INDEX 293

PREFACE

The trend of the last twenty-five years towards specialization in the field of chemistry has had a considerable impact on the methods of separation. As a result of the interaction between the progress made in physical chemistry and the demand for molecular separation in analytical and preparative techniques, a considerable number of new chromatographic processes have been developed. The introduction of gel chromatography in the early nineteen-sixties provided a completely new and efficient method of liquid chromatography which has gained ground in practically every branch of chemistry, especially in biochemistry.

The breakthrough brought about by gel chromatography was accompanied by lively literary activity: over four thousand publications have appeared since 1959. To give an idea of the rate of development, three times as many publications appeared on gel chromatography during the period 1965—1971 as those published during the first seven years of its practical application, between 1959 and 1965. An analysis of the publications by subject matter shows that interest centres particularly on the new theories and techniques of gel chromatography, the latest gel types, the introduction of molecular weight determination and gel-phase thin layer chromatography, modern automated chromatographs, regular production of special pumps, columns, etc. and their growing use, and, last but not least, commercial promotion activities for a steadily expanding market.

Over and above the new applications of the process (in the chemistry of enzymes, nucleic acids, proteins, antibiotics, in microbiology, membrane research, etc.) there has been rapid development also in the principles of the technique. New potentialities have been provided, for instance, by the development of affinity chromatography.

The routine application of the method makes its objective assessment increasingly difficult. According to a preliminary survey, gel chromatography — one or more of its several forms — is applied in more than half of all biological and biochemical activities, without any reference being made to it in the titles of the report or articles published. It is noteworthy that, in spite of its theoretical and practical significance, probably owing to the ever widening scope of application and the increasing amount of information, there are relatively few comprehensive monographs or books on gel chromatography (Determann 1969; Fischer 1969; Ackers 1970).

The dynamic spread of gel chromatography may be attributed to the fact that the technique helps solve or simplify the numerous problems caused in practice by the unusual behaviour of molecules (the common type as well as macromolecules) in the crosslinked structure of gels.

There has been a most interesting feature during the development of gel chromatography. Some of the applications (desalting, buffer change, etc.) became routine well before the theoretical bases had been sufficiently clarified. According to Professor Tiselius gel chromatography has developed at such a dynamic rate that theoretical research lagged behind the practical applications (Tiselius 1968b). This unusual fact still acts as a stimulus for researchers engaged in the mechanism of separation.

There is another factor in development which contributed to make macromolecular chemistry a science in its own right. The crosslinkage of the gels either arises from some characteristic macromolecules or creates macromolecules by polymerization reactions. It is no coincidence that the first science to apply gel chromatography has been biochemistry. The gel phase offers a medium and favourably mild conditions for the separation, fractionation, concentration, etc. of biopolymers in general — proteins, enzymes, nucleic acids, etc. — which closely resemble the strongly hydrated macromolecular biosphere of the living organism. Because of this, examination of the molecular phenomena taking place in gels has not only added a new process to those known and used in chromatography, but has also fostered the development of basic research in the applied sciences. The gel structure is considered to be the physicochemical model for macromolecular interactions and biological systems.

It is hoped that gel chromatography, both as a method and as a subject, will help in unravelling the laws and regularities of the living world.

HISTORICAL SURVEY

The basic phenomena of gel chromatography were first observed during the adsorption of ions of different sizes (Ungerer 1925). The term "molecular sieving" was first used by McBain in 1926 (cited in Porath 1962a). The crystalline crosslinkages of natural and synthetic aluminium silicates, known as molecular sieves or zeolith-permutites (Linde Molecular Sieves, Linde Air Products Co.), made it possible to separate molecules according to shape and size (Wiegner 1931; Tiselius 1934; Claesson and Claesson 1944, 1948). The phenomena observed on molecular sieves were summed up by Barrer and Brook (1953) who established and proved the correlations of adsorption and molecular size. In their monographs, Hersh (1961), Porath and Flodin (1961) studied the properties of molecular sieves in great detail.

Similar "sieving" properties were ascertained during the application of ion exchange resins (Samuelson 1944; Rauen and Felix 1948). It was found that in the structure of ion exchange resins (Wofatite, Amberlite) there is a correlation between the number of crosslinkages, the degree of swelling and the ion exchange capacity of the larger ions (Kunin and Meyers 1949; Deuel et al. 1950; Mikes 1958). This property was utilized in the separation and purification of numerous groups of compounds (amino acids, peptides, proteins, dyestuffs, etc.) (Richardson 1949, 1951; Thompson 1952; Partridge 1952). The separation by molecular weight of small uncharged molecules (sorbite, glycerin, glycol) has likewise been described (Wheaton and Baumann 1953; Clark 1958).

The experience obtained with ion exchangers directed the researchers' attention to the larger-pored polysaccharide matrices. Deuel and Neukom (1954) synthetized uncharged crosslinked galactomannane gel and used it in the desalting of colloids. Lindquist and Storgards (1955) and Lathe and Ruthven (1955) separated peptides and proteins on granulated starch particles. Later on Lathe and Ruthven (1956) established that the penetration of molecules into the gel phase depends on the structure and concentration of the gel. In an analogy with distribution chromatography they found a relationship between the molecular size and the chromatographic behaviour of the molecules. Polson (1956) observed similar regularities in agar gels. By purification of agar (Araki 1937, 1956), and the production of agarose particles free from ionic groups, a gel was obtained, which proved particularly well suited for the examination of macromolecules (Hjertén, 1961; Russel et al. 1964).

During the evolution of the principle of gel chromatography the study of electrophoresis played an important role. The analytical and preparative methods of gel electrophoresis promoted the examination of macromolecules and biopolymers (Smithies 1955; Raymond and Weintraub 1959; Davis and Ornstein 1959; Porath and Bennich 1962). The "Uppsala group" headed by Professor Tiselius have been studying the electroosmosis of natural substances and proteins in various types of gel since the nineteen-thirties. In their efforts to produce a convection-inhibiting electrophoresis carrier, the Pharmacia Factory in Sweden and the Uppsala University, between 1950 and 1960, tested polyvinyl alcohol, sorbite, mannite and several polysaccharides, including agar, starch, cellulose and dextran. The "dextran story" (Tiselius 1968a) turned the attention of the Uppsala group to dextran, a characteristic type of polysaccharide which, through steric cross-linkages, brought about the production of a new semi-synthetic gel. Also a number of random factors helped dextran gels to achieve success. In his monograph, Professor Tiselius (1968a) gives an apt description of these "random" factors, stressing, however, the need for the systematic work of a qualified and ingenious broad-minded team of researchers. These factors seemed to meet in a happy combination in the cooperation of the Biochemical Institute of the Uppsala University and the Pharmacia Fine Chemicals AB. Incidentally, dextran gel copolymerized with epichlorhydrin and packed into a column achieved surprisingly good separation even without an electric current. A closer examination of the phenomenon, published by Porath and Flodin in 1959, marked the beginning of what one may call an explosive development of gel chromatography. The dextran gels of the Pharmacia Fine Chemicals AB, produced in a wide range of molecular weights and marketed under the trade name of Sephadex, became internationally known in a very short time. The same joint research team of the Uppsala University and the Pharmacia Fine Chemicals AB clarified most of the theoretical problems. The popularity of the Sephadex gels was indebted, to no small extent, to the producers' exemplary and intensive promotion work.

The example of dextran showed clearly that the natural polysaccharides, whose properties were difficult to reproduce, were no longer suitable for the purposes of chromatography. It was proved that these materials must come from a constant source, be of a composition rigorously checked, and be produced on a large, industrial, scale. In the early nineteen-sixties the natural gel formers were replaced by semi-synthetic or synthetic polymers (Polson 1961; Lea and Sehon 1962; Hjertén and Mosbach 1962; Hjertén 1962a; Boman and Hjertén 1962). Among these, hydrophilic polyacrylamide gels produced by the copolymerization of acrylamide and methylenebisacrylamide, found widest application. The first observations were soon followed by an examination of the principles and techniques of polymerization (Curtain and Nayler 1963) and the properties of gels (Sun and Sehon 1965; Fawcett and Morris 1966). The appropriated choice of the base materials allowed the modification of several gel types (acrylate, vinyl, etc.) (Heufer and Braun 1965; Heitz et al. 1966). This trend was strongest in plastics and polymer research (industrial) where the large range of gel formers was made more variable by the use of a series of solvents of different polarities (Lengyel 1968; Udvarhelyi 1969; Muzsay 1970). The examination of gels which swelled in organic solvents began almost simultaneously with that of hydrophilic gels (Brewer 1960, 1961; Vaughan 1960, 1962), particular attention being paid to polystyrene matrices copolymerized with divinylbenzene (Cortis-Jones 1961; Langhammer and Quitzsch 1961; Morris and Morris 1963; Determann et al. 1964, 1965). Following the publications of Moore and Hendrickson, the large scale production and extensive application of polystyrene gels also began (Moore 1964; Moore and Hendrickson 1964, 1965). Similar but less intensive development took place in the field of polysaccharide-based organophilic gels (Determann 1964a; Nyström and Sjövall 1965a, 1965b, 1966a, 1966b; Sjövall and Vihko 1966a, 1966b). Under the trade name of Sephadex LH-20 an organophilic hydroxypropyl derivative of the dextran gels appeared on the market (Joustra 1967).

In the late nineteen-sixties the introduction of affinity chromatography lent new momentum to the application of gel chromatography (Cuatrecasas and Anfinsen 1971). Further new trends were the use of mixed (agarose-polyacrylamide) gel granules (Uriel et al. 1971; Boschetti et al. 1972) and the application of porous glass grains (Haller 1965; Bombaugh 1971). In the field of theory, greatest perspectives are offered by the thermodynamic interpretation of gel chromatography (Nichol et al. 1969; Albertsson 1970a; Ackers 1970; Brown 1970, 1971) and the utilization of the general hydrodynamic interrelations of liquid chromatography (Vink 1970, 1972; Snyder 1972a, 1972b; Karger 1971; Kirkland 1971a). From the relative stabilization of the theoretical bases it would seem that the first progressive period of development in gel chromatography terminated in the nineteen-seventies and any further progress must be expected through its wider application and from a more thorough study of macromolecular interactions.

ACKNOWLEDGEMENTS

The authors express their thanks to Professor László Fejes Tóth, Academician, Director, Institute of Mathematics Research of the Hungarian Academy of Sciences, for his valuable remarks on the mathematical treatment of the geometric models, and to Dr. László Szepesy, Candidate, Head of the Scientific Department of MÁFKI (The Hungarian Research Institute for Natural Gas and Petroleum) for his guidance in the hydrodynamics of gel chromatography. Thanks should go also to the Pharmacia Fine Chemicals AB for their kind permission to use documentation on the Sephadex gels.

The authors wish to extend their sincere thanks to Bio-Rad Laboratories, Richmond, California and to LKB-Produkter AB, Bromma, Sweden and authors and publishers cited in this book for rendering the documentations at their disposal.

PART I
THEORY

1. THE FUNDAMENTALS OF GEL CHROMATOGRAPHY

Several synonymous definitions are used in the literature of gel chromatography which are mostly of historical significance only and refer to the mechanism of separation. Such are, for instance, gel filtration (Porath and Flodin 1959; Strain 1960), molecular filtration—molecular sieving (Fasold et al. 1961; Hjertén and Mosbach 1962), exclusion chromatography (Pedersen 1962,) restricted diffusion chromatography (Steere and Ackers 1962), gel permeation chromatography (Moore 1964), and gel chromatography (Determann 1964a; Haller 1968). To replace the various and often misleading terminologies Determann (1964b) introduced the more rational term of gel chromatography. Also the Working Commission on Chromatography of the Hungarian Academy of Sciences proposed the use of the term "gel chromatography" in 1971. This term indicates simply and adequately that the chromatographic process concerned is, essentially, characterized by the gel phase. The molecular phenomena of the separation process take place in the liquid spaces actually constituting the gel phase and surrounding it. Accordingly, gel chromatography is a method of liquid chromatograhy which may be regarded as a special boundary variant of liquid-liquid partition chromatography.

1.1. GEL FORMATION; THE STRUCTURE OF GELS

It has long been known that some of the natural and synthetic polymers, and sometimes even inorganic compounds (silicates, phosphates, aluminium oxide, etc.), are prone to form jelly-like structures with small molecules and solvents. According to their colloid chemistry definition, gels have a semi-solid consistency, and are stable in form; they are flexible, difform systems produced by the interaction of a gel-forming compound with a solvating medium. Gel-forming substances and solvating solvents stabilize each other in the gel structure and are functional parts of one other. A characteristic property of gels is that they contain a conspicuously high percentage of solvent and little solid matter. They may lose their solvent content during drying or dehydration, and may pick up some spontaneously while swelling. Although the loss or the uptake of solvents is generally reversible, the removal of the solvating molecules may cause reversible or irreversible changes in the gel structure.

On the basis of the structural changes caused by solvation, gels fall into two groups. The structure of xerogels collapses on the removal of the dispersive medium, while the solid structure—the so-called matrix—of the steric network of aerogels remains unchanged even after the removal of the dispersive medium. An example of the former is dextran gel (Sephadex), and of the latter, silica gel.

The crosslinked structure of gels is the result of the crosslinkage of polymer molecules with straight or branched chains. Depending on the type of bond, gel structures are of two main types.

In most natural biopolymers, for instance polysaccharides, mucoproteins, mucopolysaccharides (starch, agar, latex, pectin, collagen, gelatin, etc.) gel forming — apart from the substantial properties (molecular weight, chemical structure) — is a spontaneous reversible process which depends on the temperature and concentration of the colloid solution of the polymer. The solutions of colloidal particles and macromolecules are known as sols. The sol-to-gel conversion of the gels is induced by loose secondary chemical forces, hydrogen bonds, polarization interactions and the London—Van der Waals dispersive forces. Such gel structures may be disrupted by slight physico-chemical effects (i.e. warming up), which cause sol formation. When cooled down the bonds will recombine and the sol once more solidifies into a gel.

In the second type of gel, gel formation is irreversible; the polymer chains are built up by primary covalent bonds. The colloidal-size macromolecules which are characteristic of gels are often produced by the polymerization reaction of low molecular weight monomers (acrylamide, acrylic acid, styrene, etc.). In other cases, however, as for instance in the production of Sephadex gels, the crosslinks come about by crosslinkage between high molecular weight polymer (dextran) chains. Gel structures fixed by primary chemical bonds show higher resistance to slight physicochemical effects and cannot be converted into sols by warming. The gel structure cannot be disrupted except by breakdown of the polymer molecules.

The structure of synthetic and semi-synthetic gels is generally dependent on the chemical (structural) properties of the basic substances, the relative concentration of the reagents and the solubility conditions during gel formation. The functional reactivity of monomers and their spatial orientation have a determinant effect on crosslinking. The reactions inducing crosslinkage are random processes governed by the laws of statistical probability. These processes cause voids of dimensions and geometry characteristic of the type of gel used to form the polymer chains. These voids are called the pore size of the gel. Owing to the kinetics of the crosslinking reactions, the pore sizes also follow the rules of statistical distribution. Since the gel structure is not a rigid system but a mobile, more or less fixed, network of polymer chains, the terms used in practice are average or effective pore size. The pore size is one of the most important parameters of the molecular phenomena (diffusion, permeability) of gel chromatography. Pore sizes may be influenced by the conditions of manufacture, and the solubility relationships of the initial substances and the end product.

From the aspect of pore sizes, two principal types of gel structure can be distinguished (Kun and Kunin 1964). The so-called microreticular (microporous) gels have more crosslinks in their structure, a higher dry-matter content and a lower specific solvent uptake in swelling, compared to macroreticular gels. The more uniform repetition of the crosslinks in microreticular gels produces smaller pores and renders the gel suitable for the separation of smaller molecules. Microreticular gels are obtained if the solubility of the starting substances (monomers) and the end product do not differ greatly, and if the crosslinks assume the desired gel structure only gradually — for instance by increasing the number of crosslinks between the polymer chains already fixed. The number of crosslinks, and therefore the pore size, can be readily controlled by the concentration of the bifunctional compounds which produce the bonds (i.e. with epichlorhydrin in dextran gels) and by the test conditions (catalysts, heat, etc.). Gels with fewer crosslinks, by virtue of their structure, are soft, difficult to handle and unsuitable for chromatography. Microreticular gels are mostly xerogels.

The structure of the macroreticular gels is rather heterogeneous, the spatial distribution of the matrix being uneven. Their large pores render them suitable for the permeation and separation even of macromolecules. They swell readily, and have a high solvent and a low dry-matter content. In one group of macroreticular gels fibrils of a thickness of several (100) molecules may form by aggregation of the polymer chains, often accompanied by the evolution of a microcrystalline structure (i.e. starch, agar, agarose). The structure depends on the character of the crosslinkage between the fibrils (hydrogen bridges, polarization, etc.). Flodin (1962) for instance, on the basis of minimum differences in the specific gravities, ultracentrifuged glucose bundles, characteristic of the discontinuity of the steric structure, from starch gel.

Another possibility for producing macroreticular gels is the polymerization reaction. If the monomer dissolves in the suitably chosen solvent but the polymer produced does not, then the aggregation and precipitation of the polymer chains will produce a macroreticular structure. The pore size of the gels so obtained (e. g. polystyrene) intimately depends on the quality of the solvent used for the precipitation of the polymer (Moore 1964).

Macroporous gels may be obtained also by polymerization carried out in a solution containing colloidal particles, the gel being subsequently extracted by a suitable treatment from the polymer particles. This process will leave behind voids which correspond in size to the colloidal particles (Wieland and Determann 1967).

Most macroreticular gels are aerogels which, in spite of their high porosity, are resistant to mechanical effects.

The structure of microreticular and macroreticular gels is shown in *Figures 1a* and *b*.

The solvating medium and the solvent molecules picked up during the swelling of the crosslinkages are structurally and functionally equivalent parts of the gel structure. By their solvating properties gels can be divided into so-called hydrophilic and organophilic (hydrophobic) types, according to whether they swell in water or in polar and apolar-organic solvents. Between these extreme theoretical cases a wide variety of transitional types are known. The solvating capacity of the basic substances is, in general, dependent on the gel structure. The solvent molecules pertaining to the gel structure appear in different positions and with a different energy in relation to the monomer solution or the pure solvent (Vavruch 1965; White and Dorion 1961; Clifford *et al.* 1970; Clifford and Child 1971;

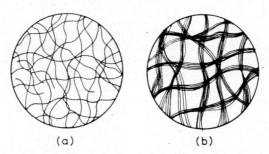

Fig. 1. The main types of gel structure
(a) microreticular; (b) macroreticular

Texter *et al.* 1975; Andrasko 1975). On the other hand, apolar blocking of the polar groups of the monomer (e.g. by methylation, hydroxy-propylization) or insertion of polar groups between the members of the apolar chains allows modification of the solvation capacity of the gels and the production of a variety of new ones.

1.2. GEL-FORMING SUBSTANCES; THEIR PRODUCTION

Although there are many substances with gel-forming properties, only a few compounds can be used for chromatography. From the point of view of chromatography these must meet the following criteria.

1. They must not react chemically with the substances to be separated or with the solvents under the experimental conditions of the chromatographic process.

2. The gel structure must have an appropriate chemical stability. The breakdown of the matrix and its spontaneous dissolution in media with different pH values must be minimal. The breakdown of the matrix can be analytically proved in gels of looser structure. For instance in the eluate of Sephadex G-200 dextran gel the anthrone reaction is positive. With continuous elution, 0.002 to 0.003% of Sephadex G-25 and G-75 gels and 0.005% of the G-200 type is lost daily (Granath 1964; Granath and Kvist 1967). The glucoside bonds of the polysaccharide gels are sensitive to strong mineral acids, while acrylamide gels containing a carboxylic acid amide are sensitive to alkalis.

3. They must be resistant to microorganisms and bacterial degradation as far as possible. The conservation of polysaccharides, which provide excellent culture media, will be discussed in Section 1.4 of Part II.

4. The gel structure must contain minimal ionic groups. Autooxidation, which is inevitable during production and storage, may produce chain-end carboxyl groups in the gels, of the order of microequivalents per gram of xerogel. The weak ion exchange properties of the gels will be dealt with in more detail in Section 1.6 of Part II.

5. When choosing the gel formers the aim should be to select a matrix of identical chemical structure which permits the production of a series which covers a broad spectrum, and enables a large variety of jobs to be carried out (e.g. Sephadex, Bio-Gel types).

6. It is essential for both theoretical and practical considerations that the gel-forming substances should yield homogeneous particles in sizes and distribution which are easy to control and of appropriate mechanical stability (see Sections 2.1 and 2.6 of Part I).

As regards origin and processing, three types — (a) natural, (b) semi-synthetic and (c) synthetic gels — are distinguished.

1.2.1. NATURAL GEL-FORMING SUBSTANCES

The first substances used for gel chromatography were natural polysaccharides and polymers. Most of these, e.g. galactomannane (Deuel and Neukom 1954), starch (Lindquist and Störgards 1955; Lathe and Ruthven 1955, 1956) and agar (Polson 1956, 1961; Killander et al. 1964; Bengtsson and Philipson 1964) are now of merely historical interest for both laboratory practice and industrial production. In the elucidation of the main interrelationships of gel chromatography the examination of starch and agar particles was of particular importance. They were, however, gradually superseded by gels of known origin and known composition, first of all by semi-synthetic dextran gels and agarose. The only gel of natural origin still in use is agarose (starch is also used but rather infrequently; Hanus and Kucera 1974).

1.2.1.1. AGAROSE GELS

Agar is probably the oldest hydrophilic gel former which is still in use. Raw purified polysaccharides (agar-agar) extracted from sea algae are widely used in the food industry and in the laboratory (bacteriology, immunology, electrophoresis).

The first workers to study the gel chromatographic properties of agar were Polson (1956, 1961) and Hjertén (1962c). Killander et al. (1964) fractionated human macroglobulins on agar particles. Immediately the first examinations proved that agar groups with ion-charge interfere with the mechanism of separation by gel chromatography. Araki (1937) stated that agar consists of two components: polar agaropectin containing sulphate and carboxyl groups, and uncharged neutral agarose. Attention turned to agarose as a substance free from ionic groups and showing superior gel-forming properties. The first method applied by Araki for purification was found to be too cumbersome for general separation practice (Araki 1937, 1956). Soon after Araki's work Hjertén (1961, 1962b) elaborated a method for the precipitation of agaropectin by cetylpyridiniumchloride.

Russel et al. (1964) described a process of agarose purification by fractionated polyethylene glycol precipitation.

Experience has shown that only properly granulated gel particles are suitable for chromatography (Hjertén 1964, 1968). Bengtsson and Philipson (1964) worked out a standard method of producing gel particles, a method which has found general acceptance in industry. This consists basically in pressing hot agarose sol through a Seitz filter, under continuous stirring

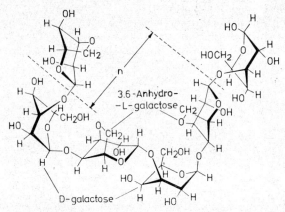

Fig. 2. The structure of agarose (Araki 1956)

conditions, into ether cooled in ice water, and when set, sieving the wet particles. For chromatographic work only purified agarose, condensed in beads of homogeneous size, is used (Hjertén 1968; Joustra 1970). Data concerning the commercially available agarose gels are collected in *Table 1*. Except for the Sagavac gels, which consist of broken granules, every agarose product consists of spheroids condensed in the form of beads. Agarose gels are marketed in swollen state, and conserved by bacterial inhibitors (usually sodium azide) in the form of aqueous suspensions. The most widely known chromatographic agarose gels are the Sepharose products (Pharmacia Fine Chemicals AB) and Bio-Gel A (Bio-Rad Laboratories). As regards the range of separation by molecular weight and by particle size, Bio-Gel A has the greatest number of fractions.

Agarose is a linear polysaccharide built up of D-galactose and 3.6-anhydro-L-galactose units.

Figure 2 illustrates the structure of the polysaccharide chains of agarose gel. Even in an aqueous solution of low concentration (less than 0.5%) agarose readily assumes the gelatinous form. Both the chromatographic properties and the pore size of the agarose gel depend on the agarose content. The lower the agarose content the larger are the molecules which are capable of penetrating into the gel structure.

By examination of the structure of agarose gels using different physicochemical methods (optical rotation, X-ray diffraction, computerized model building, NMR, etc.) it was established that the agarose chains exhibit a double-helix structure with an axial periodicity of 0.95 nm (Arnott et al. 1974; Andrasko 1975). As proved by electron microscopic examinations, the commercially available agarose particles are spheroids in the 15 to 100 μm diameter range, with a spongy surface and a structure in which the openings of the pores and of the channels can be clearly discerned (Amsterdam et al. 1975; Marsden and Wieselblad 1976). The particle structure consists of a steric network of filaments (fibres) developed from agarose chains by hydrogen bonds, including (presumably) a few hundred agarose bundles of a thickness of between 20 and 300 Å, touching at several points. In the stabilization of the bonds the conspicuously large pores may have an important share in the maintenance of the high mechanical stability and the microcrystalline structure of the fibrils (Laurent 1967; Granath and Kvist 1967). Depending on the conditions of production, agarose gels of the same dry-matter content but dissimilar properties may be obtained (Öberg and Philipson 1967).

As a natural property of their structure, agarose gels are sensitive to substances and effects which tend to loosen or disrupt the hydrogen bonds (concentrated urea and salt solutions, heat treatment), but they can be used without any difficulty in solutions containing guanidine hydrochloride (Fish et al. 1969). While a moderate concentration (approximately 2 M) of urea and an ionic compound (1 M NaCl) is not detrimental to the agarose structure, this does affect the mechanical stability of the gel beads. Agarose and the polysaccharide gels are sensitive to buffer solutions containing borate because while forming dissociable hydrogen ions, they produce a soluble complex with saccharides. Agarose gels should be used at temperatures between 0 and 40 °C, and at pH values between 4 and 9. They cannot be sterilized by drying or heating because at higher temperatures (above 50 °C) the melting of the gel bead impedes granulation and the formation of homogeneous bead sizes and destroys completely the gel structure. Agarose beads must be stored in the swollen and wet state since, if dried out, the native gel structure will not regenerate from the xerogel. On the other hand, it is an interesting property of agarose beads that, unlike other hydrophilic types, they can be treated or washed in alcohol or acetone without the risk of shrinkage. This property makes it possible for apolar contaminants (for instance lipids) adsorbed by the agarose to be removed.

Owing to their macroporous structure agarose gels can be used with advantage in the chromatography of even very large molecules, for instance nucleic acids, polysaccharides, nucleoproteines, antigens, etc. (Erikson and Gordon 1966; Kingsbury 1966; Öberg and Philipson 1967; Malchow et al. 1967; Russel and Osborn 1968; Loeb 1968; Petterson et al. 1968; Joustra 1970; Margolis 1967; Hanai et al. 1968; Höglund 1967; Demassieux and Lachance 1974). Being suitable also for the chromatography of cellular organelles and phags, they provide a transition, as it were, between the separation ranges of average-size molecules and microscopic particles. The application of agarose gels for the determination of molecular weight is dealt with in detail in Section 2.6.3.3 of Part I. With respect to *Table 1*, it should be noted that, as with other hydrophilic gels, the ranges of fractionation by molecular weight of proteins and polysaccharides deviate also with agarose gels. The exclusion molecular weights determined for proteins are generally two or four times higher than the values measured with polysaccharides (i.e. Sepharose 6B: 4 million for proteins and 1 million for polysaccharides). This tends to show that protein molecules penetrate deeper into the agarose gel structure than polysaccharides of the same molecular weight. This dissimilar behaviour may be attributed primarily to their different specific volumes and their different hydration.

An important property of agarose gels, one of considerable advantage in practice, is their flexible and reversible change of volume in the gel bed. In contrast to other gels, agarose columns do not aggregate under hydrostatic pressure, and when the pressure is released the gel bed recovers its original volume. There is no need to repack the column. In addition, to achieve optimum chromatographic properties, the pressures and flow rates (see Part II) should be determined on the basis of empirical correlations, as presented later.

To increase the mechanical and chemical stability, agarose gels with covalent crosslinks have been developed more recently (Schell and Ghetie 1968; Porath et al. 1971; Porath 1973). As for dextran gels, the crosslinks (see Section 1.2.2.1) are induced by epichlorhydrin, which links up the agarose chains in a single gel fibre, only no interfibre crosslinks forming to any significant degree. Under these conditions a crosslinked agarose gel is obtained which has the same porosity as the parent gel but substantially higher thermal and chemical stability. The crosslinked agarose beads so produced are rigid, permit high flow rates, retain their pore sizes even in other organic solvents, lend themselves to sterilization by autoclaving, and have, in fact, proved to be excellent matrices for various methods of (apolar) gel and affinity chromatography (Sepharose CL for gel filtration and affinity chromatography, Pharmacia Fine Chemicals AB, Uppsala, Sweden, 1975, and Porath 1973). For some parameters of the crosslinked agarose gels marketed under the trade name of Sepharose CL see *Table 1*.

In a similar way Laas (1975) and Porath et al. (1975) produced benzylated agarose beads, crosslinked by dibromopropanol and divinyl sulphone.

Agarose mixed with colloidal silica forms beads containing a large proportion of agarose and a minor amount of silica. High flow rates are attainable owing to their mechanical stability also at high hydrostatic pressures and these particles are also available for gel chromatography of proteins and polysaccharides (Pertroft and Hallén 1976).

1.2.2. SEMI-SYNTHETIC GEL FORMERS

Semi-synthetic gels are produced by small or large scale purification, fractionation and chemical conversion of natural base materials, primarily dextran. The best known semi-synthetic hydrophilic gels are the dextran derivatives of the Pharmacia Fine Chemicals AB marketed under the trade name of Sephadex. The Sephadex gels are most frequently used for the packing of chromatographic columns. Also the first systematic investigations into the theoretical correlations were performed on these gel types (Granath and Flodin 1961; Flodin 1962).

Table 1

Commercial agarose gels and their properties

Type	Agarose content, %	Wet particle size, μm	Fractionation range (MW million)	Manufacturer
Sepharose 6B	6	40—210	0.2— 1.5	
Sepharose 4B	4	40—190	0.1— 3.0	
Sepharose 2B	2	60—250	0.8—20	
Sepharose CL-2B	2	60—250	20*—40**	Pharmacia Fine Chemicals AB, Uppsala, Sweden
Sepharose CL-4B	4	40—190	5*—20**	
Sepharose CL-6B	6	40—210	1*— 4**	
Octyl-Sepharose CL-4B				
Phenyl-Sepharose CL-4B				
AH-Sepharose 4B (aminohexyl-)				
CH-Sepharose 4B (carboxyhexyl-)				
Bio-Gel A–0.5 m	10	150—300 75—150 40— 75	0.01— 0.5	
Bio-Gel A–1.5 m	8	150—300 75—150 40— 75	0.01— 1.5	
Bio-Gel A–5 m	6	150—300 75—150 40— 75	0.01— 5	Bio-Rad Laboratories, Richmond, California
Bio-Gel A–15 m	4	150—300 75—150 40— 75	0.04—15	
Bio-Gel A–50 m	2	150—300 75—150 40— 75	0.1—50	
Bio-Gel A–15 m	1	150—300 75—150 40— 75	1—50	
Gelarose 2%	2	—	—	
Gelarose 4%	4	—	—	Litex, Glostrup, Denmark
Gelarose 6%	6	—	—	
Gelarose 8%	8	—	—	
Gelarose 10%	10	—	—	
Sagavac SAG 10	10		0.01—0.25	
Sagavac SAG 8	8		0.25—0.7	Seravac Laboratories Ltd., Berkshire, England
Sagavac SAG 6	6	142—250	0.05— 2	
Sagavac SAG 4	4	66—142	0.2—15	
Sagavac SAG 2	2		0.5—150	

Notes: From manufacturers' brochures.
Approximate exclusion limits for *polysaccharides and **proteins.

1.2.2.1. SEPHADEX GELS

The Sephadex gels are based on dextran. Dextran is a linear polysaccharide built up of glucose units by 1.6-α-glucoside bonds. Their polymer structure includes side-chain linking with 1.8-, 1.3- or 1.4-glucoside bonds. Sephadex gels are prepared from dextran produced by sucrose culture medium from the Leuconostoc mesenteroides B 512 strain. The side-chains of dextran, obtained by fermentation, are of the 1.3-glucoside form and consist, in the main, of more than one glucose unit. From raw dextran of heterogeneous composition (with an average molecular weight between 10 and 300 million) fractions of an average molecular weight between 40 and 70,000 are isolated by repeated purifications, partial hydrolysis and alcoholic precipitation. Alkaline solutions of the dextran fractions are emulsified by the addition of stabilizers to a water-immiscible organic solvent which is then made to react with epichlorhydrin, under continuous stirring, at an appropriate temperature (40 to 70 °C). Epichlorhydrin causes 1.3-glyceride–ether bonds to form between the dextran chains. As intermediates, dextran derivatives substituted by epichlorhydrin are obtained. The chemical reactions are described by the following equations:

$$\text{Dextran—OH} + CH_2\text{—}CH\text{—}CH_2Cl \rightarrow$$
$$\underset{O}{\diagdown\diagup}$$

$$\text{Dextran—O—}CH_2\text{—}CHOH\text{—}CH_2Cl$$
$$\text{Dextran—O—}CH_2\text{—}CHOH\text{—}CH_2Cl + NaOH \rightarrow$$
$$\text{Dextran—O—}CH_2\text{—}CH\text{—}CH_2 + NaCl + H_2O$$
$$\underset{O}{\diagdown\diagup}$$

$$\text{Dextran—O}CH_2\text{—}CH\text{—}CH\text{—}CH_2 + HO\text{—dextran} \rightarrow$$
$$\underset{O}{\diagdown\diagup}$$

$$\text{Dextran—O—}CH_2\text{—}CHOH\text{—}CH_2\text{—O—dextran}$$

The crosslinking reactions cause a water-insoluble network to arise from the dextran. The schematic illustration of the structure of the crosslinked dextran gels is shown in *Figure 3*.

The gel particles are washed in water and dried, after dehydration in alcohol. Particle sizing takes place by sieving. In the early years Sephadex gels were prepared by block polymerization and crushing (Flodin 1962). The emulsification process yields dextran gels in the form of beads. When swollen, the gels consist of spheroids of nearly uniform size.

Table 2 shows the more important properties of the Sephadex G gels produced by the Pharmacia Fine Chemicals AB.

The Sephadex gels are marketed in the form of xerogel particles of defined sizes. Except for the types G-10 and G-15 every gel is available also in superfine fractions (10 to 40 μm), mainly for the purposes of thin layer chromatography and for the packing of analytical columns of increased

Table 2

Sephadex gels and their properties

Type		Dry particle diameter, μm	Water regain, W_r, g H_2O g^{-1} xerogel	Bed volume, V_d, ml g^{-1} xerogel	Specific gravity of swollen particles, d g ml^{-1}	Fractionation range (MW)	
						peptides and globular proteins	dextrans
G-10		40—120	1.0±0.1	2—3	1.24	up to 700	up to 700
G-15		40—120	1.5±0.2	2.5—3.5	1.19	up to 1 500	up to 1 500
G-25	coarse	100—300					
	medium	50—150	2.5±0.2	4—6	1.13	1 000— 5 000	100— 5 000
	fine	20— 80					
	superfine	10— 40					
G-50	coarse	100—300					
	medium	50—150	5.0±0.3	9—11	1.07	1 500— 30 000	500— 10 000
	fine	20— 80					
	superfine	10— 40					
G-75		40—120	7.5±0.5	12—15	1.05	3 000— 70 000	1 000— 50 000
	superfine	10— 40					
G-100		40—120	10.0±1.0	15—20	1.04	4 000—150 000	1 000—100 000
	superfine	10— 40					
G-150		40—120	15.0±1.5	20—30	1.03	5 000—400 000	1 000—150 000
	superfine	10— 40		18—22			
G-200		40—120	20.0±2.0	30—40			
	superfine	10— 40		20—25	1.02	5 000—800 000	1 000—200 000
Sephacryl S-200	superfine	40—105*	—	—	—	5 000—250 000	—

Notes: From the brochures: Sephadex, Gel Filtration in Theory and Practice, 1970; and Sephacryl S-200 Superfine for high performance gel filtration, 1976. Pharmacia Fine Chemicals AB, Uppsala.
* Wet bead diameter.

resolution. For routine work and for preparative columns the "fine" fractions (20 to 80 μm) are recommended, while the medium size particles (50 to 150 μm) are applicable for coarser separation jobs, to speed up the rate of flow. Finally, the coarse fractions (100 to 300 μm) are used for applications other than column chromatography: for (preparative) batch methods or to pack continuously operating basket centrifuges.

The trade names (G-10, G-100, etc.) refer to the water uptake or water regain of the xerogels (W_r = water regain, see Section 2.3 of Part I). Accordingly, the xerogel structure of the G-25 grade has a water regain capacity of approximately 2.5 ml per gram, while the G-100 type takes up 10.0 ml of water per gram.

Dextran retains the strongly hydrophilic properties produced by the polar hydroxyl groups even in the gel structure. In the production of semisynthetic gels, the substitution of cellulose or starch by dextran is both practical and theoretically significant. Owing to their favourable spatial orientation the free hydroxyl groups of dextran react with nearly equal probability in the crosslinking reaction, thus enhancing the development of a uniform crosslinked structure. The swelling and the chromatographic properties of the gels depend on the number of crosslinks. From *Table 2* it is evident that the specific volume of swelling is, by necessity, closely related to the number of crosslinks and the water regain. An increasing number of crosslinks reduces the number of polar hydroxyl groups and thereby the water regain capacity of the xerogel. The fewer the crosslinks among the dextran chains the higher will be the water regain of the xerogel

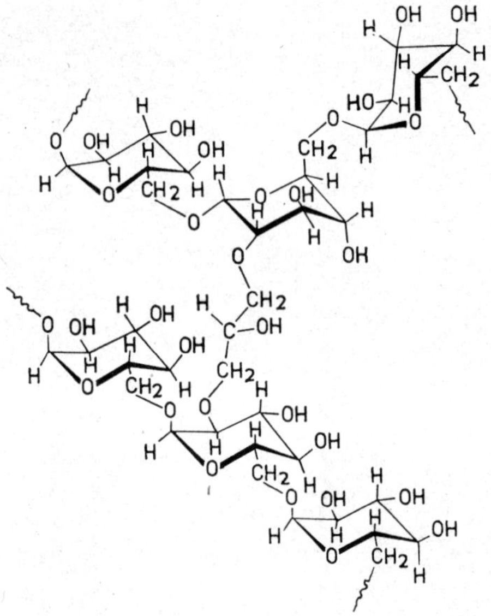

Fig. 3. The structure of crosslinked dextran gels

and the larger will its specific swelling volume become. At the same time the dry matter content and the density of the gels will decrease proportionately.

The changes in the size (diameter) of the xerogel particles while swelling can be computed from the water regain and the density of the xerogel (to a fair approximation $d = 1.64$ g/ml for all gel types), using the multiplication factor $(1 + dW_r)$ (Fischer 1969, p. 183). Electron microscopic examinations have shown (DeMets and Lagasse 1970) that the diameter of the dextran gel particles swells to twice the original dimension. This is in good agreement with practice and corresponds to an eight-fold growth in volume.

According to their mechanical and structural properties, the gels with only few crosslinks (Sephadex G-75, G-100, G-150, G-200) are called "soft or loose", while those with a greater number of crosslinks (G-10, G-15, G-25, G-50) are termed "dense" gels. Experience has proved that the soft gels are more susceptible to deformation and exhibit a higher resistance to the through-flow of fluids.

The system of crosslinks determines the pore size of dextran gels, whilst the range of their fractionation is determined by molecular weight (size). With increasing number of crosslinks, both the pore size and the limit of that molecular size (weight) which is just capable of penetrating into the gel structure will diminish (the exclusion limit). *Figure 4* shows the so-called selectivity curves, i.e. the range of fractionation by molecular weight

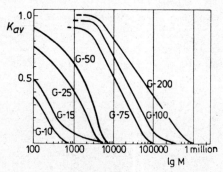

Fig. 4. The selectivity curves of Sephadex G dextran gels. The data for G-10, G-15, G-25 and G-50 refer to polysaccharides (dextran fractions), G-75, G-100 and G-200 to globular peptides and proteins

of the Sephadex G gels. It has already been mentioned for agarose gels that peptides, globular proteins and polysaccharides behave differently. *Table 2* proves that the anomalous behaviour of polysaccharides increases proportionately to the loosening of the gel structure and increasing molecular weight.

The significance of the crosslinks has been corroborated also by chemical examination of the gel structure. Oxidation of the dextran gels with periodic acid has proved that in G-25 60% of the glucose units, and in G-200 20% of these units were in a glyceride-ether bond (Granath 1964; Vándor 1965).

Dextran gels exhibit outstanding chemical properties for chromatography. Their glucoside bonds are disrupted only by strong mineral acids. Concentrated (88%) formic acid does not affect them and in dilute acids (0.02 M HCl) and dilute bases (0.25 M NaOH) they remain constant for several months (Craft 1961). Sephadex gels stand up to alkaline treatment even at 60 °C and retain their original properties in 0.1 M HCl for one to two hours. They contain a small number of carboxyl groups and have a low ion exchange capacity up to approximately $10-20\mu$ equivalent per gram of xerogel. Oxidative effects do affect them and increase the number of ionic carboxyl groups. Dry or, in neutral media, Sephadex gels can be sterilized at 110 to 120 °C for up to 30 minutes.

In addition to aqueous solutions Sephadex gels swell readily also in ethylene glycol, dimethyl sulphoxide and formamide. The G-10 and G-15 types swell even in less polar dimethylformamide. The dextran gels cannot be solvated with methanol, ethyl alcohol or acetic acid; in their aqueous solutions swelling is only partial and the pores are small.

1.2.2.2. MOLSELECT GELS

The Reanal Fine Chemicals Factory in Hungary produces and sells crosslinked dextran gels under the trade name of Molselect. Production began in 1964. Apart from some details of the licence, their preparation, denomination and properties are similar to those of the Sephadex gels.

Table 3 shows the properties of the Molselect gels of the Reanal Factory.

Table 3

Properties of the Molselect dextran gels

Type	Particle size (xerogel), μm	Water regain, W_r, g H_2O g^{-1} xerogel	Bed volume, V_d, ml g^{-1} xerogel	Fractionation range (MW)
Molselect G-10	50—100 100—320	1.0	2— 3	up to 700
Molselect G-15	50—100 100—320	1.5	2.5— 3.5	up to 1 500
Molselect G-25	50—100 100—320	2.5	4— 6	100— 5 000
Molselect G-50	50—100 100—320	5.0	9—11	500— 10 000
Molselect G-75	50—100	7.5	12—15	1 000— 50 000
Molselect G-100	50—100	10.0	15—20	1 000—100 000
Molselect G-200	50—100	20.0	30—40	1 000—200 000

Note: From the brochure: Molselect 72, Reanal Fine Chemicals, Budapest, Hungary, 1972.

Although there is little experience as yet with the Molselect gels, on the basis of their structure and their main parameters their chromatographic properties are likely to be equivalent to those of the Sephadex types. According to *Table 3* the denser Molselect gels are available in two particle sizes (coarse and fine) and the looser types in one (fine). The manufacturer guarantees a maximum concentration of ionic groups of 30 to 40μ equivalents per gram of xerogel. Otherwise the Molselect gels have the same properties as the Sephadex types.

1.2.2.3. ORGANOPHILIC DEXTRAN GELS

The separation processes in gel chromatography take place usually in polar media, i.e. in aqueous solutions. However, numerous biologically active compounds (lipids, vitamins, steroid hormones, coenzymes, etc.) are intrinsically lyophilic in character. For a change in dielectric constant of the medium or in the molecular configuration, the better part of the biological macromolecules become apolar and form an appropriately hydrophobic gel structure which swells readily in organic solvents.

Dextran derivatives which swell in organic solvents constitute an important group of semi-synthetic gels. In such applications of dextran the well known structure of the polymer and its easy-to-control chemical conversion have a decisive share (Nyström and Sjövall 1975).

1.2.2.4. THE SEPHADEX LH-20 AND LH-60

The Sephadex LH-20 and LH–60, the products of Pharmacia Fine Chemicals AB, are the best known semi-synthetic gels with organophilic and partly hydrophilic properties as well. They are obtained by the hydroxy-propylation of Sephadex G-25 and G-50 dextran gels, respectively. By blocking the polar hydroxyl groups the crosslinked structure of the dextran gel becomes partly apolar and swells even in organic solvents. Simultaneously the free hydroxyl groups produce definitely polar properties. Hydroxy-propylation slightly diminishes the pore size and the water regain of the original dextran gels. Compared to the value of 5 000 in the Sephadex G-25, the maximum molecular weight of the Sephadex LH-20 for polyethylene glycol at exclusion is approximately 4 000.

Table 4 illustrates the swelling ability of Sephadex LH-20 and LH-60 in different solvents. As seen, there is no close relationship between the po-

Table 4

Properties of Sephadex LH-20 and LH-60 dextran gels*

		LH-20	LH-60
Specific gravity (dry particles)	g ml^{-1}	1.3	—
Particle size	μm	25—100	40—120
Fractionation range	(MW)	100—4000	400—10,000

Solvent	Dielectric constant, ε_{2t} °C**	Bed volume approx, V_d, ml g^{-1} xerogel	
		LH-20	LH-60
Dimethylsulphoxide	45.0	4.4—4.6	13.4—13.8
Pyridine	12.3	4.2—4.4	13.4—13.8
Water	81.1	4.0—4.4	12.4—12.8
Dimethylformamide	36.7	4.0—4.4	12.9—13.3
Methanol	31.2	3.9—4.3	11.9—12.2
Ethylene dichloride	10.6	3.8—4.1	11.0—11.3
Chloroform (1% ethanol)	5.1	3.8—4.1	12.3—12.6
n-Propanol	22.2	3.7—4.0	11.0—11.3
Ethanol (1% benzene)	25.8	3.6—3.9	12.0—12.3
Isobutanol	11.9	3.6—3.9	10.8—11.1
Formamide	109.0	3.6—3.9	8.6— 8.9
Methylene dichloride	9.1	3.6—3.9	11.0—11.3
n-Butanol	19.2	3.5—3.8	11.0—11.3
Isopropanol	26.0	3.3—3.6	10.0—10.3
Tetrahydrofuran	7.6	3.3—3.6	9.6— 9.9
Dioxane	2.2	3.2—3.5	9.8—10.1
Acetone	21.5	2.4—2.6	5.5— 5.8
Acetonitrile	38.8	2.2—2.4	3.1— 3.3
Carbon tetrachloride	2.2	1.8—2.2	1.9— 2.1
Benzene	2.3	1.6—2.0	2.4— 2.6
Ethyl acetate	5.0	1.6—1.8	3.0— 3.2
Toluene	2.4	1.5—1.6	1.9— 2.1

* From the brochures: Sephadex LH-20 and LH-60 Chromatography in Organic Solvents, Pharmacia Fine Chemicals AB, Uppsala, 1970 and 1976.
** Data according to Dobos (1965), B. S. Nagy (1970) and Mikes (1970).

larity (dielectric constant) of the solvents and the specific swelling of the gels. The differences may be attributed to the structural features of the solvents and their polarization effects.

From solvent mixtures the xerogel absorbs the components of higher polarity first. The finding that the linear correlation between the logarithm of the molecular weight of the separated substances and their elution properties, as shown in *Figure 4*, does not hold for many cases, may refer to the specific interactions between the gel structure and the solvent molecules. For instance, in chloroform the selectivity curve of polyethylene glycol appears as a hyperbole, indicating the predominance of the adsorptive effects over separation by molecular weight.

In view of the relatively short history of Sephadex LH-20 and LH-60, further research is necessary to clarify the problem. Like dextran gels, Sephadex LH-20 and LH-60 are also marketed in the form of bead polymer xerogel particles. Their chemical stability, swelling, manipulation and properties are the same as those of the Sephadex G-25 and G-50.

1.2.2.5. OTHER ORGANOPHILIC DEXTRAN GELS

By substituting the hydroxyl groups of Sephadex G gel types, for instance by acetylization of Sephadex G-50 (Determann 1964b) or the reaction of Sephadex G-25 with aliphatic isocyanates (Heitz *et al.* 1966) further organophilic gels may be produced. The alkylation of the hydroxyl groups produces bonds of higher chemical stability. Nyström and Sjövall (1965a, b) obtained a hydrophobic gel by the methylation of Sephadex G-25, which in addition to its good adsorption ability, proved to be suitable also for gel chromatography in organic solvents. In a similar way Ellingboe *et al.* (1970) prepared Sephadex alkyl ethers with long carbon chains. These types, in spite of their promising properties, are applicable for special jobs only, and have not yet spread widely.

Quite recently, Pharmacia Fine Chemicals AB, the makers of Sephadex gels, introduced two new types of hydrophobic polysaccharide gels: (1) Sephacryl S-200 Superfine (see in *Table 2*) is prepared by covalently cross-linking allyl dextran with N,N'-methylenebisacrylamide to give a mixed, rigid dextran-acrylamide structure (Sephacryl S-200 Superfine, Pharmacia Fine Chemicals AB, 1976); (2) Octyl-Sepharose CL-4B and Phenyl-Sepharose CL-4B are derivatives of the crosslinked agarose gel Sepharose CL-4B (see in *Table 1*) containing hydrophobic n-octyl and phenyl groups, respectively (Octyl-Sepharose CL-4B, Phenyl-Sepharose CL-4B for hydrophobic interaction chromatography, Pharmacia Fine Chemicals AB, 1976).

1.2.3. SYNTHETIC GEL-FORMING SUBSTANCES

The production of synthetic gels is closely related to polymer chemistry and polymerization reactions. According to the base materials, synthetic gels fall into three categories: (a) acrylamide-acrylate copolymers, (b) styrene-divinylbenzene copolymers and (c) other mixed types of gels.

1.2.3.1. ACRYLAMIDE-ACRYLATE GELS

By homologous polymerization of the toxic water-soluble acrylamide containing a reactive double bond

$$CH_2=CH-CONH_2$$

two different products may be obtained, depending on the experimental conditions. Thermal polymerization yields a highly insoluble solid substance, while the so-called vinyl polymerization, induced by the mild activation of the double bond, yields a water-soluble linear polyacrylamide chain. For activation redoxicatalysts (for instance potassium or ammonium peroxomonosulphuric acid), initiators (e.g. tetramethylethylenediamine [TEMED]) and a regulator (β-dimethylaminopropionitrile) are used. The presence of riboflavin as sensibilizer may have a similar activating effect. If polymerization is carried out with a bifunctional acrylamide, generally N,N'-methylenebisacrylamide

$$CH_2=CH-CONH-CH_2-NHCO-CH=CH_2$$

then a water-insoluble polyacrylamide crosslinked with a gel structure will be obtained. N,N'-methylenebisacrylamide produces crosslinks between the polyacrylamide chains. *Figure 5* illustrates the crosslinked structure of polyacrylamide gels, according to Raymond and Wang (1960).

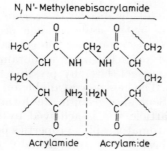

Fig. 5. The crosslinked structure of polyacrylamide gels (Raymond and Wang 1960)

Polyacrylamide gels were first used in electrophoresis (disk electrophoresis) (Raymond and Weintraub 1959; Davis and Ornstein 1959) as a carrier for the convection inhibitor. Granulated gel particles were first used for chromatography by Lea and Sehon (1962) and by Hjertén and Mosbach (1962). The first systematic examinations of the production of gels for column chromatography were carried out by Hjertén (1962a) and Curtain and Nayler (1963).

The pore size of polyacrylamide gels depends mainly on the concentration of the monomer acrylamide, but also on the ratio of N,N'-methylenebisacrylamide, which causes the crosslinks to form. By varying the ratio of the two, a wide variety of gels may be produced. Between an acrylamide concentration of 4 and 16% Hjertén (1962a) prepared gels of 5% bifunctional monomer with characteristically different pore sizes and swelling ability. Hjertén ascertained the porosity of the gels by chromatographing proteins with dissimilar molecular weights.

Sun and Sehon (1965) gave a detailed description of the interactions of the constituents and the water-regain capacity of the gels. To designate the polyacrylamide gels they suggested the use of a double group of figures indicating the percentage ratio of the monomer to the bifunctional monomer. According to this system the polyacrylamide gel marked 15—5 is prepared from a solution containing 15% of acrylamide and 5% of bisacrylamide. Sun and Sehon established that the water regain (specific swelling volume) of gels of dissimilar composition (for instance 15—5 and 10—20) may be equal, although their intrinsic structure and the size of their pores may differ widely. Fawcett and Morris (1966) proved that gels containing a relatively high amount of bisacrylamide (e.g. 6.5—15 or 8—25) have substantially larger pores than those built up from fewer bifunctional units and having approximately the same swelling ability. Using the Ogston and Laurent—Killander structural models, Fawcett and Morris proved that gels with high bisacrylamide content exhibit a macroreticular structure as seen in *Figure 1b*. This tends to prove that the structure of the polyacrylamide gels depends also on the relative amounts of their different constituents.

The carbonyl-amide groups of polyacrylamide gels produce polar-hydrophilic properties. Strong alkalis disrupt the carbonyl-amide bonds and their hydrolysis causes a rise in the ion exchange properties of the gels. For chromatography, the homogeneous, swollen gel particles proved to be best. Previously, the desired particle size was obtained by milling and subsequent sieving of the xerogel, later by pressing the wet polymer through sieves (Hjertén 1962a). The gels now available on the market are produced by emulsification (see Sephadex gels), in the form of condensed bead polymers. Homogeneous particle size is achieved by sieving the wet product. With respect to their chromatographic properties polyacrylamide gels can be reversibly dried. The best known commercially available polyacrylamide gels are the Bio-Gel P types.

1.2.3.2. BIO-GEL P POLYACRYLAMIDE GELS

Table 5 presents a survey of the trade names and most important properties of the Bio-Gel P series of gels produced by the Bio-Rad Laboratory (Richmond, California). Each of the ten different types is sold in the form of xerogel beads. The producer's catalogue specifies the range of the size of wet, swollen gel particles. The types P-2, P-4, P-6, and P-10 are available in four particle sizes, the rest in three particle sizes. There is also a superfine fraction obtainable for every type (less than 40 μm) specifically for thin layer chromatography. The Bio-Gel P types are of slightly higher polarity than the dextran gels and the time necessary for their swelling is somewhat shorter than that for the corresponding dextran gels. A compar-

Table 5

Properties of Bio-Gel polyacrylamide gels*

Type	Particle size (wet), μm	Water regain, W_r, g H_2O g^{-1} xerogel	Bed volume, $V_{\tilde{a}}$, ml g^{-1} xerogel	Fractionation range** (MW)
Bio-Gel P-2	150—300 75—150 40— 75 — 40	1.5	3.8	200— 1 800
Bio-Gel P-4	150—300 75—150 40— 75 — 40	2.4	5.8	800— 4 000
Bio-Gel P-6	150—300 75—150 40— 75 — 40	3.7	8.8	1 000— 6 000
Bio-Gel P-10	150—300 75—150 40— 75 — 40	4.5	12.4	1 500— 20 000
Bio-Gel P-30	150—300 75—150 40— 75 — 40	5.7	14.8	2 500— 40 000
Bio-Gel P-60	150—300 75—150 40— 75 — 40	7.2	19.0	3 000— 60 000
Bio-Gel P-100	150—300 75—150 40— 75 — 40	7.5	19.0	5 000—100 000
Bio-Gel P-150	150—300 75—150 40— 75 — 40	9.2	24.0	15 000—150 000
Bio-Gel P-200	150—300 75—150 40— 75 — 40	14.7	34.0	30 000—200 000
Bio-Gel P-300	150—300 75—150 40— 75 — 40	18.0	40.0	60 000—400 000

* From the brochure: Bio-Rad materials for Ion exchange, Gel filtration, Adsorption. Catalogue U/V 1970.
** For peptides and globular proteins.

ison of Sephadex G-25 and Bio-Gel P-6 shows that both the water regain and the specific swelling volume of polyacrylamide particles are higher. Another advantage offered by polyacrylamide gels is that their particles are more brittle and more resistant to mechanical effects. Owing to their dissimilar structures, the shrinkage of polyacrylamide gels produced by high salt concentration is less than that of the dextran types. This phenomenon may be attributed to the different structure of the hydrate spheres of the dextran's hydroxyl and polyacrylamide's amide groups. The difference is particularly conspicuous during adsorption, which is dealt with later (see Section 2.7.3 of Part I).

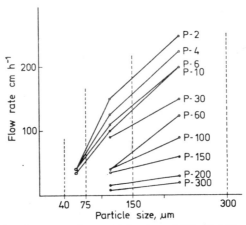

Fig. 6. The correlation between the particle size of Bio-Gel P polyacrylamide gels and the rate of flow in the bed. From the catalogue of the Bio-Rad Laboratories, Richmond, California, 1970

A further superior feature of polyacrylamide gels as compared to dextran gels is that, being synthetic, they do not enhance the growth of microorganisms and are neutral to bacterial attack.

The flow rate in gel columns closely depends on the particle size. *Figure 6* shows the flow rates recommended by the manufacturer for Bio-Gel P types of different particle size. This figure indicates also the well-known interactions according to which (assuming constant pressure) smaller particles increase the hydrodynamic resistance of the gel bed and reduce the flow rate of the moving (flowing) phase. This effect is particularly significant in dense gel structures and with small pores (P-2, P-4, P-6, P-10). The flow velocities in the soft gels with low polymer content (P-200, P-300) are lower by an order of magnitude and show less variation with particle size.

Figure 7 shows the interactions between the physical constants of the Bio-Gel P types (water regain, specific swelling volume) and the flow rate in columns packed with gels of the particle size most frequently used (75—150 μm). As seen, with increasing pore size (swelling, water regain), the flow rate in the gel columns decreases.

The ranges of fractionation in molecular weight of the Bio-Gel P types are, by and large, the same as for the corresponding dextran gels. The selectivity curves in *Figure 8* refer to globular proteins and spheroid molecules. The fractionation ranges and the limit values of exclusion molecular weights of polysaccharides and other linear molecules are slightly lower in the Bio-Gel P types than the values established for proteins.

Like dextran gels, polyacrylamide gels also adsorb aromatic compounds in a selective way. This property, bearing in mind the dissimilar chemical

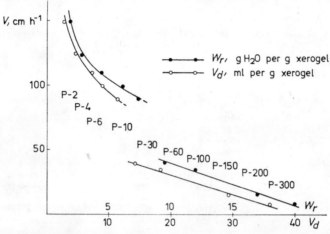

Fig. 7. The correlation between the water regain (W_r) of the Bio-Gel P polyacrylamide gels, their specific swelling volume (v_d) and the rate of flow. From the catalogue of the Bio-Rad Laboratories, Richmond, California, 1970

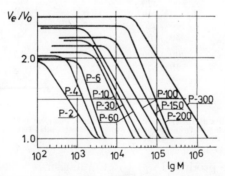

Fig. 8. The correlation between the molecular weight and the elution volume of spheroids in Bio-Gel P acrylamide gels. From the catalogue of the Bio-Rad Laboratory, Richmond, California, 1970

build-up of the gels, is surprising. In addition, it may be assumed that adsorption depends on structural factors as well. Sun and Schon (1965) observed that the more easily swelling 5—20-type gel ($W_r = 13$ g H_2O/g) binds tryptophan better than the 15—15 gel which is less susceptible to swelling, and contains also a higher ratio of dry matter ($W_r = 7$ g H_2O/g).

1.2.3.3. ACRILEX POLYACRYLAMIDE GELS

The Hungarian Reanal Fine Chemicals Factory began the supply of polyacrylamide gels, under the brand name of Acrilex, in 1968. The principal physico-chemical properties of the gels correspond to those of the polyacrylamide types previously described. Their parameters are shown in *Table 6*.

Table 6

Properties of the Acrilex* polyacrylamide and Spheron** methacrylate gels

Type	Particle size, μm	Water regain, W_r, g H_2O g^{-1} xerogel	Bed volume, V_d, ml g^{-1} xerogel	Fractionation range (MW)
Acrilex P-2	50—100 100—320	1.5	4.0	100— 2 000
Acrilex P-4	50—100 100—320	2.5	6.0	500— 4 000
Acrilex P-6	50—100 100—320	4.0	8.5	1 000— 6 000
Acrilex P-10	50—100 100—320	5.0	11.0	5 000— 15 000
Acrilex P-30	50—100 100—320	6.0	15.0	10 000— 30 000
Acrilex P-60	50—100 100—320	7.0	18.0	20 000— 60 000
Acrilex P-100	50—100 100—320	7.5	20.0	30 000—100 000
Acrilex P-150	50—100 100—320	9.0	24.0	50 000—150 000
Acrilex P-200	50—100 100—320	15.0	35.0	90 000—200 000
Acrilex P-300	50—100	18.0	40.0	100 000—300 000
Spheron 300	200—300 40— 80	—	4.5	1 000—300 000
Spheron 1000	20— 40	—	4.5	3 000—10^6
Spheron 300 BTD	40— 60	—	4.5	3 000—2 · 10^6
Spheron SE	32— 40	—	—	1 000—300 000

* From the brochure: Acrilex gel filters, Reanal Fine Chemicals, Budapest 1971.
** Spheron gels can be obtained from Lachema, Brno, Czechoslovakia. Data given from Vondruška *et al.* (1976), Čech *et al.* (1977) and Břizová *et al.* (1977).

The ten types are produced in two granular sizes which cover a narrower range (50 to 100 μm) than the Bio-Gel P types. Because of this the gel bed will be more homogeneously distributed which, in turn, allows a sharper fractionation than is feasible with the corresponding types of polyacrylamide gels.

1.2.3.4. ORGANOPHILIC AND HYDROPHILIC ACRYLATE GELS

Unlike the polar acrylamide gels, hydrophobic-organophilic gels can be obtained by polymerization of acrylic acid. Determann *et al.* (1964) produced gels from methylmethacrylate and ethylenedimethacrylate using azodiisobutyronitrile as initiator, which readily swell in organic solvents such as toluene, benzene, chloroform, and methylene chloride. It was found that the polymethylmethacrylate gel particles were suitable for the fractionation of polystyrene oligomers. By polymerizing in emulsion Determann *et al.* even succeeded in condensing beads. However, owing to the deterioration of the mechanical properties of the gels, it was no longer possible to increase the pore size by reduction of the ratio of the bifunctional monomer. Heufer and Braun (1965) produced high porosity gels of stable form by the homopolymerization of ethylenedimethacrylate in isoamylalcohol which, together with relatively low specific swelling ability, produced very favourable conditions of flow. It is assumed that these gels also belong to the organophilic type of macroreticular gel structure. The reaction kinetics of the polymerization of methylmethacrylate was studied by Smith *et al.* (1966). Recently, Břizová *et al.* (1977) used a copolymer of ethylene dimethacrylate and styrene (Spheron SE, see *Table 6*) in the analysis of trace amounts of impurities in water.

Another way of producing acrylate gels which swell in organic solvents, like dextran gels (see the Sephacryl S-200 type in Section 1.2.2.5), lies in the polymerization of suitably prepared polymer chains by crosslinking. Fritsche and Gröbe (1965) used formaldehyde for this purpose and applied the gels so obtained to the fractionation of acrylonitrile polymers.

Hydroxyethyl methacrylate copolymerized with ethylenedimethacrylate in a non polar dispersion medium yields hydrophilic acrylate gels (Vondruška *et al.* 1976, Čech *et al.* 1977). Hydroxyalkylmethacrylate gels of Spheron type (see *Table 6*) provide optimal hydrodynamic conditions in addition to a constant flow through the column.

1.2.3.5. POLYSTYRENE GELS

Although polystyrene-divinylbenzene copolymers, as the base materials of ion exchange resins, belong to the first type of crosslinked gel structure used for chromatography, it was the natural and semi-synthetic gel formers which contributed most to the development of gel chromatography. The reason for this might have been the relatively small pores and hydrophobic properties of polystyrene gels. For natural macromolecules, which are almost always polar and which are very suitable for examination in the polar phase, polystyrene gels proved to be unsatisfactory. Although, with the evolution of macroreticular structures, extreme enlargement of the pores became feasible, the organophilic properties restricted the use of polystyrene gels to the examination of hydrophobic natural and synthetic polymers (petroleum products, asphalts, plastics, etc.). In addition, adequate hydrophilic carriers (polysaccharides, polyacrylamide) were found meanwhile for the gel chromatography of the polar phase. With the development of chemical processing and plastics industries, attention turned anyway toward compounds which dissolved in organic solvents. Simultaneously, plastics chemistry, by supplying a variety of apolar gel structures, contributed greatly to the development of separation methods. To distinguish between separation by molecular size and separation other than that from the polar phase, a specific terminology (Moore 1964), that of the so-called gel permeation chromatography (GPC) has been introduced. The distinction of this process is based on the assumption that in appropriate solvents and organophilic gels the apolar molecules may better approximate the ideal separation mechanism in gel chromatography. The permeation of molecules into the gel structure depends solely on the size ratio of molecules to pores; other factors of separation, for instance the interactions between the molecules and the solvent or the molecules and the gel structure, are of little significance, sometimes negligible. Over and above these principles, whose details are open to debate, it seems that a clear distinction between gel permeation chromatography, the organophilic test conditions and the relevant instrumental analysis is the most practical idea. There is no doubt, however, that the production of polystyrene gels (and the merits of their manufacturers) had a considerable influence.

The first worker to try to separate polystyrene homologues on a polystyrene matrix copolymerized by divinylbenzene was Vaughan (1960). He later extended the range of fractionation of the crosslinked structure to molecular weights of several hundred thousands (Vaughan 1962). Cortis-Jones (1961) and Langhammer and Quitzsch (1961) proved the efficiency of polystyrene gels in adsorption and partition chromatography. Soon after, the Dow Chemical Co. began the production of polystyrene particles with a 2% divinylbenzene content, suitable for the separation of lipids and oligophenylenes (Tipton et al. 1964; Heitz et al. 1966). The systematic examination and extensive application of polystyrene gels started after the publications of Moore and Hendrickson (1965). It was Moore (1964) who proposed the use of the term "gel permeation chromatography", and went into great detail on the formation of polystyrene gels. He established that by diluting the reaction compound and by correctly selecting the solvents used for the precipitation of the gel particles — provided that the concentration of divinylbenzene is constant — the pores of the gel can be increased very considerably. According to this system he produced gels for the separation of polystyrene fractions of molecular weights between 1 000 and 8 million. During polymerization he used diethylbenzene, dodecane obtained and isoamyl alcohol for precipitation of the polymer.

More recently Makarova and Jegorov (1970), Weiss et al. (1971) and Dawkins and Hemming (1975) have studied the porous polystyrene gels obtained by anionic polymerization.

In spite of their high degree of porosity polystyrene gels are mechanically stable. On removal of the solvent their structure does not collapse completely. The polystyrene divinylbenzene matrices may be regarded as hybrids of the xerogels and aerogels. The solvent content of the gels can be easily exchanged, sometimes without the need to repack the column, by passing new solvent through the column.

Two types of polystyrene-divinylbenzene gels are known. The ones produced by the Dow Chemical Co. are marketed by Waters Associates Inc., Massachusetts, U.S.A., under the trade name of Styragel. For the properties of the eleven Styragel types, of different pore sizes, see *Table 7*. The Styragel gels are sold in the swollen state, suspended in diethylbenzene. Care should

Table 7

Chromatographic properties of the Styragel
(polystyrene-divinylbenzene) gel types*

Type of pore size, Å	Fractionation range (average MW for vinyl polymers)	Exclusion limit (MW)	Particle size, µm
60	800	1 600	50—80
100	2 000	4 000	50—80
400	8 000	16 000	50—80
1 000	20 000	40 000	50—80
5 000	100 000	200 000	50—80
10 000	200 000	400 000	50—80
30 000	600 000	1.2 million	50—80
100 000	2 million	4 million	50—80
300 000	6 million	12 million	50—80
500 000	10 million	20 million	50—80
1 million	20 million	40 million	50—80

* From manufacturer's brochure: Waters Associates Inc., Framingham, Mass., 01701 U.S.A.

Table 8

Properties of the Bio-Beads S polystyrene gels*

Type	Particle size, µm	Bed volume, V_d, ml g^{-1}	Fractionation range (MW)
Bio-Beads S-X1	40— 75	5.8	600—14 000
Bio-Beads S-X2	40— 75	6.2	100— 2 700
Bio-Beads S-X3	40— 75	5.1	up to 2 000
Bio-Beads S-X4	40— 75	4.2	up to 1 400
Bio-Beads S-X8	40— 75	3.9	up to 1 000
Bio-Beads S-X12	40— 75	2.6	up to 400
Bio-Beads SM-1**	300—500		
Bio-Beads SM-2**	300—500	2.9	600—14 000

* From manufacturer's brochure: Bio-Rad Laboratories, Richmond, California.
** Products swollen in polar solvents and water.

be taken to prevent drying out of the gel particles. Kept at 100 °C for long periods of time, the gel particles become unsuitable for use. As eluants tetrahydrofuran, trichloro-benzene, o-dichloro-benzene, toluene, m-cresol, methylene chloride and dimethylformamide are recommended. More recently Chow (1975) separated dextran fractions in aqueous solution on columns packed with Styragel 10^4, 10^5 and 10^6 Å types.

Another type of polystyrene gel consists of the Bio-Beads S produced by the Bio-Rad Laboratories in Richmond, California. These types of polystyrene gels have a considerably narrower range of fractionation molecular weight.

Two more recent representatives of the Bio-Beads S types (see *Table 8*) are the SM-1 and SM-2 which, according to the manufacturer's information, can be used even in the aqueous phase and feature an exceptionally homogeneous pore size (100 and 90 Å resp.). The Bio-Beads are marketed in the dry state. They swell readily in a variety of organic solvents: benzene, carbon tetrachloride, chloroform, dimethylformamide, chlorinated aromatic and aliphatic compounds, and are applied well to the fractionation of oils and low molecular weight styrene polymers.

1.2.3.6. OTHER COLUMN PACKINGS IN GEL CHROMATOGRAPHY

Using an appropriate method of dispersion, networks with pore sizes suitable for chromatography can be produced, theoretically, from every substance. In addition to the chromatographic media discussed above there are numerous column packings of various structures with favourable properties to choose from; these, however, are applied only in a restricted sphere at present. The majority of porous substances to be mentioned below are so-called aerogels, having mostly a solid, stable spatial structure which does not swell in solvents and is not characterized by a gel consistency, in the classic sense of the term. These aerogels are used mainly in gel permeation chromatography which takes place under unusual conditions (high pressures, elevated temperatures, in organic solvents).

Porous glass beads, equally applicable in polar and apolar solvents, provide a most interesting example for separation by molecular size. They are produced from (alkaline) borosilicate glass by heat treatment and sintering. As a result of this method homogeneous, evenly distributed, pore sizes are obtained, variable between wide limits. The pores consist of the interstices, channels and voids, of variable geometry, between the glass particles adhering one to another.

The first observations on glass particles came from Pedersen (1962) who found that in columns packed with very fine (10 to 30 μm) glass beads the larger protein molecules migrated faster than the smaller ones. The explanation for this phenomenon lies in the fact that small molecules can pass even in the liquid spaces of the smaller pores. Pore sizes, in addition to the separation of molecules of different diameters, allow also the determination of the effective molecular radii and of the fractionation of cells and cellular organella. Haller (1965) fractionated viruses on porous glass beads by using differences in their shape.

Although the phenomena observed on porous glass particles are characteristic of gel permeation chromatography, it is likely that in the mechanism of separation, apart from purely spatial effects, the slightly cationic exchange properties of the glass surface also play a role (Zhdanov *et al.* 1970; Cooper *et al.* 1971; Crone *et al.* 1975). The polar effects of the glass surface may be counteracted by a thin silicon coating (Iwama *et al.* 1971), by increasing the pH or the ionic strength, or by saturating the active points of adsorption. Protein adsorption can likewise be prevented by the polyethylene oxide coating (Hiatt *et al.* 1971) of the glass particles or by the use of alcoholic eluants or amino acid-containing buffers (Mizutani and Mizutani 1975, 1976). Since during the chromatography of apolar compounds in organic solvents the polar properties of the glass do not have an effect

(Moore 1966), porous glass particles can be used to advantage in gel permeation chromatography for high molecular weight polymers (Cantow and Johnson 1967; Ross and Casto 1968; Basedow *et al.* 1976). Barral and Cain (1968) used porous glass particles in their structural chemistry investigations.

Glass beads are ideally suited for column packing. Their high chemical resistance (only hydrogen fluoride and very strong alkalis would harm them) enable easy cleaning (e.g. with hot nitric acid or chromosulphuric acid). The particles also show extremely high mechanical stability which makes high flow velocities possible under high pressure. The material and the pore sizes are independent of temperature and enable chromatography to be carried out even at high temperatures. Porous glass beads form a homogeneous and easily reproducible packing; they do not compact during the process and make repacking and frequent calibration unnecessary. These properties are particularly suited for gel permeation and gas chromatography. Furthermore, the extreme pore sizes offer good prospects for the separation of natural macromolecules, cell fragments, etc. in the aqueous phase.

Using chaotropic buffers (KSCN), glass becomes suitable even for the purification of proteins by adsorption chromatography (Bock 1976). Glass is used in a steadily expanding field in the gel chromatography of polysaccharides (Dintzis and Tobin 1974) and of protein detergent complexes (Collins and Haller 1973; Frenkel and Blagrove 1975). Haller, for thirteen proteins, constructed selectivity curves of glass beads with different pore sizes and published an empirical equation for the correlation of the log protein molecular weight versus the elution coefficient (Haller 1973).

Table 9 shows the types and main particulars of the commercially available glass particles, together with their controlled pore sizes.

The pore diameters of the Bio-Glas products range between 200 and 2500 Å. Accordingly they are suitable for the fractionation of molecular weights between 3000 and 9 million, provided that globular beads (polystyrene) are used. The glass beads are sold in four sizes. The larger ones (150 to 300 μm) for column chromatography, the medium sizes (50 to 150 μm) for gas chromatography, beads less than 50 μm for thin layer chromatography. The specific solvent uptake of the Bio-Glas types is between 0.5 and 0.6 ml/g. The void space amounts to approximately 70% of the total gel bed volume, of which 40% is the outer volume between the glass beads and 30% the so-called inner volume of the beads bounded by the different pore sizes. The volumetric correlations of gel chromatography, to be discussed later, hold true also for the porous glass beads (see Section 2 of Part I).

Similar porous glass beads are marketed under the trade name of Controlled Porosity Glass (CPG) by the Waters Associates Inc., and the Corning Glass Works (see Haller 1965). Also these beads stand out by their extreme homogeneity and their well-defined pore sizes. They can be used with equal efficiency in polar and apolar solvents. Uniform pore sizes increase the selectivity of the column in the resolution equation (see Section 2.6.3. of Part I) whether used for their high resolution or for their sharper separation (Bombaugh 1971). However, when mixtures of highly heterogeneous molecular weight distribution are to be separated it is advisable

Table 9

Properties of porous glass particles used in column chromatography*

Type	Average pore size, Å	Fractionation (exclusion) range (MW)	
		Dextran	Polystyrene**
Bio-Glas 200	200	—	300 — 30 000
Bio-Glas 500	500	—	10 000 — 100 000
Bio-Glas 1000	1 000	—	50 000 — 500 000
Bio-Glas 1500	1 500	—	400 000 — 2 million
Bio-Glas 2000	2 000	—	800 000 — 9 million
CPG 10— 75	75	28 000	—
CPG 10— 125	125	48 000	—
CPG 10— 175	175	68 000	—
CPG 10— 240	240	95 000	—
CPG 10— 370	370	150 000	120 000
CPG 10— 700	700	300 000	400 000
CPG 10—1250	1250	550 000	1.2 million
CPG 10—2000	2000	1.2 million	12 million

* Bio-Glas Beads: Bio-Rad Laboratories, Richmond, California 94802; CPG (Controlled Porosity Glass): Waters Associates Inc., Framingham, Mass. 01701; and Corning Glass Works, Corning, N.Y., 14830. Data given according to manufacturers' brochures and Kirkland (1971, p. 259).
** In toluene.

to use beads of widely different porosities, which makes the above quoted advantages irrelevant.

The chemical composition of the high-porosity silicate particles produced by a special process (Santocel, Porasil, Merck-o-gel Si, Spherosil, Silochrom) is similar to that of the glass beads, but their adsorptive capacities, particularly if polar compounds are concerned, are better defined. One of their shortcomings is that at high flow rates they give rise to a greater zone broadening. The adsorptive effects are suppressed by additives (suppressors), for instance triethylene glycol, or chemical deactivation of the silicate particles (Bombaugh 1969a, b, c).

Silicate particles have found wide application in the field of analytical polymer chemistry (Eltekov and Nazansky 1976). Vaughan (1962) separated polystyrene fractions dissolved in benzene on porous silica gel. De Vries et al. (1967) studied silica beads sold under the trade name of Porasil. Oster et al. (1971) used Merck-o-gel Si types for gel permeation chromatography. Pacco (1971) found that the chromatographic properties of Styragel and Porasil are approximately equal.

Table 10 shows some properties of the commercially available porous silicate particles.

Spherosil particles are made in three sizes (100 to 200, 100 to 150 and 150 to 200 μm), their specific volume after solvent uptake being 1.8 ml/g.

Porous aluminium oxide gels were also found to exhibit chromatographic properties based on the separation of molecular sizes (Baba et al. 1969; Baba and Sato 1971; Sato and Otaka 1971). Kaiser (1970) experienced the same with carbon particles.

Table 10

Properties of porous silica gels*

Type	Pore size (diameter), Å	Molecular weight exclusion limits for polystyrene fraction
Porasil- 60	—	60 000
Porasil- 250	—	250 000
Porasil- 400	—	400 000
Porasil-1000	—	1 million
Porasil-1500	—	1.5 million
Porasil-2000	—	2 million
Merck-o-gel Si-150	150	50 000
Merck-o-gel Si-500	500	400 000
Merck-o-gel Si-1000	1000	1 million
Spherosil A	below 100	—
Spherosil B	100—200	—
Spherosil C	200—400	—
Spherosil D	400—800	—
Spherosil E	800—1500	—
Spherosil F	above 1500	—
Silochrom C-1	400	2 million**
Silochrom C-2	500	2 million**
Silochrom C-3	500	2 million**
Silochrom C-4	1100	6 million
Silochrom C-5	1600	20 million

* From manufacturers' catalogues and Bombaugh (1971).
Porasil: Waters Associates Inc., Framingham, Mass. 01701.
Merck-o-gel: Merck AG., Darmstadt, Federal Republic of Germany.
Spherosil: Péchiney-Saint-Gobain, France. (Marketed by Serva Feinbiochemica GmbH, Heidelberg, Federal Republic of Germany.)
Silochrom: Aerosil, Degussa, Rheinfelden (Baden).
** From Eltekov and Nazansky (1976).

A more recent trend in the examination of organophilic gel structure is the production of polymers with a phenol-resitol network (Udvarhelyi 1969). Using phenol-resitol matrices, phenolnovolak resin fractions, between the molecular weights of 500 and 2000, could be separated.

Heitz et al. (1966) first produced organophilic gels by copolymerization of vinyl acetate and butanediol-divinylether, and subsequently by hydrolysis of ester groups. The polyvinyl acetate gels are sold under the name of Merck-o-gel-OR and are especially suited for the fractionation of hydrocarbons (petroleum) (Oelert 1970).

Further column packings used in gel permeation chromatography, particularly for the separation of low molecular weight compounds, are the ethylvinylbenzene-based organophilic gels manufactured by the Waters Associates Inc. and sold as Porapak P, Q, R, S, and T (150 to 200 mesh) (Heitz 1970a, b).

The application of so-called mixed gels proved to be a promising new initiative even in affinity chromatography (Doley et al. 1976). On the basis of work by Uriel et al. the French L'Industrie Biologique Française (INDUBIO, Gennevilliers) put gel beads on the market consisting of polyacrylamide and agarose under the brand name of Indubiose AcA (Uriel et al. 1971; Boschetti et al. 1972). Agarose polyacrylamide bead polymerized gels are supplied currently by the LKB (Ultrogel AcA) (Jefferis 1975; LKB Ultrogel, Pre-swollen polyacrylamide/agarose gel beads for high-speed gel filtration, Stockholm 1975). *Table 11* indicates some of their more important parameters. The numerals refer to the polyacrylamide and agarose content of the gels, respectively.

Table 11

Properties of the LKB Ultrogel polyacrylamide/agarose gels*

Type	Acrylamide content %	Agarose content %	Fractionation range for globular proteins (MW)	Max. flow rate ml cm^{-2} hour
Ultrogel AcA 54	5	4	6 000 — 70 000	50
Ultrogel AcA 44	4	4	12 000 — 130 000	45
Ultrogel AcA 34	3	4	20 000 — 400 000	40
Ultrogel AcA 22	2	2	60 000 — 1 million	18

* From manufacturer's brochure: LKB Ultrogel Pre-swollen polyacrylamide/agarose gel beads for high-speed gel filtration, Stockholm, 1975.

In accordance with the properties of their constituents the polyacrylamide-agarose mixed gel beads are hydrophilic and swell readily. In the structure of the beads the polyacrylamide network is filled out by agarose. In low concentration (e.g. 2%) the acrylamide or agarose gels are extremely plastic, soft and difficult to handle. Not only are the mechanical characteristics of the combination of the two types better but they extend the range of the separation molecular weight of polyacrylamide up to approximately 1 million. Evidently the pore size of the polyacrylamide/agarose gel particles is determined by the structure of the polyacrylamide, bearing in mind that pure 2% agarose gels are otherwise capable of separating substances with molecular weights of even 20 to 25 million (see *Table 1*).

Figure 9 shows the selectivity curves of Ultrogel AcA gels. The slope of the curves, i.e. the range of separation molecular weight of the given type of gel — particularly if particles with higher polyacrylamide content (5%) are concerned — can be controlled to a certain extent by varying the agarose content. The reduction of the agarose concentration, compared for example with the gels 5—4 and 5—2, raises the exclusion molecular weight to 200 000 (in the case of 5—4 to 80 000; Boschetti et al., 1972, 1974). The Ultrogel AcA gel beads are spheroids of homogeneous distribution (100 to 160 µm) which, if used for the packing of chromatographic columns, allow an adequate rate of flow (approximately 10 to 20 ml/h in a column of 2×40 cm, total volume 150 ml). Owing to their agarose content the polyacrylamide/agarose gels must be used between the temperatures of 2 and 36 °C. They are sensitive to heat and to reagents prone to destroy the hydrogen bonds (concentrated urea for instance), as well as to bacterial effects. Their chemical resistance (between pH 3 and 11) is adequate. They show minimal adsorptivity. Using the calibrating proteins included in *Figure 9* the Ultrogel AcA gels can be used also for molecular weight determinations. In the recommended ranges of molecular weight the selectivity curves are practically linear. Among the proteins, lysozyme is an exception insofar as it has a higher affinity to agarose and elutes with a greater volume than would be expected.

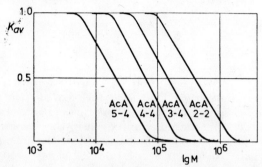

Fig. 9. The selectivity curves of the Ultrogel polyacrylamide-agarose mixed gels. The proteins used for plotting the curves (and their molecular weights) were: cytochrome C (12 400), myoglobin (17 000), ovomucoid (28 000), bovine albumin (65 000), γ-globulin (150 000), catalase (230 000), fibrinogen (340 000), ferritin (550 000), thyroglobulin (670 000) and Blue Dextran 2000 (>2 000 000)

2. THEORY

Looking back in restrospect upon the hardly ten years history of gel chromatography we may state that the evolution of its theory was influenced by several trends. Right from the outset, it applied the same techniques as used in column chromatography and even in the theoretical analogies it set out from the same methods. The basis of this theory had been the idea that the porous gel structure is of a formal geometry, and the assumption that the separation mechanism is similar to that of the liquid-liquid chromatographic methods. The separation processes which could readily be followed by, and were retraceable to, volumetric measurement yielded volumetric equations which were easy to apply in practice. These trends, including the concept of gel permeation chromatography which came to be developed later, are known by the collective term of steric-volumetric theory (Gelotte 1960; Flodin 1961, 1962; Porath 1962a, b, 1963; Squire 1964; Laurent and Killander 1964; Moore 1964; Altgelt 1970; Casassa 1971; Cameron 1971).

The second trend in the development of the theory concentrated on the examination of the molecular-kinetic phenomena of the separation processes. Investigation into the altered diffusion of the macromolecules and the colloidal-colloidal interactions in the gel network led to a new interpretation of the separation mechanism of gel chromatography (Polson 1961; Ackers 1964; Smith and Kollmannsberger 1965; Siegel and Monty 1965, 1966; Takagi 1966; Yau and Malone 1967; Di Marzio and Guttmann 1969).

Parallel with the development of the chromatographic methods and kinetic theories the examination of the hydrodynamic parameters of the gel columns received increasing attention. This, in turn, enabled a more exact formulation of the separation processes and the determination of the optimum experimental conditions of gel chromatography to be carried out (Flodin 1961; Giddings and Mallik 1966; Giddings 1967; Ackers 1970; Kelley and Billmeyer 1970; Kirkland 1971a; Snyder 1972a; Vink 1970, 1972).

Those publications which treat the processes taking place at the interface of the gel phase from the aspects of thermodynamics (osmotic pressure, entropy, etc.) represent the latest trend in research into the theory of gel chromatography (Casassa 1967; Edmond et al. 1968; Edmond and Ogston 1968, 1970; Polson and Katz 1969; Nichol et al. 1969; Ogston and Silpananta 1970; Hjertén 1970).

The realization that colloidal systems, inclusive of the gel network, cause changes in one another's solubility (Laurent 1963a, b) and thereby in the virtual volumes available to the molecules (Albertsson 1960, 1970a; Johansson 1970a, b) is closely related to the thermodynamical approach. This interpretation, at a higher level, means that recourse is taken to the principle of liquid-liquid distribution chromatography and allows thermodynamic calculation to be applied to gel chromatographic phenomena (Nichol *et al.* 1969).

The point of departure for the theory of gel chromatography is a system consisting of gel particles surrounded by the solvent volume — called the gel bed. To clarify the theoretical interrelationships the constituents and properties of the gel bed will be examined first.

2.1. THE VOLUMETRIC DISTRIBUTION OF THE GEL BED

To use gels in chromatography, gel particles must be produced. Initially the gel particles were obtained by granulation and crushing of swollen or dry (xerogel) substances.

This process, which is no longer in use, yields broken grains, heterogeneous both in size and shape. The interstices and passages between grains cause different rates of flow and turbulence in the cross-section of the column. Irregular material transport or mixing impair the sharpness of separation. The production of spherical, so-called bead-polymerized (or condensed — see agarose) gel particles solved the problem to some extent. However, eddy diffusions may occur even between spherical particles and this is one of the factors which set limits to the resolution of gel chromatography (see Section 2.6.1.5 of Part I).

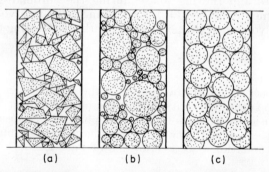

Fig. 10. The schematic structure of columns built up of
(a) broken particles, (b) heterogeneously and (c) homogeneously distributed spheroid gel particles

Figure 10 shows the schematic pattern of a gel column built up of (a) broken particles, (b) heterogeneously and (c) homogeneously distributed bead polymers. This proves that even with spheroid particles the homogeneous distribution of the sizes is of considerable importance. It will be evident that only particles of fairly equal size and spherical in shape will provide for uniform interstices. Broken particles will never achieve the same effect.

The gel particles divide the total bed volume into two distinct parts, as seen in *Figure 11*.

One of the volumes (the stationary phase of gel chromatography) consists of the solvent molecules pertaining to the gel structure. This is the so-called inner volume of the bed (V_i). The characteristics of the inner volume and its

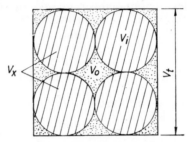

Fig. 11. The volumetric distribution of the gel bed; V_t denotes the total bed volume, V_o the outer volume of the interstices, V_i the inner volume of the gel particles, V_x the volume of the polymer network (matrix)

accessibility have already been dealt with in the section on gel structure and pore sizes; this will be reexamined in more detail in connection with the molecular-kinetic theories of the separation mechanism of gel chromatography. The other volume of the gel bed (the mobile or flowing phase of gel chromatography) consists of the interstitial or outer volume (V_o) between the gel particles and flowing phase during the chromatographic process. The total volume of the bed (V_t) is the sum of the two volumes, with the addition of the volume of the polymer network constituting the gel structure (matrix) (V_x)

$$V_t = V_i + V_o + V_x$$

In gels of looser structure the volume of the gel matrix is negligibly small compared to the other elements, but in the more compact gel types it may represent a considerable portion of the gel phase. In theoretical derivations the volume of the polymer network is considered by distribution coefficients of different formulations (see Section 2.3 of Part I).

2.1.1. THE GEOMETRIC MODEL OF THE VOLUMETRIC DISTRIBUTION OF THE GEL BED

The volumetric distribution of the gel bed, in other words the capacity of the gel column V_i/V_o (see Section 2.6.1.3 of Part I), is of considerable importance for the resolution of gel chromatography. It will be clear from *Figure 10* that in addition to the shape of the particles the volumetric distribution of the bed depends also on the steric orientation of the particles. Experience has shown that with homogeneous particle size the ratio of the inner to the outer volume is independent of the particle size, the gel material and the pore size. In what follows, theoretical derivations will assume that the gel particles are spheroids which are identical in size (homodisperse), rigid, and of stable form. These conditions are rarely met in practice, owing to the statistical distribution of the sizes and, if softer gels are concerned, to the deformation of the particles.

The problem of the volumetric distribution of the gel bed can be approximated on the basis of the random orientation of rigid homodisperse spheroids. A similar geometric model was proposed by Bernal in 1959 for the examination of the structure of fluids. It had been established earlier that with the closest hexagonal orientation the spheroids fill out the space to

$$\frac{\pi}{6}\sqrt{2} \text{ corresponding to } 74\%$$

(Coxeter 1958; Fejes Tóth 1964, 1972). However, in practice the experimentally determined space utilization of homodisperse solid spheres is considerably lower (between 59 and 65%) (Westmann and Hugill 1930; Rice 1944; Scott 1960; Bernal and Mason 1960; Susskind and Becker 1966). By a simple test the volumetric ratio of the space utilization of spheroids can be empirically established. Homodisperse glass or steel beads (bearing rollers) are packed into a 100-ml measuring cylinder and compacted by shaking or rapping. The voids are then filled up with water from a burette. The volume thus obtained is between 34 and 38 ml, which means that the volume occupied by the spheres is 62 to 66 ml. Scott (1960) found that, depending on the method of "packing", with homodisperse rigid spheroids a looser (59%) and a denser (63%) spatial structure can be achieved, in a reproducible way, by mechanical packing. Bernal and Mason (1960) used an ingenious method experimentally to determine the frequency and distribution of the spheres' points of contact (the coordination number). They established that in the looser structure each sphere contacts 6 to 8 others, in the denser one each contacts 7 to 9 spheres.

Figure 12 illustrates the distribution frequency of the points of contact or coordination numbers. An empirical examination of the coordination numbers proved that orientation may take place at least at five, and theoretically at a maximum of 12 points. According to experience the closest hexagonal of steric orientation (coordination 12) is rarely achieved in practice, and the same holds true for contact at 11 points. It should be noted here that Bernal and Mason did not take into consideration the effect of the walls of the reservoir in the determination of the coordination numbers, only of the internal regions of the heap of spheroids. In experiments performed with columns having different, sometimes flexible, walls (Susskind and Becker 1966) hexagonal orientation was observed on the walls, which favours higher coordination numbers.

It will be obvious from the above observations that the coordination numbers determined by Bernal and Mason point to the possibility of steric orientation of spheroids and that their distribution is closely related to the empirically measured values. The most frequently encountered coordination numbers and the possibility of steric orientation of spheroids suggested the idea that the volumetric distribution of the gel bed can be calculated from geometric models constructed according to given coordination numbers. For calculation purposes the geometric models were arbitrarily chosen on the basis of two criteria: (1) the steric orientation pertaining to the coordination numbers is a regular symmetic geometric model of the densest orientation of

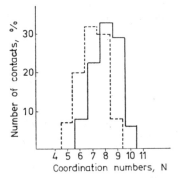

Fig. 12. The frequency of the number of contacts (coordination numbers) of homodisperse spheroids for looser (- - -) and denser (—) steric orientation (Bernal and Mason 1960)

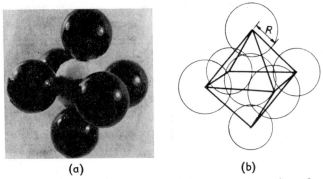

(a) (b)

Fig. 13. The geometric model (octahedron) of the steric orientation of a sphere with six points of contact

the spheroids, and (2) the geometric model pertaining to a given coordination number is a congruent volumetric unit whose repetition fills out the space continuously.

On the basis of these criteria two models may be used to calculate the volumetric distribution of the gel bed.

The steric orientation of the spheres with six contacting points can be illustrated by an octahedron (*Figs 13 a* and *b*) with spheres of radius R in its centre and its apices. The volumetric distribution pertaining to a coordination of 6 can be calculated from the volume of the octahedron and the spheres therein (spherical sections) in the following way.

The volume of the octahedron is

$$V_{oct} = 12 R^3$$

The total volume of the sphere (considering that in the volume of the octahedron one-twelfth of the spheres located on the apices take part) is

$$V_g = 2\pi R^3$$

The ratio of the volumes is

$$\frac{V_g}{V_{oct}} = \frac{\pi}{6} = 0.523$$

Accordingly, spheres oriented with the coordination number of 6 will occupy 52.3% of the space.

Figures 14 a and *b* show the regular steric orientation of nine spheres of radius R (coordination number 8). The geometric model pertaining to 8 gives the network of the hexahedron from *Figures 14a* and *b*:

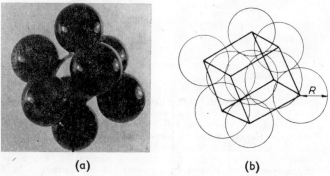

(a)   (b)

Fig. 14. The geometric model of spheres with eight points of contact (hexahedron)

$$\frac{4R}{\sqrt{3}}$$

whose volume is

$$V_{hex} = \left(\frac{4R}{\sqrt{3}}\right)^3$$

53

The volume of the spheres (one eighth of the spheres on the apices participating in the volume of the hexahedron) is

$$V_g = \frac{8\pi}{\sqrt{3}} R^3$$

and the ratio of the volumes is

$$\frac{V_g}{V_{hex}} = \frac{\pi(\sqrt{3})^3}{24} = 0.680$$

which shows that with a coordination number of 8 the spheroids fill out the space to 68%.

Illustrating the foregoing and the value of the densest hexagonal orientation (0.740) as a function of the coordination numbers according to *Figure 15*, the assumed volumetric ratio of the remaining coordination numbers

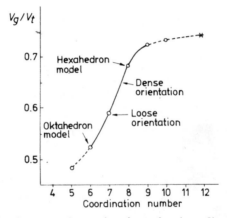

Fig. 15. The correlation between the steric orientation (coordination number) and the space utilization of homodisperse rigid spheroids (V/V_t, where V_g denotes the volume of the spheres and V_t the total volume)

The values pertaining to the coordination numbers 6 and 8 were derived by calculation from the octahedron and hexahedron models, respectively (see text). The figure marked with an asterisk shows the closest hexagonal orientation (coordination number 12) and space utilization as calculated by Coxeter (1958), namely 0.74. Close orientation: calculated = 0.649; measured 0.63 to 0.66. Loose orientation: calculated = 0.588; measured = 0.59

can be determined by extrapolation of the curve. From the frequency of the coordination numbers as determined by Bernal and Mason (1960) (see *Fig. 12*), the volumetric ratio of loose and dense steric orientations can be calculated. Weighting the volumetric ratios (V_g/V_t) pertaining to the coordination numbers according to their frequency, 58.8% is obtained for loose spheroid sets and 64.9% for the denser heaps. These values are in fair agreement with the ratios experimentally determined by Bernal and Mason (1960) and Scott (1960) and with those established by Susskind and Becker (1966), i.e. 59 and 63 to 65%, respectively. In practice the space utilization of the gel particles is somewhat higher (between 65 and 76%) than the value measured with the denser orientation of rigid spheres (63—65%) because both the heterogeneity of the grain sizes and their deformation enhance space utilization.

Concerning the mathematical derivations, stress should be laid upon the arbitrary assumption that the coordination numbers 6 and 8 correspond to octahedron and hexahedron models, respectively. However, the coincidence of the values calculated on the basis of this hypothesis and the empirical values tends to prove that the random orientation of spheroids produces groups which correspond to the steric orientation of the hexahedron and octahedron models. To verify this, further examinations are necessary. However, the geometric model of the volumetric distribution of the gel bed, supported by experience, indicates clearly that the shape, size and localization of the gel particles and the method of packing the column constitute preconditions for the success of gel chromatography, from both the theoretical and practical aspects.

2.2. THE STERIC-VOLUMETRIC THEORY OF GEL CHROMATOGRAPHY

The volumetric equations which were the first theoretical relationships of gel chromatography set out from the observation that, from columns consisting of gel particles, different substances elute in different volumes in the sequence of their molecular weights. The separation process is schematically illustrated in *Figure 16*.

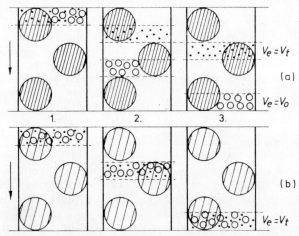

Fig. 16. Schematic illustration of the three separation processes of gel chromatography with (a) microporous and (b) macroporous gels

From among the solutes applied to the gel bed the larger ones (proteins, polysaccharides, etc.) occupy the voids between the gel particles, i.e. the so-called outer volume of the bed. Having passed through the bed they appear in the column eluate.

The smaller molecules (salts, monosaccharides, etc.) on the other hand can permeate into the inner volume of the gel particles and to elute them an amount of solution corresponding approximately to one complete volume of the column must be passed through. The elution properties of medium size molecules, which take up only part of the volume of gel particles, fall between these two extremes. According to the steric theory of gel chromatography, which is in harmony with experience, the gel structure and the pore sizes produced by the crosslinking of the polymer chains play a decisive role in the molecular separation. As evidenced by *Figure 16b*, if the pore sizes permit the diffusion of even larger molecules into the gel structure no separation will take place at all.

2.3. THE TERMINOLOGY OF GEL CHROMATOGRAPHY

Before going into the derivation of volumetric equations it is necessary to survey briefly the symbols and designations used in the theory and practice of gel chromatography.

(A) The volumetric parameters of the gel bed

V_t denotes the total volume of the gel bed and V_o the outer volume of the bed, the sum of the volumetric elements between the gel particles based on empirical and theoretical calculations (see Section 2.1 of Part I). The outer volume amounts to approximately 1/3 of that of the gel bed: $V_o = 0.33\ V_t$.

V_i denotes the inner volume of the gel bed. In the derivation of the volumetric equations (see the definition of K_d in Section 2.3) the total volume of the gel particles is regarded as equal to the maximum available in the gel phase (by electrolytes). If loose, readily swelling, large-pored gels with low dry-matter content (e.g. hydrophilic dextran gels) are concerned, the difference caused by considering the non-swelling volumetric elements rather than the true volume of the matrix is negligibly small.

(B) The volumetric parameters of the solutes

V_e denotes the elution volume, i.e. the quantity of solvent passing through the gel bed up to the appearance of the separated substance. This depends on the volumetric distribution of the gel bed, the quality of the gel and the properties of the separated substance.

$(V_e - V_o)$ denotes the net elution volume or the retention volume (V_N is the net retention volume). This is the volume occupied by the solute in the gel phase. If the gel particles do not retain the solute molecules ($K_d = 0$)

then the retention volume is equal to zero and the substance will elute in the outer volume (V_o) of the column. If on the other hand the molecules can occupy the full volume of the gel phase ($K_d = 1$) then the retention volume is equal to the total available volume of the gel phase (V_i).

V_{sep}, the separation volume, is equal to the difference between the elution volumes (see *Fig. 16*), $V_{sep} = V_{e2} - V_{e1}$. Substituting the relationship derived for the elution volume

$$V_{sep} = (V_o + K_{d2}V_i) - (V_o + K_{d1}V_i) = (K_{d2} - K_{d1})V_i$$

Accordingly the separation volume can be computed from the inner volume of the gel bed and the difference between the volumetric distribution coefficients.

V_s denotes the volume of the sample, i.e. the volume of the substances applied to the column during chromatography. In the case of optimum separation this depends on the separation volume. Obviously, two substances can only then be separated if the sample volume is equal to or less than the separation volume, i.e.

$$V_s = V_{sep} \quad V_s = (K_{d2} - K_{d1})V_i$$

This proves that the smaller the inner bed volume and the smaller the difference between the coefficients of volumetric distribution, the smaller is the sample to be used. If the difference between the distribution coefficients is considerable ($K_{d1} = 0, K_{d2} = 1$, as is the case in desalting) then the sample volume may approximate the inner volume of the gel bed. Experience has shown that best separation can be achieved if the volume of the sample is approximately one-third of the inner volume of the bed, i.e. if $V_s = 0.3\, V_i$.

Summing up the above, to separate perfectly two substances, the difference between the distribution coefficients must be at least 0.3. The sample volume has a role in the exact determination of the elution volumes and is related to the internal volume of the gel bed.

Figure 17 shows the interpretation of the elution volume when the volume of the sample is (a) negligibly small, (b) comparable to, or (c) very much larger than the volume of the column.

V_e/V_t is the relative elution volume, i.e. the elution volume referred to the outer volume of the bed. It is independent of the column size and is a parameter of the behaviour of the solutes, which can be easily calculated from the measurement findings. From its reciprocal value, the so-called retention constant (R), the relative rate of migration of the separated substances, is characterized in gel phase thin layer chromatography. Owing to the volumetric distribution of the bed (V_o), the relative elution volume depends on the method of "packing" the gel grains and the hydrostatic pressure acting on the column.

V_e/V_t is the elution volume referred to the complete volume of the gel bed. As before, it is a parameter of the chromatographic properties and is independent of the column geometry. Since V_t can be measured more exactly than V_o (V_o is measured by the chromatography of a large-molecule substance, V_t is obtained by simple volumetry), and also V_e/V_t is more exact. The column packing and the pressure difference have little effect on the value of V_e/V_t.

$V_e - V_o/V_o$ is the capacity factor of the gel bed (column) k' (see Section 2.6.1.3 of Part I for more details). It is the retention volume of the outer

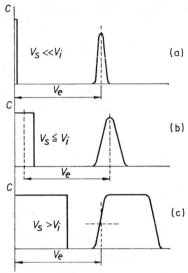

Fig. 17. The interpretation of the elution volume if the volume of the sample is (a) very much smaller, (b) comparable with, or (c) larger than the inner volume of the bed

column volume referred to unity. If the substances occupy the entire inner gel volume ($K_d = 1$, $V_e - V_o = V_i$) it denotes the capacity of the column.

K_d denotes the volumetric distribution coefficient. For its derivation see Section 2.4 of Part I. According to its definition, it occupies only that portion of the inner volume of the gel grains which is available to the molecules.

K_{av} is a more rational definition of the volumetric distribution coefficient. Laurent and Killander (1964) derived the concept of distribution coefficient from the total volume of the gel phase, including also the volume of the matrix: $(V_t - V_o) = (V_i + V_x)$

$$K_{av} = \frac{V_e - V_o}{V_t - V_o}$$

This term eliminates the errors caused in the calculation or indirect measurement of the inner gel volume, in view of the fact that if certain aerogel types (e.g. polystyrene gels) are used, not every volumetric element of the pores will participate in the solvent uptake. These volumes (although they form part of the total gel volume) do not take part in the process of chromatography. Generally, the lower the dry-matter content of the gel the smaller is the difference between K_d and K_{av}.

The correlation of the differently defined relative elution volumes and volumetric distribution coefficients are illustrated in *Figure 18*.

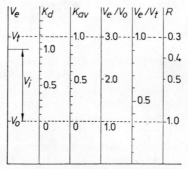

Fig. 18. The correlation between the volumetric parameters of the gel bed and those of the solutes

(C) The parameters of the gel properties

S_r, W_r denote the solvent and water regain respectively, in other words, the capacity for taking up solvent or water. The maximum amount of solvent, in ml, taken up by one gram of xerogel is a parameter which characterizes the physical properties of gels (polarity) and their structure (porosity, number of crosslinks). The solvent regain of xerogels can be determined from the difference between the weights of the dry and the swollen gel. The specific swelling volume (bed volume) is the total volume of the gel bed after the swelling of 1 g of xerogel, the portion of the total gel volume referred to 1 g of xerogel is termed the specific bed volume.

Γ M_{lim} denotes the average exclusion limit of the molecular weight — the highest average molecular weight at which the molecules become incapable of diffusion into the inner volume of a given gel. The exclusion molecular weight depends not only on the molecular structure (globular, fibrillar) but also on the physical properties of the molecules (proteins, carbohydrates — see Section 1.2 of Part I). The particle size of the gel grains is determined by the appropriate mesh number of the standard series of sieves, or by the grain sizes expressed in terms of micrometers (μm). Grain size generally refers to xerogels, less frequently to wet, swollen gel grains;

ϱ denotes the density of the solvent, in terms of g/ml;
d denotes the density of the swollen gel grains, in terms of g/ml.

2.4. THE VOLUMETRIC EQUATIONS OF GEL CHROMATOGRAPHY

Gelotte (1960), Flodin (1961, 1962) and Porath (1962a), on the basis of volumetric theories, stated that gel chromatography should be regarded as a special boundary case of partition chromatography in which, in analogy to immiscible liquids, both the stationary and the flowing phase consist of the same solvent (water, if hydrophilic gels are concerned).

The introduction of the principle of volumetric distribution was favoured by the fact that if the gel particles are in equilibrium with their environment then the concentration of the solutes in the inner and outer volumes is equal. That is, the distribution is proportional to the volumes.

In line with the volumetric distribution of the gel bed, the elution volume of solutes (V_e), in the ideal case, must not be less than the outer (V_o), nor larger than the total, volume of the column. Owing to the cumbersome nature of the measurement of matrix volume, the total volume of the solvent in the column was regarded as equal to the sum of the outer and inner volumes, i.e. $V_t = V_o + V_i$.

If in the course of column chromatography, a substance eluting with a volume of V_e occupies a volume of $V_e - V_o$ in the gel particles, the distribution constant can be written in the form

$$K_d = \frac{V_e - V_o}{V_i}$$

where K_d is the so-called volumetric distribution coefficient.

Expressing the elution volume as

$$V_e = V_o + K_d V_i$$

the basic equation of the volumetric theory can be derived.

The relationship between the elution volume and the volumetric parameters of the gel bed is illustrated by the so-called elution curves in *Figure 19*.

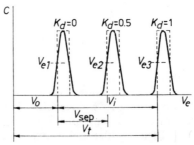

Fig. 19. The correlation between the elution volume (V_e) and the volumetric parameters of the gel bed

According to the volumetric equation of gel chromatography, the elution volume of the solutes is made up of the outer volume of the bed and the volume of the gel particles available to the gel grains, in proportion to the volumetric distribution coefficient.

The volumetric distribution coefficient is a physical constant which depends on the molecular size of the solutes (their molecular weight) and the inner gel structure (pore size). Its value may vary between 0 and 1 (see *Figs 17* and *18*).

Table 12

The volumetric distribution coefficient (K_d) of selected compounds of the Sephades G gel types most frequently used*

Compound**	K_d				
	G-25	G-50	G-75	G-100	G-200
Ammonium sulphate	0.9	—	—	—	—
Potassium chloride	1.0	1.0	—	—	—
Tyrosine	1.4	1.1	—	—	—
Phenylalanine	1.2	1.0	—	—	—
Tryptophan	2.2	1.6	1.2	—	—
Glycine	0.9	—	1.0	—	—
Cytochrome C	—	—	0.4	0.7	0.80
Ribonuclease	—	—	0.4	0.6	0.60
Chymotrypsin	—	—	0.3	0.54	0.64
Trypsin	—	—	0.3	0.54	0.62
Haemoglobin	—	—	0.1	0.30	0.58
Egg albumin	—	—	—	0.29	0.52
Bovine albumin	—	—	—	0.20	0.29
Transferrin	—	—	—	0.11	0.32
Glyceraldehyde phosphate dehydrogenase	—	—	—	—	0.31
Aldolase	—	—	—	—	0.27
Human γ-globulin	—	—	—	—	0.26
Catalase	—	—	—	—	0.26
Fibrinogen	—	—	—	—	0.03
Urease	—	—	—	—	0.20
Ferritin	—	—	—	—	0.11
α-Crystalline	—	—	—	—	0.00
Tobacco mosaic virus	—	—	—	—	0.00

* From the literature.
** In order of increasing molecular weight.

Should K_d be equal to 0, i.e. the solute molecules are completely excluded from the inner gel volume, then the elution volume is equal to the outer bed volume, i.e. $V_e = V_o \cdot K_d = 0$ indicates the limit of the exclusion molecular weight (M_{llm}) of the gel structure. *Table 12* shows that this limit depends on the composition and structure of the solutes (for instance, proteins, polysaccharides).

If $K_d = 1$ then the solute molecules may occupy the total inner volume of the gel and the elution volume is equal to the total solvent volume of the column: $V_e = V_o + V_i = V_t$. The value $K_d = 1$ is approximated only by electrolytes and the volumetric distribution coefficient rarely exceeds 0.8—0.9. K_d values higher than 1 indicate that adsorptive or ion exchange processes are associated with the gel chromatographic process.

Table 12 shows the volumetric distribution coefficient of a number of compounds, for the most frequently used Sephadex G gel types. More recently the volumetric coefficient, expressed as the total volume of the gel particles, K_{av} (see Section 2.3 of Part I and *Fig. 18*) was used to characterize the interactions of the gel and the solutes. The advantage of this technique is that it makes the determination of the inner gel volume unnecessary and permits the calculation of K_{av} from the easily measurable volumetric parameters (V_e, V_o, V_t) of chromatography.

In addition to the measurement of the elution volume of electrolytes, the inner volume of the gel particles ($K_d \approx 1$) can be computed also from the physicochemical constants of the gels. The inner volume is proportional to the quantity of the xerogel, a, and to the solvent regain of the gel (S_r, W_r)

$$V_i = \frac{aS_r}{\varrho}$$

where ϱ denotes the density of the solvent.

Regarding hydrophilic gels (ϱ_{H_2O}) this term simplifies to the formula $V_i = aW_r$. Since the quantity of xerogel packed into the column is generally unknown, the inner volume can be calculated also from the constituents of the gel bed volume and the density of the swollen gel

$$V_i = (V_t - V_o) \frac{dS_r}{\varrho \left(\frac{S_r}{\varrho} + 1 \right)}$$

or for aqueous media

$$V_i = (V_t - V_o) \frac{dW_r}{W_r + 1}$$

The density of the swollen gel is usually stated in the manufacturer's catalogue (see, for example, *Table 2* which shows Sephadex G types). This may be established also by picnometry. The density of the swollen gel may be calculated also from the properties of the gel former. According to Laurent and Killander (1964) the volume of the crosslinks of the gel (matrix) (V_m) can be determined by the partial specific volume of the polymer. Granath (1958) proved by examination that the partial specific volume of dextran is 0.61 ml. Making use of this information, Flodin (1962) found the following relationship between the water uptake and the wet density of Sephadex gels:

$$d = \frac{1 + W_r}{0.61 + W_r}$$

One of the most important factors of the volumetric equations is the outer volume of the gel bed. Over and above theoretical derivations (see Section 2.1 of Part I) the outer volume must be checked also by the determination of the elution volume of substances with appropriately high molecular weights, which are excluded from the gel particles. For this purpose proteins, polysaccharides, preferably coloured compounds as for instance haemoglobin, fine-grained China ink and the dextran derivative Blue Dextran 2000 with an average molecular weight of 2 million, marketed by the Pharmacia Fine Chemicals AB, can be used to best advantage. From the practical problems involved in the determination of the outer volume, the methodology and packing of gel columns will not be considered.

2.5. THE GEOMETRIC MODEL OF MOLECULAR EXCLUSION

In the early period of development the volumetric theories started from those relationships of gel chromatography which were easy to follow by volumetric techniques, and disregarded those interactions caused by the structure or geometry of the gels and solutes. A more thorough study of the gel structure led to the idea that the system of crosslinks acts as a physical barrier for molecules of certain defined sizes and these are thus excluded from certain volumetric elements of the gel. To verify this theory several authors have used simple geometric models.

2.5.1. THE PORATH AND SQUIRE EQUATIONS

Porath (1963) regarded the total volumes of the gel particles to be cones with a radius R, the penetration of the solutes into these cones depending on the molecular radius r. He assumed also that the column occupied by the molecules is a cone, and by comparing the volumes he arrived at the following relationship for the coefficient of the volumetric distribution

$$K_d = k \left(1 - \frac{2r}{R}\right)^3$$

in which, apart from the constant k, only the ratio of the two radii occurs. The radius of the gel cone is proportional to the swelling, i.e. to the gel regain capacity $R^3 = kS_r$, while the molecular weight of the macromolecules built up from the sequences is proportional to the square of the radius, i.e. $M = kr^2$. Substituting these relationships into the above equation he obtained

$$K_d = k_1 \left(1 - k_2 \frac{M^{1/2}}{S_r^{1/3}}\right)^3$$

In a simplified form, this equation can be used also in establishing the molecular weight, by determination of the constants a and b:

$$K_d^{1/3} = a - bM^{1/2}$$

Porath (1963) verified the validity of these equations using dextran fractions of known molecular weight and the K_d value.

Squire (1964) further developed Porath's theory by treating the extremely complex heteroplanar volumetric elements of the gels as combinations of molecular-size cones, cylinders and voids. He assumed that the separation mechanism of gel chromatography is based on regions inaccessible to molecules of the given sizes and on the variation of the molecular sizes. In the theoretical derivation Squire applied two simplifying assumptions. 1. He neglected the phenomenon of adsorption and the rate of diffusion. 2. To a first approximation he regarded proteins as uniformly hydrated spheroids.

As the starting point of the derivation he used the assumption that in a gel structure which is in equilibrium with the solvent the ratio of inner to outer volume is constant

$$V_i/V_o = \text{const.}$$

If the volume is taken to be a cone of radius R and an apex angle is assumed, one has

$$V_i = \frac{R}{3} \cdot \frac{R^3}{\text{tg } \alpha} = \text{const.} \cdot V_o$$

By the same consideration, the volume available in the gel to a protein of radius r is

$$V_{ip} = \frac{R}{3} \cdot \frac{(R-r)^3}{\text{tg } \alpha}$$

From the two above equations

$$V_{ip} = \text{const.} \cdot V_o \frac{(R-r)^3}{R^3}$$

The volume required for the elution of the proteins is

$$V_e = V_o + \text{const.} \cdot V_o \frac{(R-r)^3}{R^3}$$

Respectively

$$V_e = V_o + \text{const.} \cdot V_o \left(1 - \frac{r}{R}\right)^3 \tag{a}$$

A similar equation was obtained for a cylinder of radius R

$$V_e = V_o + \text{const.}' \cdot V_o \left(1 - \frac{r}{R}\right)^3 \tag{b}$$

or for a gap of $2R$ width

$$V_e = V_o + \text{const.''} \cdot V_o \left(1 - \frac{r}{R}\right)^3 \qquad (c)$$

For the distribution of these volumetric elements the arbitrary assumption yielding the best result is that

$$\text{const.} = 3g^3; \quad \text{const.'} = 9g^2; \quad \text{const.''} = 9g$$

where g denotes the so-called geometric factor of the volumes accessible to the molecule.

Combining equations (a), (b) and (c) in the ratio of the constant g, one obtains

$$\frac{V_e}{V_o} = 1 + g\left(1 - \frac{r}{R}\right)^3$$

In principle the equations must satisfy two criteria, namely

$$\text{if } r \xrightarrow{\lim} R, \text{ then } \frac{V_e}{V_o} \xrightarrow{\lim} 1$$

which means that the molecule is completely excluded from the gel structure, and

$$\text{if } r \xrightarrow{\lim} r_{H_2O}, \text{ then } \frac{V_e}{V_o} \xrightarrow{\lim} \frac{V_t - V_m}{V_o}$$

which means that the elution volume is equal for the solvent to the total volume of the gel bed.

Substituting into the equation his own data together with the data found in the literature, Squire stated that these two criteria were satisfied.

In the equation the radius of the spheroid proteins can be substituted by the cube root of the molecular weight. Also the radius R of the conic volumetric elements of the gel can be expressed with that molecular weight which is just not capable of penetrating into the gel (this is the exclusion molecular weight, M_{lim}). Hence

$$\frac{V_e}{V_o} = 1 + g\left(1 - k\frac{M^{1/3}}{M_{lim}^{1/3}}\right)^3$$

Since, for a given gel, M_{lim} has a constant value, the equation can be further simplified to

$$aM^{1/3} = b - \left(\frac{V_e}{V_o}\right)^{1/3}$$

This relationship measuring the constants a and b, permits a determination of the molecular weight to an accuracy of 10% without calibrating the column before each measurement. With Sephadex gels and various proteins, consideration of the equations led Squire to results which were in good agreement with the literature. He eliminated reversible adsorption with a buffer adjusted to the pH of the isoelectric point and corrected the form factor of the molecules by consideration of the hydrodynamic radius

$$r_h = \left(\frac{3M}{4N} \left(1 + \frac{W}{v\varrho} \right) \right)^{1/3}$$

where r_h denotes the hydrodynamic radius of the protein, M its molecular weight, N the Avogadro number, W the regained moisture content of the protein in g/g H_2O, v the partial specific volume of the protein, and ϱ the density of the solvent.

It should be noted that when Squire applied this equation to dextran fractions he arrived at quite different results, which he attributed to the known fact that the hydrodynamic radius of dextran is approximately 2.5 times greater than that of all other proteins of the same molecular weight.

2.5.2. THE STERIC MODEL OF LAURENT AND KILLANDER

Laurent and Killander (1964) approached the spheric interactions of gels and solutes on the basis of the steric model of the matrix. Laurent made the observation that, in their aqueous solution, polysaccharides (e.g. hyaluronic acid, dextran, etc.) show properties quite similar to those of the crosslinked gel structure (Laurent and Pietruszkiewicz 1961; Laurent et al. 1964). Depending on their molecular weights polysaccharide molecules exclude the other macromolecules from their surroundings and reduce their solubility. For instance, by increasing the concentration of high molecular weight dextran, certain proteins can be gradually precipitated from their solutions (Laurent 1963a, b).

Laurent and Killander assumed the polysaccharide chains of dextran gels (Sephadex) to be straight rigid rods, their length and location being determined by statistical probability. In the steric structure of the polysaccharide chains the volume available to spheroid molecules depends on the molecular size (r), the radius (R) of the polysaccharide fibres (cylinders) and their length (L). On the basis of the equation derived earlier by Ogston (1958) for a similar physical model, Laurent and Killander arrived at the following relationship for the coefficient of volumetric distribution

$$\lg K_{av} = - \pi L (r + R)^2$$

where K_{av} denotes the volumetric distribution coefficient referring to the total volume of the gel bed (see Section 2.5 of Part I), L the length (concentration) of the polysaccharide fibres in terms of cm/ml of gel, R the radius of the spheroid molecules.

In the verification of the equation, Laurent and Killander, to a first approximation, assumed that the rate of diffusion, the flow of the liquid and the shape of the molecules have no influence on the ideal (quasi-stationary, static) equilibrium between the inner and outer volume of the gel. For their examinations they used various cellulose hydrolysates, dextran and protein molecules. They established their equivalent spheroid radii and their gel chromatographic properties (V_t, V_o, V_e), calculating the rest of the factors of the equation (L and R) from the measurement data or taking them as already known from other investigations. Thus, for example, for the radius of the polysaccharide chains they accepted the value of $R = 7$ Å. The measured and calculated (from gel chromatography) L values of the polysaccharide chains showed good agreement and were directly proportional to the density and dextran content, respectively, of the dextran gels (Sephadex G-25 to G-200). A comparison of the calculated and measured values showed that the structure of dextran gels (Sephadex) can be regarded as a steric network made up of statistically distributed rigid fibres.

The difference between the results calculated from the equation and measured in practice was attributed by Laurent and Killander to neglect the diffusion phenomena and inaccuracies in the determination of the equivalent radii. In fact, the rate at which the equilibrium of diffusion is achieved, in addition to the size of the gel particles, their inner structure and the concentration of the polymer, depends also on the rate of diffusion of the solutes into the gel. It is a highly complicated function of the speed of free diffusion, the molecular sizes and the gel characteristics. The correction of the Ogston equation for microreticular gels is negligibly small (Fisher 1969, p. 344). Substantially greater differences are caused by variances between the forms of the molecules and the theoretically assumed spheroid. Although correction of the Ogston equation includes consideration of the shape factors also, the Laurent—Killander method is generally applied in practice. However, without theoretical verification up to now, for molecules of irregular form the Stokes radius is used with good results (Siegel and Monty 1965, 1966)

$$a = \frac{RT}{6\pi N \eta D}$$

where a denotes the Stokes radius, R the universal gas constant, T the absolute temperature, N the Avogadro number, η the coefficient of internal friction of the medium and D the diffusion coefficient.

If hydrated macromolecules are concerned, the relationship for the Stokes radius, the maximum molecular weight and use of the gel chromatographic data enable a determination of the molecular weight to be carried out

$$M = \frac{4\pi}{3} \cdot \frac{N}{v} a^3$$

where M is the molecular weight, v the partial specific molecular volume and a the Stokes radius.

According to Laurent's original assumption (1963a, b), dextran dissolved in the moving phase of the gel chromatographic material must exert a characteristic effect on the elution volume of other macromolecules (e.g. proteins). Hellsing (1965, 1968) confirmed this from his own experience. He found that the presence of dextran, polyethylene glycol, etc., causes a change in the elution properties of albumin but, owing to the shrinkage (complex coacervation, see Section 2.7.1 of Part I) of the gel particles it is difficult to ascertain clearly changes in the exclusion properties.

Fawcett and Morris (1966) performed calculations similar to those of Laurent and Killander on polyacrylamide gels and proved that in the looser gels, with fewer crosslinks, the length of the polymer chains per unit volume is proportional to the polymer concentration of the gel. They established furthermore that in networks which include numerous crosslinks, the effect of increasing concentration of the monomer participating in the reaction is merely a thickening of the chains.

2.5.3. THE LAURENT AND LAURENT ELECTRIC ANALOGUE MODEL

Among the assumptions which support the volumetric theories, the electric connection model elaborated by Laurent and Laurent (1964) claims special attention. These authors demonstrated this mechanism of gel chomatography by the analogy of the charging and discharging of capacitors connected in parallel.

The model, which is surprisingly easy to understand, enabled the plotting and study of "elution curves" to be carried out. From the theoretical aspect, the model reflects the dynamic (kinetic) idea of the separation process of gel chromatography.

2.5.4. MOLECULAR SIEVING THEORIES

On the basis of an idea proposed by Hjertén and Mosbach (1962), Hohn and Pollman (1963) — disregarding the structure of the polymer matrices — interpreted the interaction between the gel structure and the solute molecules as an operation of statistical sieving. For the relationship between the volumetric distribution coefficient and the molecular weight they used the Boltzmann distribution and obtained the following readily applicable relationship for oligomers (oligonucleotides)

$$\lg K_d = k_1 - k_2 \frac{M}{M_{monomer}}$$

where k_1 and k_2 are constants, and M and $M_{monomer}$ the molecular weights of the oligomers and the monomer, respectively.

Ackers (1967), from the analogy of the sieving model, calculated the voids available to the molecules on the basis of the normal (gaussian) distribution and arrived at a well defined linear relationship between the volumetric distribution coefficient of gel chromatography and the Stokes radii of the molecules to be separated.

2.6. THE KINETIC THEORIES OF GEL CHROMATOGRAPHY

The volumetric theories of gel chromatography set out from the macroscopic approach of the volumetric distribution of the gel bed and, using two arbitrarily chosen parameters, sought relationships which would be acceptable in practice. For the sake of simplifying the treatment, they often interpreted the properties of the solute molecules, their movement and their interaction with the gel structure by simple schematic models.

The kinetic theories of gel chromatography, on the other hand, started from a hydrodynamic examination of the stationary and flowing phases of the gel bed. Porath (1959), having established that the elution volume of the solutes is independent of the flow rate and concentration within wide limits, disregarded diffusion as a determinant factor. Polson (1961) proved at the same time that, in the case of albumin and haemoglobin of approximately the same molecular weight, the volumetric distribution coefficient is related to the diffusion coefficients rather than to the molecular weights. Others have made similar observations which showed that, apart from the simpler cases, the volumetric theories do not explain the phenomena of gel chromatography and, in order to elucidate the laws and regularities of the spheric effects, the hydrodynamic parameters of column chromatography and the molecular interactions must be resorted to.

2.6.1. THE HYDRODYNAMIC PARAMETERS OF GEL CHROMATOGRAPHY

The general hydrodynamic relationships of liquid chromatography, with certain modifications, are valid also for gel chromatography (Karger 1971). Most parameters for resolution of the columns, etc. refer to gel permeation chromatography, and, like gas chromatography, this is based on timing. In view of the peculiarities of the charge and the difficulties involved in the volumetric process (volatile solvents !), it is necessary to measure the retention time. Owing to the incompressibility of liquids with soft gels and a constant rate of flow the retention time can be substituted by the volume of the through-flow (elution or retention volume). In most relationships from theoretical derivations it is assumed that the elution profiles are symmetric and conform to the gaussian distribution and, accordingly, that the elution volume is independent of the quantity of the chromatographed sample. Experience has shown, however, that the distribution isotherms are linear and the areas below the elution curves are proportional to the quantity of chromatographed substance.

2.6.1.1. RESOLUTION IN GEL CHROMATOGRAPHY

The separation of two substances of nearly equal quantity by column chromatography may be characterized by the degree of overlap of the elution profiles, the distance between the elution peaks ($V_{e2} - V_{e1}$) and the width of the base of the elution curves (b_2, b_1). The separation or the resolution of the column (R_S, resolution) is expressed by the following equation

$$R_S = 2\frac{V_{e2}-V_{e1}}{b_1+b_2} = \frac{V_{e2}-V_{e1}}{4\sigma}$$

where V_{e1} and V_{e2} denote the elution volumes of the components, σ is the standard deviation of the gaussian distribution curve, and b_1 and b_2 denote the base width of the elution curves.

According to the gaussian curves, in the elution of two components with similar properties, $b_1 = b_2$ and $b_1 + b_2$ can be substituted by 4σ (see *Fig. 21*).

Figure 20 illustrates the way separation can be enhanced by increasing the distance between peaks (*Fig. 20b*) or by decreasing the width of the curves (*Fig. 20c*).

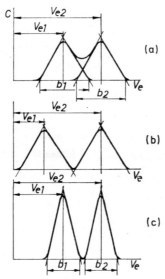

Fig. 20. The separation of two substances of equal concentration by (a) increasing the resolution of column chromatography, (b) increasing the distance of the elution profiles and (c) reducing the breadth of the elution curves

If $R_S = 1$ then the case is called 4σ separation because each fraction will have a base of width 2σ. This causes only a 2% overlap and may be regarded as satisfactory. Still better will be the separation if the resolution is $5\sigma (R_S = 1.5)$. With decreasing R_S, the overlap of the elution profile increases, and below $R_S = 0.8$ the resolution power becomes unsatisfactory.

From the above terms for the resolution power it follows that increasing elution volumes (longer columns) improve the separation, although with longer migration distance the base of the elution curves will also become broader (zone broadening, peak widening). However, the broadening of the elution curves is proportional to the square root of the length of the column (see Section 2.6.1.4 of Part I) and a longer column always provides better separation.

Using the theoretical parameters of column chromatography the resolution may be expressed in the following form (Bombaugh 1971)

$$R_S = \frac{1}{4}\left(\frac{\alpha-1}{\alpha}\right)\left(\frac{k_2'}{1+k_2'}\right)(N_2)^{1/2}$$

where α denotes the selectivity or the relative retention of the column, k' the capacity coefficient of the column, and N the theoretical number of plates (HETP), the index 2 designating the component appearing with the larger elution volume.

In the final analysis the resolution consists of three factors: selectivity, capacity and the effectiveness of the column. It seems worthwhile to study each of these factors in more detail.

2.6.1.2. THE SELECTIVITY OF GEL COLUMNS

Taking into consideration the various factors of resolution, selectivity (α) is the coefficient of the net elution volumes (see Section 2.6.1.3 of Part I), and the capacity factors (distribution coefficients)

$$\alpha = \frac{V_{e2}-V_k}{V_{e1}-V_k} = \frac{k_2'}{k_1'} = \frac{K_{d1}}{K_{d2}}$$

In gel chromatography selectivity is determined primarily by the difference between the molecular dimensions, the size of the pores in the gel structure and their distribution. One of the features which make liquid chromatography a process superior to gas chromatography is that not only the properties of the stationary phase but also the (solvent-solute) molecular interactions of the flowing phase contribute to the enhancement of selectivity.

From the equation of resolution it follows that if $\alpha = 1$ then $R_S = 0$, regardless of any other column parameter. At the same time, even very small differences in selectivity (± 0.01) cause considerable changes in resolution, particularly at values close to 1 (see the interaction between selectivity and HETP in Section 2.6.1.4 of Part I). Thus, for instance, a 10% increase in selectivity (from 1.1 to 1.2) will cause a rise in the resolution of approximately 100%, by the ($\alpha - 1$) factor.

2.6.1.3. THE CAPACITY FACTOR OF RESOLUTION

According to Section 2.4 of Part I, from the basic equation of gel chromatography the difference between the elution volume and the outer volumes is called the net elution (retention) volume (V_N):

$$V_N = V_e - V_o = K_d V_i$$

According to this definition, the capacity factor of the gel column (k') is

$$k' = K_d \frac{V_i}{V_o} = \frac{V_e - V_o}{V_o}$$

In both the stationary and the flowing phases the capacity factor expresses the ratio of the distribution of a substance, and if the substances take up the total inner volume of the gel phase ($K_d = 1$) it indicates the capacity of the column. From the above it follows that

$$V_e = V_o(1 + k')$$

It is evident that if the column has no capacity ($k' = 0$) then $V_e = V_o$, and for the resolution $R_S = 0$, this means that substances elute without separation from the outer volume of the column. For high k' values, on the other hand (for instance in gas chromatography) the coefficient $k'_2/(1 + k'_2)$ tends to 1, and the capacity factor has no influence on the resolution. In gel chromatography the k' value is small compared even to other methods of liquid chromatography (varying as it does generally between 0.5 and 2) because theoretically the value of K_d may never exceed 1.

The capacity factor can be determined experimentally from the term $(V_e - V_o)/V_o$ by measurements read directly from the chromatograms (V_e, V_o). The capacity factor indicates the ratio of the net elution volume to the outer volume of the column. It is evident from the equations that the elution volume varies proportionally with the column dimensions (V_o varies linearly with the length and quadratically with the cross-section of the column) and depends on the capacity factor of the column, i.e. the ratio of the stationary to the flowing phase. As a result the capacity factor related to the outer volume of the column (V_o) can be increased either by increasing the inner volume of the gel (V_i) or by reduction of the voids between the gel particles. It may be of interest in this connection that, according to recent examinations, if soft gels are used (Sephadex G-100, Sepharose 6B), at unchanged rates of flow the capacity and resolution of compressed columns improve considerably (Edwards and Helft 1970; Fishman and Barford 1970). Compression of the column-charge increases the HETP and, with unchanged inner volume, reduces the outer volume. Both factors unequivocally enhance the resolution of the column. In the light of these findings the practical rule, according to which soft gels must be used at hydrostatic pressures (or such flow rates) which do not cause appreciable changes either in the shape of the gel particles or the volumetric distribution of the column, calls for a thorough revision.

2.6.1.4. THEORETICAL AND EFFECTIVE HETP

The concept of theoretical plate number (N) and of theoretical plate height (HETP — Height Equivalent to a Theoretical Plate) was introduced into chromatography, on an analogy with distillation, by Martin and Synge (1941). The theory, elaborated by mathematical methods, is now routinely used in the various processes of liquid (James and Martin 1952; Glueckauf 1966a, b, c) and gas chromatography (Szepesy 1970; Leibnitz and Struppe 1970).

In practice, the HETP can be calculated with the help of two formulas. The most frequently used relationship is

$$N = 16 \left(\frac{V_{e\,max}}{b}\right)^2 = \left(\frac{V_{e\,max}}{\sigma}\right)^2$$

For liquid chromatography, James and Martin (1952) introduced the following term

$$N = 8 \left(\frac{V_{e\,max}}{\varepsilon}\right)^2$$

where ε denotes the width of the elution curve at a concentration of 0.368 C_{max}.

Figure 21 shows the parameters used in the calculation of the HETP for concentration profiles of gaussian distribution.

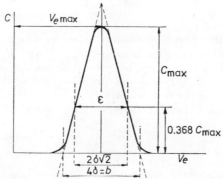

Fig. 21. The determination of HETP (height equivalent to a theoretical plate) of the gel column from the elution chart of a homogeneous substance. $V_{e\,max}$ is the elution volume measured at the point of maximum concentration profile, C_{max} the maximum concentration of the elution curve, ε the width of the concentration profile at a concentration of $0.368 C_{max}$ ($= 2\sigma\sqrt{2}$), σ the variance of the curve of gaussian distribution, and b the breadth of the base of the concentration profile ($b = 4\sigma$)

The HETP, in cm, is the coefficient of the column length (L, cm) and the theoretical plate number:

$$\text{HETP} = \frac{L}{N}$$

It will be clear from the relationship in Section 2.6.1.1 of Part I that the resolution of chromatographic columns is proportional to the square root of the HETP. In turn, from the formula for HETP, it follows that resolution, too, is proportional to the square root of the column length (V_e being proportional to V_t). Accordingly, doubling the column length would increase the resolution power by 1.4. However, the zone width and the time taken for chromatography would also increase simultaneously.

More recently, to express the effectiveness of chromatographic columns, or more exactly, to express the theoretical plate number corrected by the square of the capacity factor, the so-called effective plate number (N_{eff}) has been used:

$$N_{\text{eff}} = N \left(\frac{k'}{1+k'} \right)^2 = 16 \left(\frac{V_e - V_o}{b^2} \right)^2 = \left(\frac{V_e - V_o}{\sigma} \right)^2$$

The effective plate number, in addition to the zone width (σ), takes into consideration the column capacity also. Substituting the effective plate number into the equation of resolution, the product of two relatively independent parameters is obtained:

$$R_S = \frac{1}{4} \left(\frac{\alpha - 1}{\alpha} \right) (N_{\text{eff}})^{1/2}$$

Figure 22 indicates the relationship between column selectivity and the effective plate number at a resolution of $R_S = 1$ and $R_S = 1.5$, respectively. *Figure 22* shows that the lower the selectivity of the column the greater is the effective plate number required for adequate separation. At the same time, slight changes in the low selectivity values go with great differences in the effective plate number. With increasing selectivity the effective plate

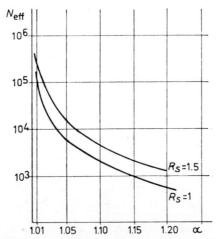

Fig. 22. The correlation between column selectivity (α) and the effective plate number (N_{eff}) in liquid chromatography at $R_S = 1$ and $R_S = 1.5$ resolution power, respectively

number necessary for separation decreases, and the resolution power of the column becomes increasingly insensitive to variations in the effective plate number. The above relationship for the resolution of column chromatography (in terms of α and N_{eff}) provides the answer to the question of whether a concrete problem of separation can, in practice, be resolved. As a rule of thumb, in liquid chromatography an effective plate number of between 500 and 750 is sufficient for most separation jobs.

2.6.1.5. THE INFLUENCE OF THE FLOW RATE ON HETP

Linear chromatographic systems distinguish two types of factors which may have an effect on ideal transport processes (Giddings 1965): (1) the so-called longitudinal diffusion acting along the column, and (2) the factors influencing the speed at which the diffusion equilibrium between the moving and the stationary phases is restored. These are caused partly by local disequilibria and partly by mixing, because of irregular (turbulent) flow. In liquid chromatography the longitudinal diffusion is only of secondary importance, while the effect of hydrodynamic factors arising from column packing and improper localization of the gel particles is substantial. Flodin (1962), on the basis of experience gained with ion exchangers (Helfferich 1959), found that, for substances excluded from the gel phase, deformations in the elution profiles are caused mainly by irregularities of flow which are produced by the charge of the gel grains. For substances penetrating into the gel, the irregularities may be the result of local disequilibria, since the flow rate has no effect whatsoever on diffusion which takes place in the gel phase (Cazes 1966).

The influence of the flow rate is expressed implicitly by the variations of the HETP. *Figure 23* shows the correlation between the HETP and the

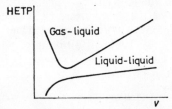

Fig. 23. The correlation between the HETP and the rate of the moving phase (v) in gas-liquid and liquid-liquid chromatography

rate of the flowing phase in gas-liquid and liquid-liquid partition chromatography. It is evident that in liquid chromatography the HETP is very much less dependent on the flow rate since diffusion in liquids is lower by orders of magnitude than in gases (10^4 to 10^5 times). In general, an increasing flow rate reduces the effectiveness of the column (Van Deemter *et al.* 1956; Heitz and Coupek 1968; Waters *et al.* 1969; Little *et al.* 1969).

The relationship between the HETP and the flow rate is shown also by the following simple empirical formula which is readily applicable in practice (Waters et al. 1969)

$$\text{HETP} = Av^n$$

where A denotes the material constant characterizing the column packing, v denotes the rate of the flowing phase, n denotes the value of the power in liquid chromatography, namely a value between 0.3 and 0.6.

For the determination of the n, values pertaining to two different HETP and flow rates must be measured:

$$\frac{\text{HETP}_1}{\text{HETP}_2} = \left(\frac{v^1}{v_2}\right)^n$$

In his monograph Giddings (1965) goes into considerable detail on the extremely complex flow phenomena of liquid chromatography and proposes the following expression to equate the effect of the diffusion coefficient (D_M), the grain diameter (d) and the flow rate (v)

$$\text{HETP} = \left(\frac{1}{2d\lambda} + \frac{D_M}{vd^2}\right)^{-1}$$

where λ denotes the so-called correction factor of packing, approximately 0.5.

Recent research into the chromatographic process aims at the elucidation of the correlations between the flow phenomena and the resolution of the gel column. This subject, however, falls beyond the scope of this book. The following authors have dealt with the problem: Kirkland 1971a; Ackers 1970; Bly 1970; Bly et al. 1971; Snyder, 1972a, b; Casassa 1971; Cameron 1971; Kelley and Billmeyer 1970; Svensson 1966; Winzor 1966; Giddings 1967, 1970; Bombaugh et al. 1969b; Bombaugh and Levangie 1969c; Snyder 1969; Deelder 1970; Cheetham and Winzor 1970; Hibberd et al. 1970; Scott, 1971.

2.6.1.6. THE HETP IN GEL CHROMATOGRAPHY

Flodin (1961, 1962) was the first to prove that the HETP value varies in gel chromatography under the influence of the particle size and the flow rate. Making use of the Glueckauf (1955b) relationship ($\varepsilon = 2\sigma\sqrt{2}$) he measured the HETP values for different Sephadex gels in model substances. His data are compiled in *Table 13* which shows certain general regularities for the influence of the particle size and the flow rate. The size of the gel particles influences the column resolution by restoring the equilibrium of diffusion (see the Giddings equation). Smaller particles are associated with lower HETP values and higher column effectiveness. Increased flow rates, owing to the disequilibrium of diffusion and to greater irregularities (turbulence), cause a drop in the HETP value. Added to this is the molecular size which, although indirectly, exerts its effect through the diffusion rate. With larger molecules (for instance uridylic acid, see *Table 13*) the HETP values are higher and vary more readily with the flow rate. The HETP values of small molecules (for instance, hydrochloric acid) are essentially lower, and small

Table 13

Height Equivalent to Theoretical Plate (HETP) for the Sephadex G gel types
(Flodin 1961, 1962)

Sephadex type and solvent (size of column)	Solute	Gel particle size, μm	Flow rate, ml h^{-1}	HETP, cm
Sephadex G-25 Water (2×65 cm)	Uridylic acid	150—250	10	0.039
			24	0.072
			51	0.149
			90	0.262
			130	0.414
			190	0.549
		75—120	10	0.0102
			24	0.0182
			51	0.210
		40— 60	24	0.011
Sephadex G-25 Water (2×67 cm)	Hydrochloric acid	150—250	14	0.105
			24	0.069
			48	0.044
			70	0.045
			96	0.051
		40— 60	14	0.023
			24	0.026
Sephadex G-25 Water (3.5×39 and 1.8×49 cm)	Glycine	150—300	120	0.064
		40— 75	18	0.011
Sephadex G-25	Glucose	40— 75	30	0.040
	Cellobiose	40— 75	30	0.039
	Cellotetrose	40— 75	30	0.033
Sephadex G-50	Isoleucine	75—150	24	0.031
Sephadex G-75	Lysozyme	75—150	36	0.36
Sephadex G-100	Albumin	40— 75	160	0.08
Sephadex G-200	Albumin	75—150	75	0.62
			50	0.48
			25	0.23
		60— 75	25	0.08

molecules, even at higher flow velocities, diffuse rapidly enough to attain equilibrium. On the other hand, at low flow rates the HETP value of small molecules increases as a result of longitudinal diffusion.

Smith and Kollmansberger (1965) observed the same effect of flow rate and molecular size (benzene and p-dibromo-benzene) on Styragel in tetrahydrofuran.

2.6.1.7. CALCULATION OF THE COLUMN DIMENSIONS ON THE BASIS OF THE HETP

If the material (molecules to be separated, type of gel, coefficient of the volumetric distribution) and the hydrodynamic parameters (grain size, flow rate) of the chromatographic process are known, the smallest allowable dimensions of the gel bed (column length) can be calculated. In case of normally distributed (gaussian) concentration profiles the HETP values re-

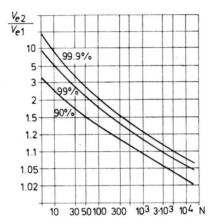

Fig. 24. The correlation between the coefficient of the elution volumes and the HETP for a 90%, 99% and 99.9% separation of the same quantity of two substances (the corresponding elution volumes are V_{e1} and V_{e2}) with concentration profiles conforming to gaussian (symmetric) distribution (Glueckauf 1955)

quired for the 90%, 99% and 99.9% separation of two components of equal quantity and of V_{e1} and V_{e2} elution volume can be obtained from the curves in *Figure 24* (Glueckauf 1955a). According to the volumetric equations

$$V_{e1} = V_o + V_i K_{d1}$$
$$V_{e2} = V_o + V_i K_{d2}$$

and hence

$$\frac{V_{e1}}{V_{e2}} = \frac{V_o/V_i + K_{d1}}{V_o/V_i + K_{d2}}$$

where V_o/V_i denotes the geometric constant of the gel bed whose value (according to Flodin 1961) may vary between 0.25 and 0.50, depending on the type of the gel. For adequately packed Sephadex columns $V_o/V_i = 0.30$.

If V_{e1}, V_{e2}, K_{d1} and K_{d2} are known, by making use of the HETP, the column dimensions necessary for effective operation can be calculated on the basis of the following examples.

1. Desalting of protein on a Sephadex G-25 column (particle size 150 to 250 μm; HETP = 0.1 cm)

$$K_{d\ \text{protein}} = 0 \qquad K_{d\ \text{NaCl}} = 0.9$$

$$\frac{V_{e\ \text{NaCl}}}{V_{e\ \text{protein}}} = \frac{0.3 + 0.9}{0.3} = 4$$

A comparison of the above with the curves in *Figure 24* shows that, in the case of a 99% desalting, the value of N is between 10 and 15. To be effective, the column (HETP = 0.1 cm) must accordingly be at least 1 to 1.5 cm high.

2. Separation of saccharose and glucose on a Sephadex G-25 column (particle size 40 to 60 μm; HETP = 0.03)

$$K_{d\ \text{saccharose}} = 0.7 \qquad K_{d\ \text{glucose}} = 0.8$$

$$\frac{V_{e\ \text{saccharose}}}{V_{e\ \text{glucose}}} = 1.1$$

For a 99% separation of saccharose and glucose, the value of N is approximately 3 000. To keep the column size within rational limits, a gel with such a particle size should be chosen so that, at a moderate (not excessively low) flow rate, the smallest HETP value is produced. Assuming the above parameters, in order to separate the two substances a column of 90—100 cm height is required.

2.6.2. THE THEORY OF RESTRICTED DIFFUSION: ACKERS' INVESTIGATIONS

In the gel structure the restricted diffusion of the molecules to be separated provides one of the foundations on which the kinetic theories of gel chromatography rely (Steere and Ackers 1962; Ackers and Steere 1962). The idea of restricted diffusion is based on the observation that macromolecules diffuse slower in agar membranes of different concentration than in their free solutions, and that the rate of diffusion depends on the molecular size (Ackers and Steere 1962). Assuming the gel pores to be cylindrical-conic these workers derived the restriction of diffusion from the size (steric factors) and from friction (friction factors) of the molecules. The correlation of these factors, in the case of free diffusion, is described by the Stokes radius (see Section 2.5.2 of Part I). The effect of the gel structure they calculated by means of a suitably chosen model.

Ackers and Steere (1962) calculated the effective pore size of the gels from the restriction of the diffusion of macromolecules of known radii, making use of the Renkin equation (Renkin 1955), elaborated for porous models similar to gels.

The Renkin equation is

$$\frac{A_R}{A_0} = \left(1 - \frac{a}{R}\right)^2 \left[1 - 2.104 \frac{a}{R} + 2.09 \left(\frac{a}{R}\right)^3 - 0.95 \left(\frac{a}{R}\right)^5\right]$$

where A_0 denotes the total cross-section of the pore surface, A_R the effective cross-section available to diffusion, a the Stokes radius of the molecule, and R the effective pore radius $\left(1 - \frac{a}{R}\right)$, the so-called Ferry's steric factor.

The effective pore sizes for given gel types, determined on different macromolecules, showed good agreement and verified the researchers' assumptions.

Ackers (1964) applied the principle of restricted diffusion to gel chromatography, also. Comparing the equilibrium and the dynamic states, he demonstrated the role of fluid flow and the molecular interactions. For an equilibrium model he used a gel bed consisting of swollen gel particles. Using a protein solution of known quantity and known concentration, from the relationship

$$K_{d\ \text{equil.}} \cdot V_i = \frac{Q_t}{C_0} - V_0$$

he calculated the equilibrium distribution coefficient of the bed ($K_{d\ \text{equil.}}$, Q_t is the total amount of the solute, and C_0 the solute concentration outside the gel particles). This term yields the same formula as the volumetric equation in Section 2.4 of Part I, provided that the coefficient Q_t/C is substituted by the elution volume. Ackers compared this relationship with the dynamic model of column chromatography, in which the coefficient of dynamic distribution ($K_{d\ \text{dyn}}$) can be calculated from experimentally measurable data in a similar way:

$$K_{d\ \text{dyn.}} V_i = V_e - V_0$$

When comparing the two models, the equality

$$V_e - V_{o\ \text{dyn.}} = \frac{Q_t}{C_0} - V_{o\ \text{equil.}}$$

was met only when Sephadex G-25 and G-100 gels were used. In gels with a larger number of crosslinks, owing to interactions between the molecules and the gel structure (adsorption), the protein distribution coefficients measured on the column yielded higher values than in the equilibrium (stationary) system. In looser gels, on the other hand (Sephadex G-200, granulated agar), because of the heterogeneous and irregular pore sizes, an inverse and more defined difference was observed. In *Figure 25* the K_d values calculated from the equilibrium and the dynamic models are compared for Sephadex G-100 and G-200, on the basis of Ackers' investigations (1964).

On the basis of his findings, Ackers supposed that in macroreticular gels the elution volume of large molecules is determined by the molecular diffusion. Since the diffusion of macromolecules into the gel phase takes place at a slower rate, these elute faster than smaller molecules. Linking up the elution properties with the Stokes molecular radii, as found in the Renkin equation, and the effective dimensions of the gel pores, he arrived at the

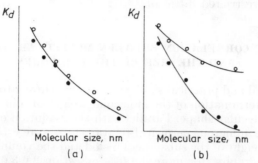

Fig. 25. A comparison of the values of the volumetric distribution coefficient (K_d) in equilibrium and in the dynamic state as function of the size of the spheorid molecules, on (a) Sephadex G-100 and (b) Sephadex G-200 gel (—o— equilibrium state, —●— dynamic state); (Ackers 1964)

following relationship for the volumetric distribution coefficient of gel chromatography:

$$K_d = \frac{V_e - V_o}{V_i} = \left(1 - \frac{a}{R}\right)^2 \left[1 - 2.104 \frac{a}{R} + 2.09 \left(\frac{a}{R}\right)^3 - 0.95 \left(\frac{a}{R}\right)^5\right]$$

The validity of this equation was confirmed partly by the substitution of molecules with known Stokes radii and accepted elution data, the results showing fair agreement with previous measurements; and partly by calculation of the Stokes radii, making use of the volumetric distribution of column chromatography and the known pore sizes, the results being in good agreement with those obtained by other methods (free diffusion, roentgen diffraction, electron microscopy). Further analysis of the equation enabled a proper interpretation of the distribution coefficient to be carried out, depending on the pore-size distribution and the molecular size.

The theory of restricted diffusion is more or less in contradiction with practical experience, in that the elution volume is independent of the flow rate. To explain this fact, Ackers assumes that the gel particles are surrounded by a solvent layer which does not take part in the flow, but whose thickness depends on the flow rate and influences inversely the elution volume. This theory, proposed by Ackers, seems to be supported by examinations of the loosely and closely bound solvent content of the gel structure (see Section 1.1 of Part I; Clifford *et al.* 1970; Clifford and Child 1971).

Smith and Kollmansberger (1965), by chromatography of aliphatic and aromatic hydrocarbons, confirmed the role of restricted diffusion. They established that, unlike other separation processes (e.g. gas or ion exchange chromatography), the diffusion coefficients pertaining to different flow rates are inversely proportional to the elution volumes.

Yau and Malone (1967), and Di Marzio and Guttman (1969) have discussed the theory of the separation mechanism of gel permeation chromatography, on the basis of restricted diffusion.

2.6.3. THE CORRELATION BETWEEN MOLECULAR DIMENSIONS AND THE SIZE OF THE GEL PORES

The theoretical and practical significance of gel chromatography is proved elegantly by determination of the molecular weight of solutes, the molecular size and the molecular shape. Parallel with the evolution of volumetric and kinetic theories, numerous empirical relationships were established between the molecular weight of the substances tested and the volumetric parameters of gel chromatography (for more detail see Section 2.6.3.6 of Part I). Most important are the investigations which provide new information on the molecular structure of the gels or the materials to be separated.

2.6.3.1. MOLECULAR WEIGHT DETERMINATION BY GEL CHROMATOGRAPHY: ANDREWS' INVESTIGATIONS

On the basis of the pioneering works by Granath and Flodin (1961), and Whitaker (1963), the determination of the molecular weight of solutes has become one of the most widely used analytical applications of gel chromatography. Andrews and his working group (Andrews 1962, 1964, 1965, 1970; Andrews et al. 1964; Downey and Andrews 1965) laid the empirical findings on a common basis and by gel chromatographic examination of some thirty proteins founded, as it were, the method and representation of molecular weight determination. Experience has shown that, on Sephadex G-200 dextran gel, a linear correlation exists over a wide range between the elution volume of proteins and the logarithm of their molecular weights. Towards globular proteins and polypeptides free from carbohydrate content, the G-75 and G-100 gels react in a similar way. Gel chromatography confirmed the statement that the partial specific volume of globular proteins (even that of fibrous molecules) conform to the gaussian distribution and that their average value is 0.73 ml/g. Accordingly, as far as their shape, density and gel chromatographic properties are concerned, most globular proteins show a regular response. The linear portion of the Andrews' selectivity curve *(Fig. 26)*, in agreement with the results of occasional electron microscopic examinations (e.g. *E. coli*, phosphatase, apoferritine, myoglobin, etc.), showed that, when dissolved, most proteins and enzymes assume a spheroidal shape.

Molecular weights determined by gel chromatography are found to deviate from literature values. This fact called attention to anomalies in the protein structure or the protein composition. The findings of Andrews *(Table 14)* prove that the molecular weight of glycoproteins determined by gel chromatography differs from the data given in the literature, depending on their carbohydrate content. This may be attributed to higher hydration of the carbohydrate chain and the elongated-expanded molecular structure. It should be noted here that, notwithstanding its high carbohydrate content, ceruloplasmin exhibits ideal globular protein properties. Human γ globulin,

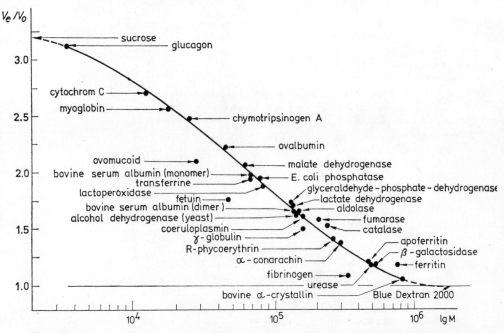

Fig. 26. The correlation between the volumetric parameters of gel chromatography (relative elution volume, V_e/V_0) and the logarithm of the molecular weight on a Sephadex G-200 column (Andrews 1965)

in spite of its molecular weight, which is the same as that of ceruloplasmin, elicits different elution properties. This fact makes it possible for the two proteins to be separated by gel chromatography. In contrast to the value of 65 000 quoted in the literature, Andrews obtained a value of 71 000 for the molecular weight of albumin, which caused him to assume that the structure of the molecule is slightly elongated. As regards the form of the proteins, *Table 14* indicates that the deviation of the molecular weight of γ globulin–fetuin–fibrinogen (taking into account their carbohydrate content) increases in the same sequence, by the axial coefficient of the molecules (see Section 2.6.3.2 of Part I.

Using the relationship between the diffusion coefficient and the Stokes radius (Section 2.5.2 of Part I), Andrews represented the elution volumes of the proteins tested as a function of the reciprocal of the diffusion coefficient (*Fig. 27*).

Table 14

Relationship between the gel chromatographic behaviour and carbohydrate content of selected glycoproteins (Andrews 1965)

Glycoprotein	Carbohydrate content, %	Molecular weight		Difference, %
		Literature	Gel chromatography	
γ Globulin	2.2	160 000	205 000	+ 30
Fibrinogen	3.1	330 000	733 000	+120
Egg albumin	3.5	45 000	41 000	− 10
Transferrin	3.9	68 000	74 000	+ 10
Ceruloplasmin	7.1	160 000	155 000	0
Thyroglobulin	8.1	670 000	1 330 000	+100
Fetuin	19.2	46 000	115 000	+150
Ovomucoid	21.0	28 000	55 000	+100

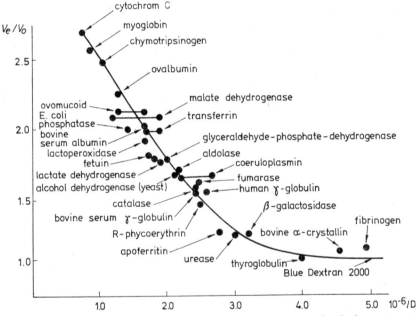

Fig. 27. The correlation between the volumetric parameters of gel chromatography (relative elution volume V_e/V_o) and the reciprocal of the diffusion coefficient ($10^{-6}/D$) on a Sephadex G-200 column (Andrews 1965)

The behaviour of proteins which deviate from the linear relationship was attributed by Andrews to the difference between the free diffusion of the molecules and their restricted diffusion in the gel phase. Similar deviations may be caused by the differences in the value of the diffusion coefficients. *Figure 27* shows that the greater the molecular weight of the proteins the greater is the deviation from the linear correlation. This observation confirms the assumption that, if the gel structure (pore sizes) and the protein sizes are comparable, molecular interactions cause decisive changes in the diffusion properties.

Some of the main results of Andrews' investigations are as follows:

(a) Gel chromatography over the linear portion of the selectivity curves allows the determination of the molecular weight of proteins, enzymes and macromolecules at an approximately 10% accuracy, provided that appropriate standards are used.

(b) Linear correlation between the molecular weight and the volumetric parameters exists, mainly in the case of carbohydrate-free globular proteins; the anomalous properties may provide further information in an examination of the structure and shape of the molecules.

(c) Gel chromatography is particularly important in the isolation and purification of enzymes and in the preliminary, orientative, determination of their molecular weight. The detection of enzymes by specific reactions

Table 15

Physicochemical and gel chromatographic parameters of enzymes and proteins frequently used for molecular weight determination*

Protein	Molecular weight, $M \cdot 10^5$	Diffusion coefficient, $D \cdot 10^7$	Stokes radius, Å	K_d Sephadex	
				G-100	G-200
Cytochrome C	0.124	13.0	17.0	0.70	0.80
Ribonuclease A	0.136	—	19.2	—	0.68
α-Chymotrypsin	0.225	—	22.8	0.54	0.64
Trypsin	0.240	—	24.1	0.54	0.62
Haemoglobin	0.320	—	24.0	—	0.58
Egg albumin	0.450	7.8	28.0	0.29	—
Bovine albumin	0.670	5.9	35.0	—	0.29
Malate dehydrogenase	0.790	5.2—8.1	—	—	0.38
Transferrin	0.880	5.3—5.8	40.0	0.11	0.32
Glyceraldehyde phosphate dehydrogenase	1.170	5.0—5.5	43.0	—	0.31
γ-Globulin human	1.400	3.8	52.2	—	0.26
Aldolase (yeast)	1.470	4.6	—	—	0.27
Catalase	2.250	4.1	52.0	—	0.26
Fibrinogen	3.250	2.0	107.0	—	0.03
Urease	4.900	3.3	61.0	—	0.20
Thyroglobulin	6.700	2.5	—	—	0
Ferritin	7.50	—	79.0	—	0.11
α-Crystalline	7.9—8.4	2.2	—	0	0

* From Andrews (1964, 1965, 1970) and Determann (1969).

allows the examination of even minute quantities of raw preparations. Such circumstances render enzymes particularly well suited to serve as reference points when plotting selectivity (calibration) curves. *Table 15* shows the physicochemical constants and gel chromatographic parameters of a number of more frequently used standard enzymes and proteins.

2.6.3.2. THE ROLE OF THE MOLECULAR SIZE AND SHAPE IN GEL CHROMATOGRAPHY: SIEGEL AND MONTY'S INVESTIGATIONS

Although in disagreement with Andrews' findings, Siegel and Monty (1966) assigned much greater importance in the separation mechanism of gel chromatography to the size and shape of the dissolved molecules, and to the phenomena of diffusion related therewith. Siegel and Monty, having performed gel chromatography on a series of proteins, established that their elution properties depended on the Stokes radii rather than on the molecular weight. According to their assumption, the molecular weight and the Stokes radii are related only in proteins with nearly the same friction coefficient and nearly equal partial specific volume. They determined the molecular weight and the friction coefficient of proteins (f/f_0) by density-gradient ultracentrifugation. The Stokes radii were calculated from the results of gel chromatography, comparing them with the values in the literature. The molecular weights and friction coefficients were calculated from the following equations

$$M = 6\pi\eta \, Na \, S(1 - \nu\gamma)$$

$$f/f_0 = \frac{a}{\left(\dfrac{3\nu M}{4N}\right)^{1/3}}$$

where S denotes the coefficient of sedimentation, ν the partial specific volume of the molecule, η the inner friction coefficient of the medium, γ the specific gravity of the medium, a the Stokes molecular radius (see Section 2.5.2 of Part I), and N the Avogadro number.

Siegel and Monty performed their tests on a Sephadex G-200 gel column with proteins of nearly equal friction coefficient and nearly equal partial specific volume. *Table 16* shows the more important physicochemical constants of the standard proteins used.

Table 16

Molecular parameters of the standard proteins*

Protein	Stokes radius, Å	Sedimentation coefficient, $S \cdot 10^{13}$	Molecular weight, $M \cdot 10^3$	f/f_0
Fibrinogen	107	7.9	330	2.34
Ferritin	79	65.0	1300	1.13
Urease	61	18.6	483	1.19
Catalase	52	11.3	250	1.25
Alcohol dehydrogenase (yeast)	46	7.4	150	1.28
Bovine albumin	35	4.3	65	1.30
Haemoglobin	24	2.8	32	1.14
Cytochrome C	17	1.9	12.4	1.09

* From Siegel and Monty (1966).

To calculate the results of gel chromatography use was made of the volumetric equations and the physical constants of the Sephadex G-200 gel:

$$K_d = \frac{V_e - V_o}{V_t - V_o - V_x}$$

$$K_{av} = \frac{V_e - V_o}{V_t - V_o} = \frac{K_d(V_t - V_o - V_x)}{V_t - V_o}$$

The volume of the gel matrix can be calculated from the specific volume of the gel bed (which for Sephadex G-200 is 30 ml/g) and from the density of the xerogel (with Sephadex G-200, $d = 1.65$ g/ml):

$$V_x = \frac{V_t}{30 \cdot 1.65}$$

Accordingly, the available volume of the gel beads for Sephadex G-200 was $K_{av} = 0.96\, K_d$.

Figure 28 shows that the Stokes radii of the proteins are in close relationship with the measured values of gel chromatography, irrespective of whether these are compared with the $(K_d)^{1/3}$ value of the Porath equation or with the $(-\lg K_{av})^{1/3}$ parameter of the Laurent—Killander theory (see Section 2.5.1 of Part I). Similarly an unequivocal correlation was found with the Stokes radii quoted in the literature when the gel chromatographic findings were substituted into the Ackers equation. No such correlation could be established in connection with the molecular weights or with the Squire equation, because, according to the experience of Siegel and Monty, the $(V_e/V_o)^{1/3}$ factor does not exhibit linear change with the Stokes radius.

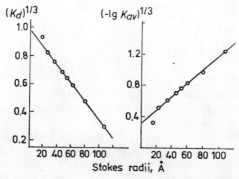

Fig. 28. The correlation between the Stokes radii of proteins and the volumetric parameters of gel chromatography on the basis of the Porath $(K_d)^{1/3}$ and the Laurent—Killander $(-\lg K_{av})^{1/3}$ equations (Siegel and Monty 1966).

The investigations of Siegel and Monty provided an interesting explanation for the anomalous behaviour of certain proteins under gel chromatography. The reverse elution of fibrinogen and ferritin, apart from their widely different molecular weights (see *Table 16*), is caused not only by the form factors (coefficients of friction) but also by their dissimilar hydration degrees and the extremely small partial specific volume (0.59 ml/g) of ferritin. Similar differences may be caused by the adsorptivity of the proteins, as, for example, in the case of cytochrome C. These examples underline the need for caution in selecting the standard proteins, and the potentialities of gel chromatography in solving problems of structural research.

Siegel and Monty also verified their findings in another way. They substituted the results obtained during the gel chromatography of standard catalase of known diameter ($r = 199$ Å) into the basic hydrodynamic equations given above, then calculated the Stokes radii and the friction coefficients of the proteins on the basis of the Ackers' equation. The results obtained are shown in *Table 17*. This table shows also the molecular weights calculated according to the same method.

Table 17

Molecular parameters of standard proteins measured by gel chromatography and calculated from the Ackers' equation*

Protein	Stokes radius, Å	Molecular weight, $M \cdot 10^3$	f/f_0
Fibrinogen	110	360	2.35
Urease (monomer)	61	480	1.18
Catalase	52	240	1.27
Alcohol dehydrogenase	46	140	1.34
Bovine albumin	35	62	1.34
Haemoglobin	24	28	1.20
Ferritin	78	2100	0.92
Cytochrome C	10	8.9	0.75

* From Siegel and Monty (1966).

From a comparison of *Tables 16* and *17*, it will be evident that, by the gel chromatographic measurement of the volumetric distribution coefficient, knowing the molecular radius of catalase, the Stokes radii calculated from the Ackers' equation are in fair agreement with the values quoted in the literature. The anomalous behaviour of ferritin and cytochrome C has already been mentioned. Although Siegel and Monty treated the problems of shape and structure in a rather one-sided way, they proved beyond doubt that a correlation between the molecular weight and the gel chromatographic properties exists only if the molecules are of a similar structure, of equal hydration, and of similar Stokes radii.

2.6.3.3. THE CORRELATION BETWEEN THE PORE SIZE AND THE CHROMATOGRAPHIC PROPERTIES OF GELS: THE DETERMANN EQUATIONS

A characteristic trend in the development of gel chromatography has been the combination of the different theoretical interpretations and practical observations. Anderson and Stoddart (1966a, b), assuming that the radii of the Porath cones and the effective pore radii as calculated by Ackers are identical, arrived at the following logarithmic relationship, confirmed by experience and used in practice: $K_d = k_1 - k_2 \lg M$. Takagi (1966) came to a similar conclusion for the chromatography of linear polyethylene fractions on Styragel.

Taking into consideration the combination of the theories and practice, Determann put the logarithmic correlations into a concrete framework of numerical equations. This work was promoted by the fact that most biological macromolecules, enzymes and proteins are tested on Sephadex gels whose properties are easily reproducible under standard laboratory conditions, and hence allow the introduction of uniform reference points in gel chromatography.

Determann and Michel (1966), mainly for hydrophilic dextran gels (Sephadex), established theoretically that there is a general correlation between the wet, swollen density of the gels and the number of pores per unit volume. With unperturbed proteins having nearly the same partial specific volume, it may be assumed that the molecular weight is proportional to the cube of the molecular radius, and the average volume of the individual pores is nearly equal to the macroscopic volumetric parameters of the gel chromatographic process. The conditions chosen by Determann and Michel put onto a practically common denominator the theories based on exclusion (Laurent and Killander 1964) and on restricted diffusion (Ackers 1964) and yielded the following formula which is identical to the empirical equations (see Section 2.6.3.4 of Part I)

$$\lg M = M_0 - (k_1 - k_2 d) \frac{V_e}{V_o}$$

where d denotes the density of the gel.

Within the limits of empirical deduction (Sephadex G-75, Sephadex G-200), this equation did prove of general validity, but for the denser gel types (below Sephadex G-50), it was no longer adequate. Making use of standard proteins and enzymes, Determann measured the constants of the equation and arrived at the following relationship:

$$\lg M = M_0 - (6.062 - 5.00 d) \frac{V_e}{V_o}$$

The fact that any relationship of this kind can be derived between the molecular weight and the gel type (density) on both the theoretical and practical basis indicated that, at least for some types of dextran gel, the separation mechanism is identical and that a kinetic-volumetric interaction between the geometric microstructure of the gels and the molecules to be separated is a realistic assumption, as regards the mechanism of gel chromatography.

Determann (1969), by collating the works of some thirty authors with his own findings, determined the constants of the equations relative to various Sephadex gels and the exclusion molecular weights of the proteins. His equations, which can be used with advantage in practice, have been compiled in *Table 18*.

Table 18

Relationship between volumetric parameters in gel chromatography and the molecular weight of solutes for different types of Sephadex gel*

Sephadex type	Equation	Exclusion limit (Mw) for globular proteins
G-200	$\lg M = 6.698 - 0.987 \dfrac{V_e}{V_0}$ $\lg M = 5.731 - 2.16 \cdot K_d$	500 000
G-100	$\lg M = 5.941 - 0.847 \dfrac{V_e}{V_0}$ $\lg M = 5.070 - 1.35 \cdot K_d$	125 000
G-75	$\lg M = 5.624 - 0.752 \dfrac{V_e}{V_0}$	75 000
G-50	$\lg M = 5.415 - 0.864 \dfrac{V_e}{V_0}$	35 000

* From Determann and Michel (1966).

Figure 29 shows that the Determann equations formulated with the standard enzymes and proteins, in the molecular weight ranges corresponding to the gel types, are approximately linear. It should be noted, however, that for the points of reference Determann, too, chose globular proteins with similar friction coefficients and similar specific partial volumes. The properties of ferritin, urease and fibrinogen deviated from the equation (Sephadex G-200); for these proteins the Determann formula yielded correct results only if the logarithm of the Stokes radii rather than the molecular weight was substituted into it. From a critical study of Siegel and Monty's (1966) observations Determann (1969) stated that, owing to their extremely high molecular weights, ferritin, fibrinogen and urease fall outside the linear fractionation range of Sephadex G-200 and that their aggregation and carbohydrate content are not known, accurately enough.

Unlike dextran gels, the polyacrylamide gels, apart from some special fields, as for example the examination of ribonucleases (Anderson and Carter 1965) and globulins (Alvord *et al.* 1966; Inman and Nisonoff 1966), have rarely been used so far for molecular weight determinations. Ward and Arnott (1965) demonstrated in the case of glycoprotein hormone that, owing to the greater space requirement of the carbohydrate content of the molecule, the logarithm of the molecular weight is not proportional to the elution volume, even for polyacrylamide gels. Battle (1967), working with fourteen proteins in a simple set-up, found linear correlation similar to the Andrews and Determann isotherms on Bio-Gel P-100, P-200 and P-300. His examinations confirmed that, because of its dissociation, haemoglobin cannot be used as a standard protein and that glycoproteins also show anomalous behaviour on polyacrylamide gels.

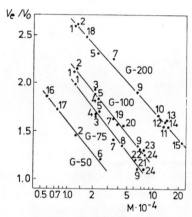

Fig. 29. A graphic illustration of the Determann equations. The correlation between the relative elution volume (V_e/V_o) and the molecular weight of proteins, on Sephadex G dextran gels. The proteins and their molecular weights pertaining to the reference points were as follows: 1. cytochrome C 13 000; 2. ribonuclease A 13 600; 3. trypsin inhibitor (soya + bean) 21 500; 4. α chymotrypsin 22 500; 5. trypsin 24 000; 6. chymotrypsinogen A 25 000; 7. pepsin 35 500; 8. albumin 45 000; 9. bovine serum albumin (monomer) 6 700; 10. glyceraldehyde-phosphate dehydrogenase 117 000; 11. bovine serum albumin (dimer) 134 000; 12. aldolase (yeast) 147 000; 13. human γ globulin 140 000; 14. alcohol dehydrogenase (yeast) 147 000; 15. catalase 230 000; 16. kallikrein inhibitor 6 500; 17. trypsin inhibitor (lima beans) 8 400; 18. methemoglobin 17 000; 19. peroxidase-1 40 000; 20. α hydroxysteroid dehydrogenase 47 000; 21. glycerophosphate mutase 64 000; 22. malate dehydrogenase 79 000; 23. enolase 80 000; 24. creatine phosphokinase 81 000

In the determination of molecular weights and molecular sizes above the fractionation range (maximum 300 000) of dextran and polyacrylamide gels, the introduction of agarose gels was a significant event (Hjertén 1962c; 1968; Locascio *et al.* 1969; Marrink and Gruber 1969; Joustra 1970; Lehmann 1970). Agarose gels of different concentrations permitted the determination of the "virtual" molecular weight of excessively large molecules, e.g. catalase (Lynn 1971), lipoproteins (Margolis 1967; Hanai *et al.* 1968; Maggi *et al.* 1968; Papenberg *et al.* 1970; Bowden and Fried 1970; Quarfordt *et al.* 1972), nucleic acids and nucleoproteins (Kingsbury 1966; Erikson and Gordon 1966; Öberg and Philipson 1967; Loeb 1968; Petrovic *et al.* 1971), polysaccharides (Malchow and Lüderitz 1967; Bathgate 1970), enzymes (Pagé and Godin 1970), macroglobulins (Killander *et al.* 1964; Russel and Osborn

1968), viruses (Pettersson et al. 1968), phages (Höglund 1967; Yoshinaga and Shimomura 1971), cell fractions and particuli (Tangen et al. 1971).

Figure 30 shows the selectivity curves of the Sepharose 2B, 4B and 6B gels containing 2, 4, and 6% agarose, respectively (Pharmacia AB, Uppsala), versus the selectivity curves of the Sephadex G-200 dextran gel. These curves prove that the calibration points plotted for dextran fractions are very different from the values determined for proteins.

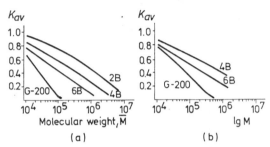

Fig. 30. Calibration curves for Sepharose 2B, 4B and 6B agarose (and Sephadex G-200 dextran) gel columns for molecular weight determination. The selectivity curves for the polysaccharides (a) were determined by dextran fractions of known molecular weight. The base points of the calibration curves plotted for proteins were: ribonuclease, ovalbumin, transferrin, glucose oxidase, thyroglobulin and α crystalline (from Beaded Sepharose 2B, 4B, 6B; Pharmacia Fine Chemicals, Uppsala, 1969)

An interesting and novel application of agarose gels in molecular weight determinations is the chromatography of proteins and polypeptides in guanidine-*hydrochloric acid* media. Contrary to expectation, the structure of the agarose gels built up with hydrogen bonds does not become disrupted by guanidine-hydrochloric acid; simultaneously, the reagent causes the proteins and polypeptides to assume the same (random coil) configuration and eliminates the form factors which render the determination of molecular weight difficult. In this manner Sepharose 6B enables very sensitive determinations in the molecular weight range from 1 400 to 80 000 in 6 M guanidine-hydrochloric acid to be carried out (Bryce and Crichton 1971; Heinz and Prosch 1971; Klaus et al. 1972).

2.6.3.4. DETERMINATION OF THE MOLECULAR WEIGHT BY THIN LAYER GEL CHROMATOGRAPHY

The use of gel particles in thin layer gel chromatography (for more details see Part III) is a quick and efficient method for the determination of the molecular weight of various macromolecules and proteins (Determann 1962; Johansson and Rymo 1962, 1964; Morris 1964; Andrews 1964; Determann and Michel 1965; Hanson et al. 1966; Wieland and Determann 1967; Aspinall and Miller 1973). Determann and Michel (1965), in the same way as in column chromatography, also measured the constants of the equations formulated for molecular weight determination by thin layer gel chromatography. As standard protein and point of reference they used the coloured cytochrome C molecule and introduced the so-called cytochrome retention factor, R_{cyt}. On Sephadex G-200 superfine gel they arrived at the following relationship for the determination of the molecular weight:

$$\lg M = 3.024 + 1.092\, R_{cyt}.$$

The high adsorption of cytochrome C on Sephadex gels (Siegel and Monty 1966) makes this protein unsuitable to act as the base point for any method of molecular weight determination. Also the experimental conditions (buffer, pH, ionic strength, etc.) which determine adsorption are still unclarified.

A fact of general validity for all gel types is that the logarithm of the molecular weight is proportional to the reciprocal value of the migration distance d (Radola 1968a, b; Jaworek 1970a, b; Wasyl et al. 1971):

$$\lg M = k\,\frac{1}{d}$$

This relationship naturally holds only for the combination of such molecules and gels in which the molecule may occupy the inner volume of the gel phase and where the molecular weight falls within the linear portion of the selectivity curve of the gel. Since, in the course of molecular weight determination by thin layer chromatography, fully excluded and fully permeable (marker) molecules and standard compounds of dissimilar migration rates are run in parallel, a comparison is possible with any (standard) molecule in the following way

$$\lg M_x = \lg M_{st}\,\frac{d_{st}}{d_x}$$

where M_x and M_{st} denote the molecular weights of the unknown and standard compounds, and d_x and d_{st} are the distances to which the unknown and the standard compounds, respectively, migrate from the point of application.

In globular proteins, a similar fair agreement exists between the layer chromatographic properties and the logarithm of the effective (Stokes) radii (Wasyl et al. 1971) which enables the determination of the molecular dimensions to be carried out to an accuracy of 5 to 10%.

2.6.3.5. THE EXAMINATION OF POLYDISPERSE SYSTEMS

The determination of the molecular weight of polymer mixtures which consist of several individual constituents having similar physicochemical properties is one of the more important applications of gel chromatography (Nilsson and Nilsson 1974; Berger 1974, 1975). The average molecular weight of polydisperse systems may be characterized by the average molecular weight referred to the molecular number (M_n), or by the average molecular weight by weight ($\overline{M}$)

$$M_n = \frac{\Sigma n_i M_i}{\Sigma n_i}\,; \qquad \overline{M} = \frac{\Sigma n_i M_i^2}{\Sigma n_i M_i}$$

where n_i and M_i denote the molecular number and the molecular weight, respectively, of the components found in the ith fraction.

The average molecular weight referred to the molecular number is expressed by the coefficient of the sum of the total quantity of the materia and hte molecular numbers. The number of molecules and the molecular weights are generally determined by end-group analysis and by measurement of the osmotic pressure. The average molecular weight by weight considers every molecular size in the ratio of its weight determined in the sample.

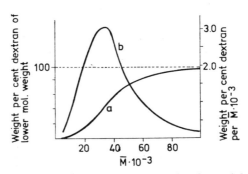

Fig. 31. Gel chromatographic examination of the molecular weight distribution of the Rheomacrodex (Pharmacia AB, Uppsala) dextran fraction ($\overline{M} = 43\ 900$; $\overline{M}_n = 32\ 700$). Column chromatography was applied to measure the dextran content and the molecular weight of the separated fractions. The differential curve of the molecular weight distribution (a) indicates the percentage of the material introduced to the column which will be eluted up to a defined value of the molecular weight. Its integral curve (b) shows the weight per cent distribution of the dextran fractions having dissimilar molecular weights, calculated for 10^3 molecular weight units

This parameter is determined generally from measurement of light dispersion and viscosity. If the sample is heterogeneous, then M_n is always greater than $\overline{\mathrm{M}}$. The coefficient $\overline{\mathrm{M}}/\mathrm{M}_n$ indicates the degree of inhomogeneity of polydisperse systems. Its value for homogeneous materials is 1, for sharply distributed polymer fractions between 1.0 and 1.20, and for technical polymer mixtures between 10 and 1,000.

Making use of the information obtained from the gel chromatography of polydisperse systems (the quality, the molecular number, viscosity, the refractive index, the average molecular weight of the materials measured in fractions) one arrives at the characteristic (integral and differential) molecular weight distribution curves according to *Figure 31*. In practice, the elution (differential) curves are plotted without the collection of fractions, merely using the continuous recording of the material properties. To convert them into integral curves, a computer program is used. To determine the molecular weights, the columns are calibrated with a series of standard polymer fractions of sharp distribution (dextran, polystyrene), corresponding to the material, solvent, etc., being examined. About the nature of a polymer, the mechanism of polymerization, the modality of the polymer, its branching, decomposition, the effect of the additives, etc., only the molecular weight distribution curves yield a realistic picture. These data were previously plotted by lengthy, cumbersome and laborious methods. Gel chromatography simplified the examination of natural and synthetic hydrophilic and apolar polymers to a very great extent.

2.6.3.6. THE EMPIRICAL EQUATIONS AND GENERAL CORRELATIONS OF THE MOLECULAR WEIGHT DETERMINATION

Parallel to the steric-volumetric and molecular-kinetic theories, many attempts were made to formulate the empirical equations of gel chromatography. *Table 19* surveys the empirical correlations obtained between the chromatographic parameters and the molecular weight of the solutes in various applications, with substances of different polarity and with solvents. It will be evident that the empirical equations express basically the same logarithmic correlation between the molecular weight and the elution parameters. Many authors verified the validity of the models derived by Porath (1963) and Ackers (1964), in various gel chromatographic applications (Auricchio and Bruni 1964; Carnegie 1965a, b; Brewer 1965). Some authors, for example Moore and Hendrickson (1964, 1965), related the molecular properties to the elution volume on the basis of the carbon atom number (the length of the C—C chain) of the polymers to be separated. Others again recommended the use of the hydrodynamic volume (Grubisic-Gallot and Benoit 1969; Le Page *et al.* 1968), or in the case of apolar small molecules, the molar volume (Cazes 1966, 1967), as the parameter of molecular size. Meyerhoff (1965a, b), partly on the basis of theoretical considerations, in addition to the steric effects and the viscosity number, also took into account the molecular interactions. His equation yielded satisfactory results primarily for apolar polymers of different configurations (vinyl, styrene, polymethyl methacrylate, cellulose trinitrate).

The theory of molecular weight determination needs certain complementary remarks, in several respects. A precondition for the application of gel chromatography is the existence of an unequivocal correlation between the space requirement and the molecular weight of the substances examined. The correlations of volumetric-hydrodynamic parameters and the molecular sizes sometimes suggest more complex shape and structural factors than the molecular weight calculated from the chemical formula. The theoretically derived and the empirical equations refer to macromolecules of nearly the same partial specific volume and of a rigid closed structure. It is assumed that the aromatic and apolar groups of the macromolecules are intra-structurally located and the adsorptive, ionic, etc. interactions of the gel are negligible. Apart from a few exceptional cases (e.g. fibrinogen) the shape factors of the natural proteins are not essentially different from the space requirements assumed in the theoretical derivations.

Their globular shape and molecular weight render most simpler proteins and enzymes particularly well suited to demonstrate the correlations of gel chromatography. On the other hand, the filament-shaped nucleic acids have a substantially lower exclusion molecular weight on the various gels than proteins of the same molecular weight. Also, polysaccharides have a larger space requirement and an elution volume smaller than their molecular

Table 19

Empirical relationships between the volumetric parameters and the molecular weight of solutes in gel chromatography

Equation	Solute	Solvent	Gel type	Author(s)
$V_e/V_t = k \cdot \lg M$	Polysaccharides	Water	Dextran	Granath and Flodin (1961a, b)
$V_e - V_0 = k \cdot \lg M$	Paraffins	Toluene	Rubber	Brewer (1961)
$K_d^{1/3} = k_1 - k_2 \cdot M^{1/2}$	Proteins, enzymes	Water	Dextran	Wieland et al. (1963)
	Oligostyrenes	Chloroform	Acrylate	Determann et al. (1964)
	Globular proteins	Water	Dextran	Auricchio and Bruni (1964)
	Oligopeptides	Phenol acetic acid–water	Dextran	Carnegie (1965a, b)
$V_e/V_0 = k \cdot \lg M$	Proteins	Water	Dextran	Whitaker (1963)
$\ln K_d = k \cdot M$	Dinucleotides	Water	Dextran	Hohn and Pollmann (1963)
$V_e = k \cdot \lg M$	Proteins	Water	Dextran	Andrews (1964)
$K_d = k_1(\lg M - k_2)$	Peptides; hormones	Water	Dextran	Sanfelippo and Surak (1964)
$V_e = k \cdot \lg (\text{chain length})$	Polyethylene; polyethers	Apolar	Styrene	Moore and Hendrickson (1965)
$V_e = k_1 - k_2 \lg (M^{1/2} \cdot \eta^{1/3})$	Cellulose nitrate; polystyrene; polymethacrylate	Tetrahydrofuran	Styrene	Meyerhoff (1965a, b)
$k_1 - k_2 \cdot \lg V_e = M$	Polysobutylenes	Cyclohexane	Synthetic rubber	Brewer (1965)
$(k - 1.234)^2 + (y - 1.234)^2 = 1.577$ $y = \left(\dfrac{2.13 \cdot M^{0.413}}{P}\right)^2$	Proteins; polysaccharides	Water	Porous glass beads (CPG)	Haller (1973)

weights indicate (Ward and Arnott 1965; Anderson and Stoddard 1966b). The polysaccharide content causes a considerable change in the gel chromatographic behaviour of the glycoproteins (see Section 2.6.3.1 of Part I). When assessing the results, the sensitivity of the macromolecules to denaturizing effects must also be taken into consideration.

It may be stated in general that, through structural changes, the environmental effects increase the space requirement of the macromolecules. The effect of guanidine-hydrochloric acid has already been mentioned in Section 2.6.3.3 of Part I. In a 5 M urea solution, albumin divides into two fractions of which the denatured product seems to have the higher molecular weight (Ward and Arnott 1965; Olesen and Pederson 1966). Acylation causes an extremely large increase in the space requirement of albumin (Habeeb, 1966). Thermal denaturation influences the gel chromatographic behaviour of proteins by modifying the tertiary structure (Selby and Maitland 1965; Leach and O'Shea 1965). The reduced elution volume of albumin under heat treatment is regarded as one of the experimental proofs supporting stericvolumetric theories (Přistoupil 1965; Přistoupil and Ulrych 1967; Ulrich and Přistoupil 1971). Similar phenomena are caused by chemical conversion, alkalic treatment, the performic acid oxidation, etc. of proteins (Lynn 1971). For instance, the denaturation of ribonuclease can be closely followed by changes in the virtual molecular weight (De la Llosa *et al.* 1966). This is not merely a technical problem since, on the non-linear portions (at either end) of the selectivity curves of the gels, molecular weight determination becomes inaccurate and may cover up differences arising from fine-structural changes. The results of molecular weight determination may be also affected by aggregation, by the association of similar or dissimilar molecules and ions, or by complex formation, etc. These factors are borne out by examination of the experimental conditions and by the dependence of gel chromatography on concentration. The study of the elution profile of chemically reacting, or reversibly associating systems has done a great deal to solve similar problems (Winzor and Scheraga 1963; Ackers and Thompson 1965; Winzor and Nichol 1965; Winzor 1966; Gilbert 1966; Nichol *et al.* 1969; Cheetham and Winzor 1970; Ackers 1970). The effect of the polarity of the solvents and of the gel structure will be dealt with in more detail in connection with adsorption.

Summing up, it may be stated that the determination of the experimental conditions in accurate detail and the consideration of the above outlined molecular and structural interaction are indispensable prerequisites for the determination of molecular weight by gel chromatography.

2.7. THE THERMODYNAMIC ASPECTS OF GEL CHROMATOGRAPHY

On the basis of molecular-kinetic theories the thermodynamic examination of gel chromatography sets out from the assumption that, in near-equilibrium states, the transition of solutes from free solution to the gel phase can be calculated from the affinity differences related to the gel and the solvent, and from the changes in the enthalpy and entropy of the gel chromatographic process. Accordingly, the thermodynamic state of the solutes in the gels undergoes a change in relation to the liquid space of the flowing phase, and the mechanism of separation is basically similar to the principle of liquid-liquid partition chromatography. Brönsted (1931) proved that the equilibrium of the partition of the various phases depends on the molecular weight of the substances which take part in it

$$K_d = \frac{C_1}{C_2} = e - \lambda \frac{M}{kT}$$

where C_1 and C_2 denote the concentration of the substances in the different phases, M the molecular weight, k the Boltzmann constant, T the absolute temperature, and λ a constant which characterizes the substances and the phases.

If low molecular weight materials are concerned, the equation shows the tendency towards uniform partition between the phases ($C_1 = C_2$). With increasing molecular weights, however, the equilibrium shifts, to an increasing degree, towards one of the phases. Brönsted's equation was in several respects suitable also for the interpretation of certain gel chromatographic phenomena, such as the exclusion of macromolecules and the uniform distribution of the small molecules. The relative insensitivity of the method to variations in the temperature and in the flow velocity, and the close correlation between the molecular sizes and the elution volume, directed attention to the mechanism of steric exclusion. However, the formal correlations between the molecular sizes and the volumetric parameters could not explain the molecular interactions to a sufficient degree. Increasing the dry-matter content of the gels and reducing the pore sizes, increasingly well defined adsorption, polarization, ion exchange, etc. effects were observed which depended on the temperature and the rate of flow.

The thermodynamic theories call for a more detailed study of the molecular phenomena of gel chromatography and examination of the interactions between the gels and the solutes from the aspects of (1) the distribution of immiscible polymer solutions, (2) the osmotic properties of the gel phase, and (3) the effect of adsorption and temperature (concentration).

2.7.1. SIMILARITY BETWEEN THE DISTRIBUTION EQUILIBRIUM OF GEL CHROMATOGRAPHY AND IMMISCIBLE-PHASE POLYMER SOLUTIONS

The similarity between gel chromatography and the immiscible polymer solutions is one of the starting points on which thermodynamic theories rely. According to previous observations, dilute aqueous solutions of differently charged, or neutral, macromolecules, while forming complex coacervates, divide into two (or more) immiscible phases. In two of his monographs, Albertsson (1960, 1970a) gave a detailed description of the formation and properties of two-phase polymer systems. The aqueous solutions of synthetic polymers (e.g. polypropylene glycol, polyethylene glycol, polyvinyl alcohol, methylcellulose, hydroxypropyldextran, dextran fractions, dextran sulphate, natural macromolecules, proteins, nucleotides, cells and subcellular particles, viruses, etc.) may form immiscible phases which permit their preparative separation (Albertsson and Baird 1962; Hofsten 1966; Walter et al. 1965, 1967, 1968a, 1969; Karlstam and Albertsson 1969; Griebel and Smith 1968; Albertsson 1965; Alberts 1967; Albertsson and Nyns 1959, 1961; Iverius and Laurent 1967; Tiselius et al. 1963; Walter and Sasakawa 1971; Albertsson 1970b; Pettijohn 1967; Albertsson 1967; Philipson 1966). Further literature is available in the publication entitled *Dextran Fractions* of the Pharmacia AB, Uppsala 1971. The division of the phases and the distribution of the solutes (the polymers proper) can be described thermodynamically by the free energy and entropy variations of mixing. The changes in energy arising from the various interactions can be controlled by the molecular size and the charge, by the polarity of the medium, the concentration, the quality and the charges of the ions (Albertsson 1960, 1965; Walter et al. 1968b; Johansson 1970a, b; Edmond and Ogston 1970; Albertsson et al. 1970a, b). As for the correlation between the coefficient of interphase partitition and the molecular weight of the substances participating in it, the Brönsted equation given earlier in this section holds valid. The thermodynamic reasons for the phase separation were studied in detail by Edmond and Ogston (1968, 1970).

The analogy between the mechanism of gel chromatography and the incompatibility of the polymer solutions was verified also by Laurent (1963a, b). Depending on their hydration energy and molecular weight, polysaccharides (dextran) and proteins cause changes in one another's solubility. Plasma proteins may be precipitated by the addition of dextran or polyethylene glycol (Iverius and Laurent 1967).

The similarity of immiscible polymer solutions and gel chromatography is reflected most clearly in the theory proposed by Heitz and Kern (1967), according to which the structure of macroporous gels may be regarded as sterically fixed polymer solutions whose concentrations decrease as they proceed towards the inside of the pores. This explanation, from the aspect of the steric-volumetric theories, may be interpreted in such a way that the voids (outer volume) among the gel particles are not constant but depend on the degree of solvation of the polymer network of the gel. When the various (macro)molecules interact with the gel structure they cause a change in the partner's solvation process. Ogston and Silpananta (1970), with respect to the intrinsic osmotic properties of gels, demonstrated that the inner volume of the Sephadex gels varies according to the quality and concentration of the solutes, and that, in materials with extremely high osmotic activity (e.g. polyethylene glycol), a complex coacervation can be observed in the gel particles with a phase contrast microscope. This effect, arising from fixation of the polymer structure, will manifest itself in the dissimilarity of the volumes accessible (elution volume) to the solutes. Apart from extremely large molecules, the flexibility of the polymer chains and the dynamic equilibrium of the solvation process do not prevent, theoretically, penetration of the macromolecules into the gel pores. However, owing to the incompatibility of macromolecules, some of the solvent molecules in the pores are not accessible to the solutes.

With regard to the role of macromolecular interactions, Hellsing (1965, 1968), by chromatographing albumin on Sephadex G-200 column in solutions of different polymers (dextran, polyethylene glycol, polyvinyl alcohol, Ficoll), observed that changes take place in the elution volume.

Ogston and Wells (1972), using sulpho-ethyl-Sephadex, modelled the osmotic properties of the polysaccharide-polyanionic structure of the gel-like cartilage tissue.

2.7.2. EXAMINATION OF THE OSMOTIC PROPERTIES OF GELS

Edmond et al. (1968) found that the osmotic properties of gels provide a possibility of studying the mechanism of gel chromatography from thermodynamic aspects. They established that if the outer osmotic pressure is increased by a completely excluded macromolecule, the inner volume of the Sephadex gel will decrease. The effect of molecules which occupy only part of the gel structure is weaker since the penetrating materials also contribute to the raising of the inner osmotic pressure. Substances of very high osmotic activity also modify the solvated structure of gels (Ogston and Silpananta 1970).

Nichol et al. (1969), taking into consideration the osmotic equilibrium of the gel particles, regarded the gel polymer network to be a semipermeable film, and derived an equation for the membrane model which corresponds to the Gibbs–Donnan equilibrium and is in good agreement with experimental results. Comparing their data with the findings of Edmond et al. (1968), they came to the conclusion that the Gibbs–Donnan equilibrium plays a significant role in influencing the behaviour of charged macromolecules, and that gel chromatography may be regarded as analogous with the system of liquid-liquid partition.

Polson and Katz (1969) examined the osmotic interactions between the gel phase and albumin molecules on agarose particles and derived the following theoretical relationship

$$G = \frac{1}{M} \cdot \frac{K_d}{1} - 1$$

where G denotes a constant related to the interaction of the polymer fibres of the gel and the Brownian movement of the proteins, M denotes the molecular weight of the protein and K_d denotes the volumetric distribution coefficient for the gel chromatographic process.

Polson and Katz, using globular proteins of known molecular weights, found that the value of G is constant ($G = 3.28 \cdot 10^{-5}$). The K_d values of the fibrillar molecules, lower than expected, were attributed to the more intensive interaction of the fibriform molecules and the gel polymer fibres.

2.7.3. THE INFLUENCE OF ADSORPTION AND TEMPERATURE IN GEL CHROMATOGRAPHY

Gel chromatography is well known as the method used for separation by size of large molecules. However, with the increasing number of crosslinks and the increasing density of the gel structure, further molecular interactions come to the fore which decisively determine the mechanism of the separation of small molecules. From the point of view of the elution volume, the processes for the separation of small molecules fall into three groups: (1) when during normal elution the separation corresponds to the molecular sizes, (2) when, in aromatic, aliphatic or electrostatic adsorption the elution volume is larger than expected, and (3) when the phenomenon of ionic exclusion causes a reduction in the elution volume.

The examination of the different (mainly dextran) gels has shown that the polymer chains exhibit a considerable capacity to adsorb aromatic, alicyclic and heterocyclic compounds (Gelotte 1960; Granath and Flodin 1961; Glazer and Wellner 1962; Marsden 1965; Carnegie 1965a, b; Gelotte and Porath 1966; Janson 1967; Eaker and Porath 1967; Lindquist 1967; Sweetmann and Nyhan 1968, 1971; Brook and Housley 1969a, b; Khym and Uziel 1970; Simkin 1970; Brook and Munday 1970a; Wasternack and Reinbothe 1970; Ehrlich et al. 1971; Crone 1971; Ziska 1971; Determann and Lampert 1972; Wasternack 1972a, b; Williams 1972; Brown 1972). Determann and Walter (1968) proved that the aromatic adsorption of dextran gels is localized to crosslinks of the ether type.

Marsden (1965), over and above aromatic adsorption, demonstrated that numerous aliphatic compounds, mainly the n-alcohols of non-ideal properties, for example urea, thiocarbamide, formamide, etc., show a behaviour during gel chromatography which is in contrast to their molecular size and to the steric theory. Many observations have proved that carbohydrates and polysaccharides are bound to dextran gels (Öbrink et al. 1967; Brown and Andersson 1971; Goodson et al. 1971; Bertonière et al. 1971; Carter and Lee 1971; Martin et al. 1972). Streuli (1971a, b) found that the sorption phenomena evolved from the combinations of the π-electron and hydrogen bonds, depending on the properties of the solvent, of the substances to be separated and of the gel structure.

Similar experience was gained during the gel chromatography of ions and inorganic salts (Porath 1964; Saunders and Pecsok 1968; Neddermeyer and Rogers, 1968, 1969; Ortner and Spitzy 1968a, b; Yoza and Ohashi 1969; Ueno et al. 1970a, b, c; Ogata et al. 1971; Yoza et al. 1971a, b). Pecsok and Saunders (1968) established that while cations are bound to the gel by a weak physical adsorption, anions (for example, to the amide groups of polyacrylamide gels) are bound by hydrogen bonds. Non-sorbent ions and molecules on gels are partially excluded from the inner volume owing to the strong hydration of the gel matrix. With low ion concentration the K_d values (particularly with dextran gels) decrease considerably, according to the Donnan equilibrium, on account of the weak ion exchange groups of the gel and the exclusion of the anions. At higher temperatures the K_d values were lower. Yoza et al. (1971b) determined a series of endothermically and exothermically reacting ions, which made possible the introduction of thermal detectors in the gel chromatography of inorganic substances.

Whitaker ascertained as long ago as 1962, that, except for a few proteins, under normal laboratory conditions the elution volume of proteins is independent of temperature. Marsden (1965) came to the same conclusion. With respect to gel adsorption, he demonstrated also that compact gels, with a large number of crosslinks, and substances of low molecular weight are greatly dependent on temperature. Leach and O'Shea (1965) examined variations of the elution volume of nine proteins in the temperature range between 25 and 40 °C. They found that the influence of temperature decreases with increasing molecular weight. Temperature influences the volumetric parameters by causing changes in the molecular structure and in the space requirement. Selby and Maitland (1965) established that variations in the gel structure (swelling) are also determinant factors. In the case of apolar gels (Styragel), on the other hand, the elution volume was found to be independent of the temperature (Moore and Handrickson 1965). Lampert and Determann (1971a) observed that, as regards the temperature-dependent changes in gel porosity, dextran and polyacrylamide gels show contrary behaviour. They found, at the same time, considerable changes on Sephadex LH-20 which they attributed to the disruption of the hydrogen bonds (oxonium structures). Öbrink et al. 1967), from the chromatography of a high molecular weight polysaccharide fraction on Sephadex G-200, established that the separation mechanism is created primarily by steric effects and that temperature has little effect, through causing changes in the flexibility of the polymer chains.

Brown (1970, 1971) demonstrated by thermodynamic calculations that, owing to weakening interactions between the gel and the solvent and between the solvent and the solute on Sephadex G-15 dextran and hydroxyethylcellulose gels, the distribution coefficient increases with rising temperature. He experienced the contrary with polyacrylamide (Bio-Gel P2) gel and found unchanged linear isotherms in polystyrene gels (Poragel 60). Following the investigations of Brown in 1972, he concluded that the value of the distribution coefficient of low molecular weight substances depends on energy and steric factors that can be expressed in terms of the changes in enthalpy (ΔH^0) and entropy (ΔS^0):

$$K_{av} = e^{-\Delta H^0/RT} \cdot e^{\Delta S^0/R}$$

From the point of view of thermodynamics, the steric mechanism of separation is characterized by a preponderance of the entropy factor, which is independent of temperature. In the structure of macroreticular gels (Kirret et al. 1966), numerous substances, including most proteins, show the same behaviour. In the above relationship the temperature-dependence of the distribution coefficient is expressed by the enthalpy factor which, in turn, can be described by the interactions between solute and gel (adsorption). The polarity of the solvent and its interaction with the solute or the gel structure is of similar significance. In his experiments Brown found considerable differences between the dextran and hydroxyethyl cellulose, or polyacrylamide types of gel. In the former, temperature causes a reduction of the interactions between the gel and the solvent and between the solvent and the solutes (exothermic behaviour). In addition, two effects come into play during chromatography. On the one hand desolvation of the gel enhances gel–solute interactions and augments the accessibility of the gel surface, while on the other it reduces the solubility of the solute. Both effects promote sorption. Although higher temperatures also reduce interactions between the gel and the solute, this reduction is of a much smaller degree. In contrast to this, higher temperatures increase the interactions of gel and solvent in polyacrylamide gel (endothermic behaviour). At higher temperatures, on the other hand, sorption effects tend to become weaker. It is assumed that there are essential differences also in the localization of water molecules in the structure of dextran and polyacrylamide gels. The more polar the gel structure (e.g. dextran) the lower the sorption effect, owing to the inhibitory effect of the solvent. The effect of strong electrolytes (NaCl) on the sorption phenomena of gel chromatography is characteristic. On dextran gels the sorption effect (distribution coefficient) increases because the salt tends to cause desolvation of the gel matrix. In the case of polyacrylamide gels, on the other hand, NaCl reduces the distribution coefficient, suppressing the influence of the amide groups. These phenomena characterize the dissimilar separation mechanism of the two gel types.

In dextran gels solvated water represents the deactivating layer which reduces interaction between solutes and the hydroxyl groups of the gel structure. The effect of the polarity of the solvent is clearly demonstrated by dimethylformamide, with which the chromatography of cellodextrin on hydroxyethylcellulose gel is completely insensitive to temperature. Temperature in such cases does not cause any appreciable change in the degree of gel solvation. Also the distribution coefficient is insensitive to temperature in the cellodextrin-dimethylformamide—polystyrene gel system.

Similar tests performed on cellulose based gels indicate the effect of solvents and temperature (Chitumbo and Brown 1973a, b; Brown and Chitumbo 1975).

These observations confirm the well-known rule of adsorptive chromatography according to which the sequence of elution and the effectiveness of separation are determined by the relative polarities of the solutes, the gel interface and the solvent. It may be stated in general that, in the chromatography of polar substances, lower polarity of the solvent enhances the sorption effect. With apolar substances the reverse holds true.

2.7.4. THE INFLUENCE OF THE CONCENTRATION OF SOLUTES IN GEL CHROMATOGRAPHY

It follows from thermodynamic derivations that, theoretically, gel chromatography cannot be independent of the concentration of the solutes. In practice, however, to detect any appreciable change in the elution properties, very high concentrations are required. The determinant factor is mainly the viscosity of the sample, which may render the flow conditions irregular and increase zone-broadening to such an extent that it changes the elution profile completely and deteriorates the resolution. The effect of viscosity may be compensated by using an eluant of similar viscosity.

In theory the effect of concentration lies in two factors: (1) the distribution isotherms are non-linear and dependent on concentration, and (2) the solutes undergo molecular changes (association, dissociation, complex formation, etc.; Baghurst et al. 1975). The influence of concentration on the volumetric parameters of gel chromatography was studied by means of molecular weight determination. Andrews (1964) observed a reduction in the molecular weight of haemoglobin and β-lactoglobulin when these were diluted, an observation which may be attributed to the associative equilibrium of the subunits. Similar examples were quoted by Determann (1969) on concentration-dependent aggregation and in connection with enzyme inhibitor—enzyme substrate interactions. Winzor and Nichol (1965), while examining ovalbumin, trypsin and serum albumin on Sephadex G-100 gel, found that the elution volume depends on concentration. They proved that concentration-dependent distribution isotherms are non-linear. They could not detect a similar effect with the dextran 500 polysaccharide fraction. Winzor (1966) studied in detail the effect of concentration on the zonal and frontal variants of gel chromatography. Polson and Katz (1969) established an increase in the K_d value at very high concentrations.

PART II
METHODS AND TECHNIQUES

1. GEL CHROMATOGRAPHIC METHODS

The discussion of the methods and techniques of gel chromatography follows the logical sequence of practice: the aspects of selection, treatment, packing into columns or other uses of the gels are first discussed, then types, pore size, chromatographic processes and equipment and the theories of chromatography are considered from the point of view of the physico-chemical properties of the solutes and the actual job in hand. Apart from theoretical research, gel chromatography is applied mostly at some intermediate stage of a chemical operation (purification, separation), which presupposes a certain degree of familiarity with the chemical composition, molecular weight, stability, solubility, etc. of the substances investigated. In the lesser but theoretically more important part of the applications, gel chromatography serves to detect the presence of similar substances or the characterization of the properties (molecular weight, molecular size, molecular distribution, aggregation, complex formation, etc.) of the (isolated) substance in its pure form. Accordingly, gel chromatographic methods fall into two categories.

Group separation is the collective name for those operations in which the molecular weights of the separated compounds differ widely and contain other components as well (for instance protein desalting, the separation of peptides, proteins, oligosaccharides and polysaccharides, etc.). Relatively coarse separation may take place on large samples in short columns, or simply in a centrifuge tube (see the section on the batch process).

Fractionation is carried out under precisely determined conditions, using the small differences between the molecular sizes to distinguish between compounds of related properties or for their identification (e.g. the fractionation of oligosaccharides and albumin oligomers, molecular weight determinations, molecular distribution investigations, etc.). In fractionation the sample volume is limited and the optimum adjustment of the hydrodynamic parameters of the long and slim columns (packing, grain size, flow rate, etc.) are of particular importance for resolution. There is no sharp difference between the two processes, the choice generally depending on the type of job in hand.

As regards its purpose, gel chromatography may be preparative or analytical. Group separation, as the name implies, is preparative in character, fractionation being performed mostly for analytical purposes.

1.1. SELECTION AND PREPARATION OF GELS

The selection of the gel depends first and foremost on the solubility of the substance to be tested. For substances which dissolve readily in water (polar) or in organic solvents (apolar), hydrophilic and hydrophobic (organophilic) gels, respectively, should be used. Column packings absorbing polar and apolar solvents alike (e.g. Sephadex LH-20, porous glass beads, Bio-Beads SM-1, SM-2) may also be used. Such packings are particularly advantageous if the substances have intermediate properties whose separation may be enhanced or modified by stressing the polar or apolar character.

Another aspect considered when selecting the gel is the purpose of the chromatographic process and the molecular weight of the substances to be separated. For group separation a pore size should be chosen in which the low molecular weight components fill out the entire gel bed ($V_e = V_o + V_i$) and the large molecules are excluded from the inner space of the bed ($V_e = V_o$). To speed up the process small-pored gels are preferable in which a high flow rate can be achieved and which are easier to control. The minimum pore size is determined by the components which have the lowest molecular weight. Thus, for example, for the desalting of proteins even the densest gel types (Sephadex G-10, G-25, Bio-Gel P-2, P-6) may theoretically be used. However, with increasing polymer content of the gels, macromolecular adsorption becomes increasingly manifest. For such applications therefore Sephadex G-50 should be chosen which offers another, additional, advantage: water adsorptivity, or more specifically a specific column capacity which is two times longer than that of Sephadex G-25. For the separation of, for example, peptide fragments with molecular weights between 3 and 30 000, or proteins with a molecular weight of 300 000, the use of gels with smaller pores than those of Sephadex G-75 should be avoided because they would exclude even peptides from the inner volume of the gel phase.

The choice of the gel type depends also on the volume of the sample and the substances to be separated. For group separation (e.g. proteins and amino acids) relatively large samples are required (maximum 0.3 V_i; see Section 2.3 of Part I). If the aim is to recover the macromolecules with a minimum of dilution, denser gels should be used which by their low specific swelling restrict the volume of the column (V_o) and also that of the sample. If, on the other hand, the aim is to collect the amino acids in a relatively small volume then a porous gel should be chosen on which the degree of fractionation, i.e. zone broadening, is less.

In contrast to the above, the gel type for fractionation should be so selected that the nominal or expected molecular weight of the substances falls in the middle of the separation molecular weight range of the gel (see Section 2.6.3.1 of Part I). The steeper the slope of the selectivity curve of the gel and the smaller the molecular weight range covered by the distribution of the pore size, the more effective will the fractionation be. However, if more complex, multi-component (polydisperse) systems are examined, uniform broadening of the fractionation range might become necessary. Granath and Kvist (1967), for instance, used a mixture of Sephadex

G-100 and G-200 for the determination of the molecular weight distribution of dextran products (in the 10 000 to 150 000 range).

Extremely large molecules, above all other factors, determine the gel type applicable. Lipo-, glyco- and nucleoproteins, in general, compounds with molecular weights above 300 000, viruses and cell organella must be chromatographed on agarose gels of appropriate pore size. Gerlich et al. (1970) demonstrated, that, on Sephadex, agarose and cellulose gels, particles larger than 400 nm become irreversibly adsorbed.

For the various applications of gel permeation chromatography a wide variety of synthetic organophilic, wetting or swelling gels are known. The experimental conditions and the solvent qualities, temperature, etc. are more important in the choice of the gel than the properties of the solutes. In view of the specificity of the problem and the available copious literature, only a few relevant surveys will be quoted (Muzsay 1970; Kirkland 1971a, b, c; Zweig and Sherma 1972; Heitz 1975).

1.2. PARTICLE SIZE

The importance of the particle size in gel chromatography has been discussed in Section 2.6 of Part I. In loose porous gels the flow properties of the bed are determined by the particle sizes (see Section 2.3.2 of Part II; Porath 1972). In general, the smaller gel particles improve separation whereas excessively fine or excessively coarse particles affect the resolution of the gel chromatographic process. Fine particles clog the column and reduce the rate of flow, whilst with particles larger than optimum size the fluid film becomes thicker and varies between wide limits. The diffusion of the solutes shows local variations and takes a longer time than with thinner fluid films. Equilibrium states are achieved at a slower rate and unevenly.

The optimum particle size depends on the degree of resolution required for proper separation. For group separation, the preparative separation of large volumes and industrial separation processes even coarser particles may be used. The particle sizes generally applied in gel chromatography are between 50 and 100 μm on average but for column chromatography or for thin layer chromatography, which call for high resolution, very fine particles are needed. The commercial gel types are available in preseparated fractions. The best average values of the Sephadex dextran gels (xerogel particles) are as follows: coarse 170 μm, medium 100 μm, fine 50 μm, superfine 30 μm (Fischer 1969). The optimum size of the gels sold in the wet and swollen state (agarose, Styragel) is approximately 150 μm. The homogeneity and distribution of the particle sizes can be checked by a simple microscope with an ocular micrometer.

A precondition for chromatographic methods is the narrow (theoretically homogeneous) distribution of the particle size in the stationary phase (Ekman et al. 1976). The finer the particles the more difficult it is to achieve homogeneous distribution. The gel particles produced in a plant or in the laboratory show a heterogeneous distribution. Sieving is the most frequent method of producing homogeneously distributed particles. Sieving is used in the first place for the grading of dry, non-aggregating xerogels which are easy to fluidize. Wet swollen gel particles are fractionated mainly by sedimentation. In compliance with the international standard specifications (DIN, US Standards), sieving takes place on a suitable series of sieves. *Table 20* shows a comparison of the particle sizes used in chromatography (μm) and the number of meshes per square cm in the sieves.

Table 20

Conversion of particle sizes from imperial to metric

Mesh	µm	Mesh	µm	Mesh	µm
30	500	80*	170*	200	74
40	360	100	140	250	62
50	290	120	125	280***	52***
60	250	150**	105**	325	44
70	210	170	88	400****	38****

* Coarse,
** Medium,
*** Fine,
**** Superfine particle size.

Another method for the separation of particle sizes is sedimentation. Under laboratory conditions, particles of different sizes are separated by the so-called static sedimentation process. In a simpler form of static sedimentation a thin suspension of swollen gel particles is allowed to settle in a tall vessel (measuring cylinder) then, after a certain defined length of time (5 to 15 minutes) the supernatant with the finer particles is decanted. Repeating this operation three to five times the fine fragmented particles, liable to clog the column, can be removed.

The particles of homogeneous size settle with a sharp boundary surface, leaving the supernatant quite clear. Different particle sizes can subsequently be separated by fractionated settling, depending on the time required for their settling.

Figure 32 shows the principle of continuous or counterflow settling (fluidization). Gas (compressed air, see *Fig. 32a*) or suspension *(Fig. 32b)* flowing counter to the settling gel particles — counter to gravitation — sweep away the smaller particles, keeping the larger and heavier ones floating, or allowing them to settle. *Figure 32b* shows the process of counterflow settling in a simple laboratory set-up (Hamilton 1958). Filtered tap water is made to flow across the gel suspension in the separating funnel (1), the flow rate of which is retarded by the shape of the vessel. The particles swept away with the flow may be collected in a Büchner funnel (2). The separation of particles of the desired size can be sensitively controlled by the rate of flow.

The Alpine Multi-Plex Zickzacksichter apparatus shown in *Figure 33* utilizes the aerodynamic differences caused by different particle sizes. It is suitable for the fractionation of xerogels, on both an industrial and a laboratory scale. Its operation is schematically illustrated in *Figure 34*.

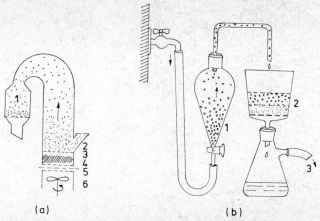

Fig. 32. Continuous or counterflow settling methods applied in the fractionation of gel particles by size
(a) fluidization by air stream: 1. collecting sieve; 2. batcher; 3. supporting sieve; 4. drying bed; 5. locks to control the flow rate; 6. fan; (b) separation of swollen gel particles under laboratory conditions: 1. separating funnel; 2. Büchner funnel; 3. vacuum line

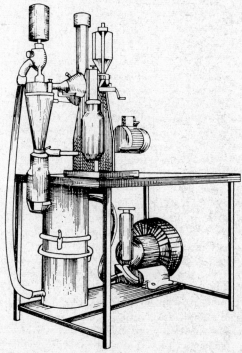

Fig. 33. The Alpine Multi-Plex Labor Zickzacksichter 100 MZR apparatus for the continuous separation of particles

111

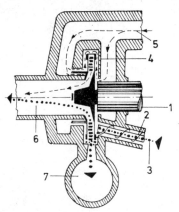

Fig. 34. The principle of the operation of the Alpine Multi-Plex Labor Zickzacksichter 100 MZR. The particles (3) introduced by the batcher (2) are centrifuged out through the labyrinth (zigzag) clearances of the vane (4) mounted on the rotating shaft (1). An air stream, counter to the sense of centrifugation (5), is produced in the labyrinths which carry the lighter particles to the outlet throat (7) and subsequently to the sieving-collecting system. The heavier and larger particles (6) can be collected along the periphery of the vane. The fractionation of the particles by size can be controlled by changing the rate of vane rotation and the velocity of the air flow

1.3. SWELLING

In practice the settling of gel particles and their separation by size take place almost simultaneously with swelling. The higher the specific solvent absorption of the gel structure (S_r, W_r: see Section 2.4 of Part I) the longer the time required to reach complete swelling. However, by accelerating the absorption of the solvent molecules (for instance over a hot water bath), the swelling time can be cut considerably. Hot swelling has another advantage: heat treatment destroys the pollutant microorganisms and removes dissolved air from the gel suspension.

If cold swelling is applied, the suspension must first be deaerated at low pressure. *Table 21* indicates the time required for the cold and hot swelling of different Sephadex gels. The swelling capacity of the finest soft gels (superfine Sephadex G-75, G-200) is slightly less than that of other products of the same porosity (40 to 120 μm). Swelling is dependent neither on the (low) ionic strength of the solution, nor on the quality of the ions and the pH of the medium (within the limits of gel stability). Depending on their porosity the polyacrylamide gels (Bio-Gel P, Acrilex) need approximately half the time for swelling as compared to dextran gels.

Depending on the structure of the polymer and the quality of the solvents, the swelling of the organophilic gels shows great variety. Many gel chromatographic column packings, for instance most aerogels (glass and silicate beads, Porapak, etc.), do not swell at all when taking up solvents, they only become wet.

The application of partially swollen gels is a field relatively less known and explored. The literature and references report on the selective reduction of the pore sizes. To cause polysaccharide (Sephadex) gels to swell acetic acid–pyridine–water (Porath and Lindner 1961) or phenol-acetic acid–water (Synge and Youngson 1961) mixtures were used.

Table 21

Minimum swelling time for Sephadex and Bio-Gel P gels at room temperature and on a boiling water bath

Type of Sephadex	Minimal swelling time (hours)		Type of Bio-Gel	Minimum swelling time (hours)	
	25 °C	100 °C		25 °C	100 °C
G-10			P-2		
G-15			P-4		
G-25	3	1	P-6	2—4	—
G-50			P-10		
G-75	24	3	P-30	10—12	—
			P-60		
G-100					
G-150	72	5	P-100	24	—
G-200			P-150		
LH-20	3	—	P-200	48	—
LH-60			P-300		

The swelling of xerogels takes place in the following manner: the particles selected for chromatography, of suitable size, are mixed with a solvent quantity equal to 10 to 20 times the swollen volume — a dilute (0.1—1%) saline or buffer solution — and allowed to stand (preferably over a hot water bath) until perfect swelling has been achieved. If hot swelling is used great attention must be paid to the gel temperature. Columns must be packed at the same temperature at which the chromatography is performed.

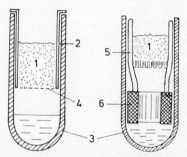

Fig. 35. Centrifuge filter inserts for determination of the specific swelling volume and solvent retention of gels

1. gel particles; 2. metallic filter insert; 3. centrifuge tube; 4. fine meshed (400) metal on polyamide filter; 5. filter cup with sintered glass filter layer; 6. distance piece (hard rubber ring)

When swollen, the gel suspension is stirred repeatedly, allowed to settle, the supernatant decanted and replaced with fresh solvent.

It is rather difficult to ascertain the gel losses during swelling and washing. The quantity of xerogel (packed into the column) may be calculated from the specific volume of swelling (see Section 2.4 of Part I). The specific swelling volume is a well-known property of commercial gels which, together with the solvent retention capacity, can be determined in the following manner. A defined volume of gel, made to swell in distilled water or pure solvent corresponding to its polarity (and allowed to settle until it reaches the equilibrium state), is washed and centrifuged for 20 minutes at 3000 rpm in the filter insert illustrated in *Figure 35* (weighed when empty). The filter insert is weighed once more, then dried in a vacuum to weight stability, and weighed again. The parameters sought for can be calculated from the weight of the dry gel, the quantity of solvent regained and the equilibrium volume.

1.4. PRESERVATION AND STERILIZATION

Most of the gels used in practice (primarily polysaccharides) provide outstanding culture media for microorganisms which, by disruption of the gel structure, may completely change their chromatographic properties. In otherwise resistant gels (e.g. polyacrylamide) proliferating bacterial or fungal colonies present similar risks, endangering the substances to be separated. Gel suspensions and columns when not in use must therefore be sterilized or kept in a solution which inhibits the proliferation of microorganisms. For the sake of completeness, mention should be made of the organophilic gels, which being indifferent to the aspect of bacterial decomposition, can be safely kept in organic solvents. In these (e.g. Styragel), the only risks are the water content of the solvents or, if dry, some irreversible structural alterations.

To preserve hydrophilic gels, phenol, cresol, toluene, formalin, butyl-alcohol and chloroform were recommended earlier (Flodin 1962). Today, according to unanimous opinion, such compounds are unsuitable for inhibiting bacterial growth because they are not only ineffective, except in high concentration, but they also cause shrinkage of the gel particles in addition; they cause most proteins to precipitate because they deteriorate the chromatographic apparatus and deform the plastic parts of the equipment. Swollen Sephadex (Molselect) gels can be sterilized without risk of deterioration by autoclaving at 110 °C for 40 minutes. When heated to 120 °C in the dry state, they burnish and decompose. Still greater attention and caution are necessary if agarose gels are used because they cannot be sterilized by heat treatment (autoclaving) and must be kept cold, in a preservative solution.

What follows is a description of the properties of a number of slightly fungicidal bacteriostatic agents which have proved their worth in gel chromatographic practice.

The gel-preserving agent most frequently used is a 0.02—0.1% solution of sodium azide (NaN_3) which above 270 nm shows practically no optical absorption but hampers the detection of nucleic acids (254 nm). It readily dissolves in water, does not react with carbohydrates and proteins, and has no influence on serological reactions. With heavy metallic salts it forms insoluble (explosive) compounds, and interferes with the fluorescent tracing of proteins and with the anthrone reaction.

Chloretone [trichlorobutylalcohol $Cl_3C-C(OH)(CH_3)_2$] can be used with weak acid solutions only. A concentration of 0.01 to 0.02% ensures a satisfactory bacteriostatic effect in gel chromatography. In alkaline media, however, the compound decomposes above 60 °C. It does not, however, interfere with anthrone reaction and has the lowest optical absorption of all preservative agents.

Merthiolate (Merfen, ethylmercury thiosalicylate), in a 0.005 to 0.01% solution, may be used against sporadic forms of bacterial infections. Its shortcoming is that it is effective in a weakly acid solution only. With heavy metal salts it is prone to precipitate and deactivate the $-SH$ groups. It readily dissolves in water, does not interfere with the anthrone reaction and causes no changes in the chromatographic properties of proteins and carbohydrates.

Hibitane (chlorohexidine), in a 0.002% solution, is a potent bacteriostatic agent but has little effect against fungi. In a higher concentration it precipitates in the presence of chloride and sulphate ions and decomposes when heated. It has a high degree of optical absorption.

Phenyl mercury salts (acetate, borate, nitrate), in 0.001 to 0.01% weakly alkaline solutions, are other effective bacteriostatic agents. In the presence of halogen and nitrate ions these salts are prone to precipitate and with reducing agents to decompose. Like merthiolate, they enter into reaction with compounds containing the $-SH$ group. The optical absorption in the ultraviolet range is considerable.

The above bacteriostatic agents have a haemolytic effect, they affect the walls of the erythrocytes (and other cells) and their membrane systems. In view of the highly sensitive biological objects and biochemical applications of gel chromatography, before chromatography the preservative agents must be removed from the gels, preferably simultaneously with the adjustment of the equilibrium state of the column.

1.5. PURIFICATION AND DRYING

Temporarily, swollen gel particles can be stored in a refrigerator, in a solution of one of the above-mentioned bacteriostatic agents. Sodium azide is best for this purpose, owing to its low molecular weight and minimal side-effects. If longer storage is intended or if the impurities are hard to remove, the gels should be purified in polar non-aqueous solvents and dried. However, this is permissible only with gels whose structure is suitable for a reversible conversion of lyogel to xerogel (aerogel). Agarose gels, for instance, cannot be so treated without deterioration of their structure, unless in a 0.1 M NaOH/1 M NaCl solution or in acetone, followed by a thorough washing in water and subsequent regeneration. The first step in the purification of other hydrophilic gels is repeated washing in water to remove salts and other water-soluble substances. This operation, associated with the homogenization of the particle size, should be performed in the same way as that indicated in *Figure 32b*. For drying, the mass exhausted through the Büchner funnel should be equilibrated with alcohol solutions of increasing concentration (30, 50, 70, 90%), which will cause a gradual shrinkage and water loss of the gel particles. When partially shrunk the gels may also be stored in 60—70% alcohol, which facilitates their repeated swelling. After treatment in alcohol the gels should be washed in diethyl ether, and dried at 60 to 80 °C, with thorough ventilation.

1.6. THE MOVING PHASE OF GEL CHROMATOGRAPHY

One of the most characteristic features of gel chromatography is the chemical identity of the solvent molecules which constitute the stationary and moving phases. Only part of the solvent volume of the stationary phase — depending on the gel type (approximately 20% if hydrophilic gels are concerned) — is bound so closely to the polymer network as to become inaccessible for the non-adsorbed molecules or ions (Pecsok and Saunders 1968). The flowing (moving) phase has a dual effect. On the one hand it determines the degree of solvation of the gel, its pore size, and the ratio of the inner to the outer volumes; on the other hand the dielectric properties, the chemical effect, the ionic strength, etc. of the medium determine the solubility and shape (solvation), and the secondary, tertiary structure, etc., of the solutes.

Depending on the gel properties, the moving phase may be a polar or an apolar solvent. The swelling of Sephadex LH-20, shown in *Table 4*, is only slightly related to the dielectric properties of the solvents. The relatively low specific swelling volume of formamide ($D_{25°} = 109$) and the large specific swelling volume of pyridine ($D_{25°} = 12.3$) are attributable to molecular interactions. In the new, more apolar, trimethyl-silyl-Sephadex gels there is a more regular relationship between the dielectric properties of the solvents and the solvent regain of the gels (Tanaka and Konishi 1972). Streuli's (1971a, b) investigations proved an interaction existing between the moving phase and organophilic gels.

The majority of the gels used in practice (dextran, polyacrylamide, agarose) have strongly hydrophobic and polar qualities and their moving phase consists almost exclusively of the aqueous solutions of salts, ions and buffers. The anomalies experienced during elution in distilled water confirm the significance of the composition of the moving phase. In solutions of low ionic strength the exclusion of negatively charged small-molecule compounds, so-called ion exclusion, finds explanation in the effects of the weak ionic (carboxyl, sulphate) groups of the polymer chains (Gelotte 1960; Ames et al. 1961; Hohn and Schaller 1967).

In aqueous solutions the polymer chains exhibit weak cationic exchange properties (Gelotte 1960; Miranda et al. 1962). Eaker and Porath (1967) concluded from their experiments that on Sephadex G-10 gel the pH and salt content of the buffer cause considerable changes in the elution of the amino acids. Ion exclusion depends also on steric factors, i.e. the electrically polarized layer of water molecules located close to the polymer chains. According to another assumption the dissimilar rate of the penetration of ions is caused by the asymmetric distribution of the charge in the cross-linkages, and from the buffer ions a double layer may be formed which determines the diffusion of other ions (Porath 1967). It is known that the elution properties and the separation of alkali metal and alkali earth metal ions depend on the quality, size, complex forming properties, adsorption, etc. of the anions (Neddermeyer and Rogers 1968, 1969; Ogata et al. 1971; Zeitler and Stadler 1972).

Thermodynamic examinations of gel chromatography verify, even theoretically, the role of the ion milieu (Nichol et al. 1969). Apart from adsorption, discussed previously, only uncharged substances can be chromatographed in distilled water without similar side-effects. In the course of the elution of charged macromolecules in distilled water a Donnan effect, similar to that for membranes and ion exchangers, takes place, the essential feature of which is that the macromolecule is not capable of diffusing against the small counterions into the gel structure. The difference in potentials retards the molecule, thereby increasing its elution volume.

Ion exclusion, weak ion exchange properties of the hydrophilic gels and the Donnan effect of the macromolecules can be suppressed by even a low ionic strength ($\mu = 0.02$ to 0.4). To increase the ionic strength, a solution of neutral salts (NaCl) is used. If the salt concentrations are high (above 1 M NaCl), the partial loss of the hydrate shell of the gels and a gradual decrease of the volume of the gel bed are observed. Polyacrylamide gels are relatively less sensitive to the desalting effect of high salt concentrations (see Section 2.7.3 of Part I).

To adjust the appropriate pH of the moving phase, various buffers may be used in gel chromatography, between pH values of 3 and 10, at temperatures below 50 °C. Borate and buffers containing hydroxyl ion cannot be used because they form complexes with the polysaccharides (gels), and they destroy the hydrogen bonds. When selecting the buffers it is advisable to chose volatile compounds which can be removed with ease by lyophilization (e.g. ammonium acetate, ethylenediamineacetate). The pH of the buffers is generally so chosen that it renders the (macro)molecule maximally soluble and stable, reduces intermolecular effects to a minimum, whilst being at an adequate distance from those isoelectric points most favourable for adsorption and aggregation of the molecule. Depending on the charge (pH) of the solutes, the following buffers are recommended as counterions. Cationic buffers: alkyl amines, aminoethyl alcohols, ammonia, ethylenediamine, imidazole, tris-(hydroxymethyl)-methylamine, pyridine, veronal (below pH 7.5). Anionic buffers: acetate, citrate, phosphate, glycinate, barbiturate, veronal (above pH 7.9). If structural, enzymatic or other reasons do not justify the presence of alkali earth metals (Ca, Mg) or trace elements (Zn, Cu, Fe, Mn, Co) then the stability of the macromolecules can be increased by 0.01 to 0.001 M EDTA (ethylenediaminetetraacetic acid disodium salt). For the examination of enzymes buffer solutions composed of tris-

(hydroxymethyl)-methylamine malate may be used with advantage (Culling 1963; Dawson *et al.* 1969).

Among the additives dissolved in the moving phase the agents enumerated in Section 1.4 of Part I, for conservation of the so-called solubilizing compounds, antioxidants and viscosity-enhancing substances are most frequently used.

The best known solubilizer is urea. This is used in concentrations between 3 and 12 M to keep various (structural) proteins in solution, or for the dissociation of the hydrogen bonds of complexes or aggregates, e.g. α crystalline (Bloemendal *et al.* 1962). Guanidine-hydrochloric acid may also be used for this purpose (see Section 2.6.3.3 of Part I for more detail). Another large group of solubilizers includes various (cationic, anionic and neutral) detergents of which the best known are sodium dodecylsulphate, sodium deoxycholate and Triton X-100 (for more details see Pharmacia AB *Separation News*, November 1972). In addition to the solubilizers, antioxidants — primarily mercaptoethanol or dithiothreitol — are used for the protection of the tertiary structure of the macromolecules (disulphide bridges, etc.) (Anfinsen and Haber 1961; Cleland 1964). The composition of the moving phase is generally so chosen as to ensure minimal viscosity. In the examination of highly viscous samples the error caused by zone broadening can be corrected by increasing the viscosity of the moving phase (Fischer 1969). For this purpose large molecular weight (approximately 50 000) dextran or Ficoll fractions are used in adequate concentration.

2. THE METHODS USED IN COLUMN GEL CHROMATOGRAPHY

The equipment and methods of gel chromatography are relatively simple. Most processes can be implemented with the not too expensive basic equipment of column chromatography, as used in liquid chromatography. *Figure 36* shows the simpler (a) and more complex (b) apparatus for gel chromatography. Various accessories — pumps and valves, electronic automatic fraction collector, detectors and recording systems — may facilitate and speed up the work, and simplify the operations but they raise

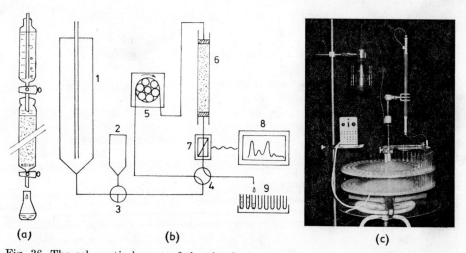

Fig. 36. The schematic layout of the simple (a, c) and complex (b) instrumentation of gel chromatography

In (b): 1. solvent reservoir; 2. sample; 3. cock; 4. two-bore cock; 5. peristaltic pump; 6. column; 7. sensor detector UV, RI); 8. recorder; 9. fraction collector

the costs involved. Even so the costs of gel chromatography will not exceed the investments required for other modern instrumental methods (gas chromatography, spectroscopy) of analysis.

The most important part of the equipment is an appropriately chosen and well-prepared chromatographic column. The criteria to be met in building up the gel column and in its handling are based essentially on the general principles of liquid chromatography, relating to the sizes of the stationary and moving phases, their volumes, the charge, the mixing spaces, the conditions of flow, etc. (Flodin 1961; Gelotte 1964a; Rothstein 1965; Gelotte and Porath 1966).

2.1. THE SELECTION OF THE CHROMATOGRAPHIC COLUMN

The selection of the chromatographic column is determined by (a) the type of job (analytical, preparative or industrial separation), (b) the type and particle size in the stationary phase (soft or rigid gel, porous glass beads, etc.), (c) the quality and composition of the moving phase (corrosive, volatile, organic, viscous, high density, etc.), and (d) the pressure, flow and temperature conditions prevailing in the course of chromatography. In the past (often still today) gel chromatography made use of the simplest known solutions of liquid chromatography, i.e. an open column packed with the stationary phase. Soft, plastic gels and recent methods call for special types of columns, however.

2.1.1. COLUMN TYPES AND THEIR APPLICATIONS

In its simplest form the chromatographic column consists of a vertically fixed glass or plastic pipe, of uniform diameter or tapering downwards, in which glass-wool, glass beads or a layer of quartz sand, sometimes a built-in sintered glass filter or some other suitable filter insert (Witt plate, filter paper) supports the stationary phase, and permits passage of the moving phase. To reduce the mixing volumes (see the relevant section) both the filter layer and the zone joining the moving phase to its outflow should be kept as small as possible. To stop the flow a cock or clamp (Mohr) is provided. From among the simple columns illustrated in *Figure 37*, the one marked (d) should be the choice of preference, owing to its favourable properties and simple lay-out. If the gel particles are fine, they might clog the filter pores and reduce or completely stop the liquid flow. With superfine gels this defect can be remedied by the application of a coarse-grained gel layer, a few mm thick, deposited over the filter insert, which is neutral from the point of view of separation. Another special column, produced individually, is the one shown in *Figure 37e*. This is a microanalytical column which prevents drying out of the gel bed when the moving phase is drained from the reservoir (Patrick and Thiers 1963). Carnegie and Pacheco (1964) built up their own special microcolumns from gel particles over the object plate of a microscope. Ogston and Wells (1970) elaborated a method of osmometric molecular weight determination which consists of measuring the volumetric changes in a single particle of Sephadex gel in a liquid-flow specially induced under the microscope. More recently very thin (2 mm diameter) long columns have been introduced for analytical gel chromatography (similar to those applied in gel permeation and gas chromatography) (Catsimpoolas and Kenney 1972).

In spite of their numerous practical features, the said columns can only be used for simple jobs (desalting, group separation, buffer change) or for informative tests. For systematic work, in order to ensure the reproducibility of the test conditions, precision-type columns are required which have a uniform cross-section and yield sharp zones, and which provide direct contact between the gel bed and the filter inserts. A further criterion is that

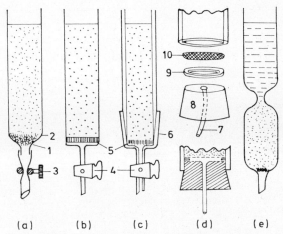

Fig. 37. Simple chromatographic columns

(a) glass tube with constriction: (1) with glass wool plug at the lower end, (2) with fine glass beads, and (3) with clamping cock; (b) with stopcock (4) and sintered glass filter plate, (5) built-in; (c) the cleaning of the columns and the filter can be facilitated by ground joints (6); (d) home-made column with minimal mixing volume: above the rubber stopper (8), with polyethylene tubing (7) inside a 1-mm thick polyethylene ring (9) and a close-meshed polyamide filter (10) is applied with a close fit in the column; (e) microanalytical column with constriction in the middle

the columns should be built up of mechanically stable, possibly transparent, elements which resist heat, chemical effects and deformation, and which are easy to manipulate and wash. Many researchers have examined ways and means of building up optimally operating continuous and intermittent columns (Roubal and Tappel 1964; Fox *et al.* 1969; Ledvina 1971). Porath and Bennich (1962) constructed a column for recycling gel chromatography in which the gel bed is surrounded by movable pistons fitted with filter plates and gaskets. This column permits even a reversal of the flow.

Figure 38a shows the Porath—Bennich column in its modified version, marketed by the LKB Co. The precision columns produced by Pharmacia Fine Chemicals AB, available in a wide choice of lengths between 15 and 100 cm and diameters between 0.9 and 5 cm, operate along a similar principle. *Figure 38b* illustrates the ingenious chromatographic columns of the Whatman Co. in which a flexible piston, in a close fit in the column, replaces the packing rings in the teflon joints. This lay-out provides a vacuum at the point where the filter insert joins the discharge pipe.

Precision-type chromatographic columns are made usually of boron silicate glass or transparent plastics, the joints, fittings, pistons, lines, etc.

of synthetic compounds being resistant not only to acids, alkalis and corrosion but even to organic solvents (plexiglass, polyoxymethylene, polytetrafluoroethylene, silicon rubber, teflon, Vyon, Tygon, etc.). Special care is taken also to prevent the substances from coming into contact with the metal parts. If the columns are large and if they operate at high temperatures and under high pressures, or if organic solvents or organophilic gels

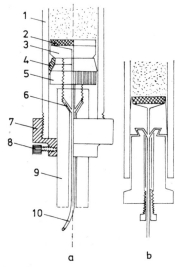

Fig. 38a. The layout of the 4200-type chromatographic column of the LKB Produkter AB (Sweden)

1. plexiglass column; 2. porous plastic filter plate; 3. conic joint piece; 4. rubber gasket; 5. pinch screw; 6. joint and conic fitting of the plastic tubing; 7. threaded lid; 8. piston rod fixing screw; 9. plexiglass piston rod; 10. plastic tubing

Fig. 38b. The layout of the Whatman-type precision column with teflon piston

are used, the columns may be made of metal, stainless steel or ceramics. To prevent clogging the factory-made columns are provided with fine-pored (less than 40 μm), thin (1—2 mm), easy-to-replace plastic (porous polyethylene, polypropylene, teflon) filter inserts.

2.1.2. COLUMN SIZES AND PROPORTIONS

The dimensions and proportions of the gel chromatographic columns are determined by the job in hand. For preparative gel chromatography, the thin, long analytical types are unsuitable and shorter and larger-diameter gel beds are used. The molecular weights of proteins were determined by Whitaker (1963) on Sephadex G-100, using a column of 1.1 cm diameter and 192 cm length. For examination of the cyclic oligoamide content of polyamide, Kusch and Zahn (1965) used an extremely long (4 by 500 cm) column. For group separations, the desalting of proteins, and buffer changes in large-molecule substances, short and wide columns should be used (10—20 cm long, 2 to 4 cm diameter) (Porath and Flodin 1959; Flodin 1961). In industrial applications columns with diameters between 10 and 20 cm, 1 m high, of 100 to 2 500 litres capacity are not infrequent (e.g. the Sephamatic system of the Pharmacia AB, Uppsala; see also Samuelsson et al. 1967a, b; Jansson 1971; Porath 1972). The optimal proportion for laboratory columns is between 1 to 5 and 1 to 100. The theoretical plate number and the resolution increase as the column becomes longer but the dimensions cannot be increased beyond a certain limit. Owing to their packing and the difficulty of maintaining proper flow conditions, the plastic soft gels (Sephadex G-200, Bio-Gel P-300, agarose gels) are not used in columns with diameters larger than 1 to 5 cm and longer than 1 to 2 m. Resolution can be enhanced by the connection of several shorter columns in series or, if a single column is used, by repeated passage of the sample, i.e. the recycling technique. Gel beds of large diameter (90 cm) divided into several shorter (50 cm) sections have proved their worth, particularly in industrial methods of protein chemistry (Porath 1972). Column capacity increases quadratically with the diameter. However, in large diameter columns homogeneous packing of the gel bed becomes increasingly difficult, and the same holds true for the uniform and sharp application of the sample.

The method of calculating the requisite and effective dimensions of the gel column was dealt with and demonstrated with practical examples in Section 2.6.1.7 of Part I. This calculation presupposes knowledge of the distribution coefficients of the substances and the hydrodynamic parameters (theoretical plate number, selectivity, capacity coefficient) of the gel bed, from preliminary examinations. Snyder (1972a, b) deals in detail with the determination of the optimal conditions (including the dimensions of the column) in liquid chromatography.

For simpler operations (e.g. desalting), the size of the gel bed is determined by the volume of the sample. Here is an example to illustrate the method of calculation for desalting by gel chromatography. Ten ml of protein solution are to be desalted on Sephadex G-25 (coarse) gel whose constants are as follows: W_r, 2.3 g H_2O xerogel; density of the swollen gel (d), 1.11. According to Section 2.3 of Part I the ratio of the sample to the inner gel volume is given by the relationship $V_s = 0.3\ V_i$ which yields an inner gel volume of 33 ml. According to the equation $V_i = aW_r$, $a = 14$ g xerogel will be needed for desalting. From the combination of the equations

$$V_i = aW_r \text{ and } V_i = (V_t - V_o)\frac{dW_r}{1 + W_r}$$

one obtains

$$V_t - V_o = \frac{a(1 + W_r)}{d}$$

Substituting into this relationship the known data, one has $V_t - V_o = 42$ ml. Since for Sephadex gels $V_t = 3\ V_o$, $V_o = 21$ ml and $V_t = 63$ ml, the criteria of the above equation are obviously satisfied. This case, therefore, indicates that desalting should be done on a gel column of 2 by 20 cm in which, after the passage of 21 ml column volume, the protein will dilute to a maximum of 15 ml.

2.1.3. THE WALL EFFECT AND ITS ELIMINATION

It is well known that, owing to the adhesion and adsorption of the solvent molecules located close to the wall, a so-called flow profile evolves in the cross-section of liquids in flow. By delaying (tailing effect) the material transport, the flow profile affects both the sharpness of the zones and the resolution of the chromatographic process. The slimmer the column and the greater the wall area in relation to the mass of the gel bed, the higher will the wall effect be. This problem is negligible if preparative sizes are concerned, but it becomes rather serious in long and slim analytical columns. If it is necessary to diminish the wall effect (Bathgate 1970) the inside of the column may be treated with dichlorodimethylsilane. For the same purpose, and to improve the flow properties and reduce adsorption, silanized porous glass beads and column packings may be used with advantage (Cooper and Johnson 1969; Masamichi and Terutake 1971; Sachs and Painter 1972).

2.1.4. THE MIXING VOLUME AND ITS REDUCTION

Those spaces of the chromatographic columns which do not form part of the gel bed but, by a mixing effect arising from eddies or diffusion, may alter the composition of the liquid introduced into or discharged from the column, are called the "mixing or dead" volumes. The mixing volume distorts the shape of the gradients of concentration or pH and decreases the sharpness of the separation of fractions. Therefore this volume should be reduced to the greatest possible extent. An important criterion that modern chromatographic columns must meet is that the sum of the inevitable mixing volumes (filter plates, piston bores and conic spaces) should be less than 0.1% of the useful volume of the column. In addition, the mixing volume includes the sum of the volumes of the pipes, cocks, valves, pumps, flowing-cell cuvettes, fraction collecting syphons, etc. for transferring the solvent. From the point of view of chromatography one distinguishes between mixing volumes ahead of the column (up to the introduction of the moving phase) and behind it. If the composition of the solvent introduced into the column remains unchanged during the chromatographic process then the mixing volume ahead of the column will affect the volume of the sample and the sharpness of the feed only. The sample should therefore be introduced to the filter insert directly, or to the stretch of the inflow closest to the column (as is known, in gas or gel permeation chromatography the sample is introduced directly into the stationary phase). The heads of the modern chromatographic columns are provided with separate inlet bores for the moving phase and for the sample (Sephadex Laboratory Columns, K9, K15, K25; Pharmacia Fine Chemicals AB, 1969). The mixing volumes which evolve behind the column may cause substantially greater errors. One of the major shortcomings of columns a, b, and c in *Figure 37* lies in the relatively large size of the filter bed and the dead volume beneath it, which, however, may be reduced by a filling of glass beads. In the latest columns and chromatographic apparatus it is the volume of the piping and that of the flowing-cell cuvettes used in the automatic sensors which must be guarded against. The smaller the volume of the chromatographic column and of the fractions the greater the influence of the mixing volume. To reduce the mixing volume, the inlet and outlet pipes of the column should be as short and as small in diameter as possible. For analytical examinations it is advisable to measure the volume and the effect of the dead volumes by model tests. From the above it follows that, for instance, the circulating pump should be located ahead of the column in the circulation system.

2.2. INSTRUMENTS AND AUXILIARIES USED IN COLUMN CHROMATOGRAPHY

Based on the principles illustrated in *Figure 36*, the instrumentation of column chromatography falls into two groups: (a) the accessories required for simple column chromatography (pipes, cocks, joints, Mariotte flasks, solvent reservoirs) and (b) the auxiliaries required to speed up and facilitate the more complex automated operations (pumps, sensors and recorders, fraction collectors, electronic valve systems, gradient mixers, etc.). This division reflects the method and the trend towards a gradual enlargement of the basic instrumentation.

2.2.1. TUBING AND JOINTS: THE SAFETY LOOP

Like the column joints, the tubing carrying the solvents and the solutions of the separated substances is made of materials which resist heat, chemical effects (acids, alkalis, oxidants), organic solvents, pressure and mechanical stresses. The materials best suited for this purpose are polyethylene, polyvinyl chloride (Tygon, PVC), polytetrafluoroethylene (Teflon), synthetic and silicon rubbers. *Table 22* shows the properties and the sources of the plastic tubing in widest use. Since polyethylene, polyvinyl chloride and teflon tubing, if under sustained mechanical stresses (compression), are prone to suffer lasting deformation, they cannot be used, for example, in peristaltic pumps. For this purpose specially flexible high-quality silicon rubber tubes (Esco, Rau-sik) are available. Teflon exhibits exceptionally high resistance to solvents, heat and pressure but, being brittle and hard, it is difficult to handle. However, if heated cautiously over a flame, it softens up and becomes suitable for drawing even into capillary tubes, in a similar manner to glass. If there are no specific requirements to meet, polyethylene and PVC (Tygon) tubes are best for gel chromatography. In the commercially available columns (see *Fig. 38a* and *b*) the tubes are connected by specially designed joints. This arrangement calls for conic tube ends. *Figure 39*

Table 22

Commercial connecting tubings and their properties

Designation	Manufacturer	Application	Properties (composition)
Polyethylene Tubing Cat. No. 3039	LKB Produkter AB, Bromma, Sweden	Tubing, nipples	Thermoplatic, elastic, deformable, resistant to mild chemical effects, translucent
PVC Tubing (Tygon) Cat. No. 3094			
Viton Tubing Cat. No. 3095			Synthetic rubber, black
Teflon (PTFE)	The Fluorocarbon Co., South Clementine, California, U.S.A.	Tubing, nipples, pistons	Thermoplastic at higher temperature, rigid, resistant to chemicals and organic solvents
Rau-Sik*	Rehau-Plastics GmbH, 8673 Rehau, Federal Republic of Germany	Peristaltic pumps	Stable up to 120—180 °C, elastic, resistant to organic solvents, translucent, crosslinked polysiloxane
Silicon Rubber Tube Cat. No. 989— 12781 and 12782	LKB Produkter AB, Bromma, Sweden	Peristaltic pumps	Translucent silicon rubber
Silescol Translucent			Thermostable up to 120—180 °C, chemically inert translucent silicon rubber
Vinescol 23	ESCO (Rubber) Ltd., Great Portland St., London, U. K.	Specially for peristaltic pumps	Thermostable up to 200 °C, resistant to oils, conc. sulphuric acid (!), organic solvents, black synthetic rubber
Butyl XX			Chemically resistant, pressure-tight black synthetic butyl rubber
Escoplastic MS			Chemically resistant, translucent

* Marketed by LKB Produkter AB, Bromma, Sweden.

shows the facilities used for shaping the tube ends and the operation proper. The easiest way to connect the tubing is to insert flexible polyvinyl chloride pipes, having thicker walls and smaller cross-sections, between each pair. At higher pressures the pipes may be connected via an intermediate piece in the manner shown in *Figure 40*.

The safety loop illustrated in *Figure 41* is an important part of the gel columns operating under hydrostatic pressure. It works on the principle of communicating vessels and prevents complete draining of the liquid from the gel bed. The drawback of the safety loop, if located behind the column, is that it increases the mixing volume. The pipe should therefore be as short as possible and the fractional collector should be located immediately beside the column. The length of the safety loop should be so determined that drainage takes place at a higher level than that of the gel

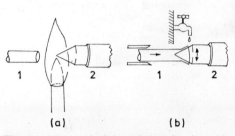

Fig. 39. The conic joint of the plastic tubing
(a) using the special conic metal tool (2) heated over a flame, the tip of the plastic tubing (1) is extended, and (b) cooled down to set

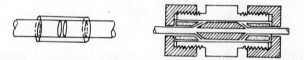

Fig. 40. Fitting the plastic tubing used for chromatography with a special joint produced for higher pressures (LKB)

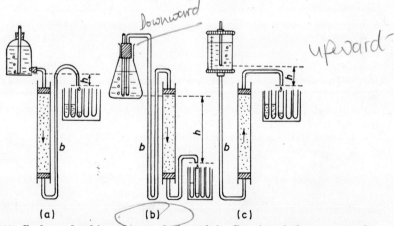

Fig. 41. Mariotte flasks and tubings for regulation of the flow in gel chromatography. Flow in the direction of gravity (a, b) and in reverse (c). The correct location of the safety loop b before the column in (b), (c) or after the column in (a) will control the hydrostatic pressure difference h

127

bed. If loose soft gels, highly sensitive to pressure differences, are used the optimum pressures are very low (a few cm of H_2O) and a safety loop is indispensable. This, if the relative positions of the Mariotte flask and the column are fixed, ensures the easy and sensitive variation of the hydrostatic pressure (level difference).

2.2.2. SOLVENT RESERVOIRS AND MARIOTTE FLASKS

In the simpler methods of gel chromatography, solvent reservoirs or Mariotte flasks are used which provide an appropriate hydrostatic pressure, to maintain a constant flow in the column. The easiest way to store the solvent is to utilize the column space above the gel bed for this purpose, but the method has a shortcoming in that the volume is limited and a drop in the liquid level would alter the flow rate. This simple technique is therefore restricted to operations which are quick, which take place in a small volume and which are relatively insensitive to the test conditions. With smaller solvent volumes a drip funnel (see *Fig. 36a*), similar to a Mariotte flask, is mounted on the column to ensure a constant rate of flow.

In columns of larger capacity and using softer gels, the accurate adjustment of a constant flow rate is essential. For this purpose Mariotte flasks of a different lay-out are used. As will be evident from *Figure 41* the hydrostatic pressure difference, adjusted by the level of the column outlet (safety loop), is independent of the quality of eluant in the vessel. The cylindrical Mariotte flasks sold by Pharmacia AB (*Fig. 41c*) with conic joints can be used also for filling up the gel bed (Sephadex Gel and Eluant Reservoirs; R9, R15, R25, R50 and R100).

2.2.3. PUMPS

For operations which call for a high degree of precision and constant flow velocity and which take a long time, or which use chromatographic equipment in a closed system (for instance in recycling gel chromatography), Mariotte flasks operating on the principle of gravitation are unsuitable. For reasons of principle and practice, liquid flow pumps must be provided. For experiments on a laboratory scale (a) piston and (b) peristaltic pumps are applied. *Table 23* shows some parameters of the commercially available micropumps.

Figure 42 illustrates a number of laboratory-scale micropumps of the more characteristic types.

Piston pumps produce higher pressures. Their drawbacks are intermittent operation and the ensuing hazard of pulsation. Another similar shortcoming is the risk of mixing in the piston space and contact between the liquid and lubricants, or the metallic surfaces. Since peristaltic pumps can produce the pressures and outputs required for gel chromatography they should preferably be used. They offer additional advantages insofar as the liquid in peristaltic pumps comes into contact with the tubing only, and in most types, by altering the tubing diameter and the transmission ratio

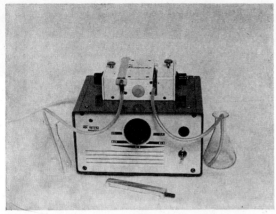

Fig. 42. Laboratory micropumps
(a) Ma-63 Mikrotechna (piston type, made in Czechoslovakia; (b) Peripump — a peristaltic finger-type pump produced by MTA Kutesz, Hungary); (c) Hiloflow (peristaltic finger-pump made by F. A. Hughes, U.K.); (d) LKB 4912A (a peristaltic roller pump made by LKB, Sweden); (e) Perpex 10 200 (a peristaltic roller pump made by LKB, Sweden)

Table 23

Laboratory pumps for gel chromatography and their specifications

Type	Manufacturer	Design	Capacity	Channels
Hiloflow pump	F. A. Hughes Ltd., U. K.	Peristaltic finger type	1—125 l h^{-1}	1
Peripump	MTA, Kutesz, Hungary	Peristaltic finger type	12—100 ml h^{-1}	2
4912 A	LKB Produkter AB, Bromma, Sweden	Peristaltic roller type	0—390 ml h^{-1}	2
Perpex pump 10200	LKB Produkter AB, Bromma, Sweden	Peristaltic roller type	changeable gear boxes 4—180 ml h^{-1}	1
Vario Perpex pump 12 000	LKB Produkter AB, Bromma, Sweden	Peristaltic roller type	0.6—400 ml h^{-1} adjustable	1
Multi Perpex pump 2 115	LKB Produkter AB, Bromma, Sweden	Peristaltic roller type	5—5 000 ml h^{-1} adjustable, changeable gear boxes	4
Miniflow	LKB Produkter AB, Bromma, Sweden	2 pistons	3—150 ml h^{-1}	1
MA-63	Mikrotechna, Czechoslovakia	1 piston	0—580 ml h^{-1}	1

the flow rate can be varied between wide limits. Powerful synchronous motors ensure a uniform rate of flow. Peristaltic pumps deliver the liquid according to two principles, roller and finger. *Figure 43a* shows the operation principle of rotary (roller) pumps, *Figure 43b* that of finger-type peristaltic pumps. Peristaltic pumps reduce the mixing volume to a minimum. By arranging the pump behind the column (suction), it can be used to control the rate at which the sample is fed. Multichannel pumps, in combination with a multibore cock (sample-eluant) should be connected ahead of the column because such a lay-out prevents air from being sucked into the column through a defective joint.

A flexible tubing of appropriate quality is one of the most significant accessories of peristaltic pumps. Silicon rubber proved best for the purpose (see *Table 22*). However, the tubing must be replaced after every fifth or seventh use.

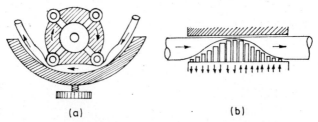

Fig. 43. The operational principle of peristaltic pumps. The fluid in the tube is delivered (a) by a roller or (b) by fingers which perform a peristaltic movement

2.2.4. THERMOSTATING THE COLUMN: HEATING AND COOLING MANTLES

In most applications of gel chromatography, e.g. in biochemistry and the study of proteins and macromolecules, fluctuations in room temperature do not affect the results to any appreciable degree. For more accurate determinations (for instance, the measurement of molecular weight) or in order to diminish the interaction of the gel structure and the solutes (for instance, in the chromatography of ribonucleic acids at a temperature of 40 °C) the columns must work at constant temperature. Owing to the susceptibility of sensitive macromolecules to denaturation and to enzymatic and bacterial decomposition, etc. or when considering the reaction kinetics of the processes involved, gel chromatography is often performed in a refrigerator or in a cold room (at temperatures between 0 and 4 °C). The Recychrom system of the LKB Produkter AB, for instance, is sold complete with a box which can be thermostated between -20 and $+40$ °C. The complete equipment for gel permeation chromatography (for instance, the Waters GPC), including the columns, tubing, pumps, sensor (refractometer), the systems for storage and batching of the sample, etc., is located in a closed system and kept at constant temperature. The solvent is introduced into the system across a heat exchanger. For the gel chromatography of small molecules, ions and inorganic compounds, thermostatic control is practically indispensable.

Gel columns can be thermostated by means of heating-cooling mantles in which a liquid of an ultrathermostat or of another heat-stabilizing system is circulating. The temperature must be kept constant, within ± 1 °C. The thermostating mantles, arranged as in the Liebig condenser, are made of the same material as the columns and are detachable and form an independent part of the equipment (LKB Recychrom system).

2.3. PACKING AND CHECKING THE COLUMN

An appropriately prepared gel bed packed into the column is of primary importance in gel chromatography. The method of packing decisively determines the hydrodynamic parameters, the uniformity and rate of flow, and the degree of zone broadening, and, through all these, the resolution of the chromatographic process. When packing a column, consideration should be taken of the specific properties and swelling of the gel, as well as of the theoretical information and practical experience available regarding the homogeneity of the grain size and the steric arrangement of the gel grains.

2.3.1. THE METHOD AND CONDITIONS OF PACKING

Only completely swollen gel particles, in the equilibrium state, should be used to pack the column. Subsequent swelling of the particles in the column may clog the path of flow and cause the destruction of the column. The equipment, the gel suspension and the solutions should be at the temperature of the environment (for instance, taking into account whether the gel was stored in a cold room or in a refrigerator). The gel suspension must be deaerated at low pressure before being packed into the column. It is of particular importance that the gases should be removed from the micropores of the wetted, but not swollen beads (glass, silica, etc.). The filter layers should be treated in the same way. The filters in the piston are generally fine-pored disks from which air must be removed under vacuum while immersed in the eluant (e.g. in a suction flask). The pistons prepared in this way should be kept immersed in the solvent until used. The outlet pipe of the piston should be closed until it is inserted into the column. The filters (glass-wool, glass grains) in the simple types of column should be positioned under a solvent layer, a few cm thick. To remove air bubbles, a slight vacuum should be produced by a tap-water-operated vacuum pump in the closed column. This method has proved advantageous also in checking the joints of the piston and the pipes in the latter types of column.

The column chosen for chromatography is mounted in the vertical position. The vertical alignment can be checked with a plumb-line or with a level. *Figure 44* shows that, in obliquely positioned, long, thin columns, owing to the asymmetry of the convective flow of the suspension, the material sediments unevenly. This alters not only the flow cross-section but also enhances zone broadening during the chromatographic process. In wider and shorter columns this effect will diminish. Long, thin columns are generally more difficult to pack uniformly and reproducibly whereas compact brittle gels pack easily and quickly in wide and short columns. The packing of soft, plastic, loosely structured gels calls for great attention and care.

Several methods are known for ensuring even packing of the gel bed. According to the simple process described by Flodin (1961) and Gelotte (1964a) the closed column is filled up to approximately one-third with

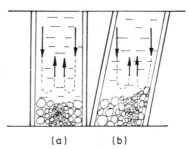

Fig. 44. Convective flows and their effect on the sedimentation of the gel particles (a) in vertical and (b) in oblique columns

solvent. Then a 5 to 10% (vol.) suspension of the gel is poured, in a thin jet, alongside a glass rod, over the latter. After the spontaneous sedimentation of a 1—2 cm thick gel layer, the flow in the column is slowly started up, the gel suspension being continuously replenished. While the gel is settling, care must be taken to prevent the gel particles separating with a sharp boundary before the bed has attained its final volume. Piecemeal packing of the gel layers and visible sedimentation zones are clear indications of the unevenness of packing. In such cases the packing must be repeated. If soft gels are used, to increase the flow rate, coarse rigid gel grains (e.g. Sephadex G-26) should be deposited to a few mm thickness over the bottom of the column.

During sedimentation of the gel, convection currents of the pattern shown in *Figure 44*, visible to the naked eye, tend to evolve and cause the finer grains to settle along the column's axis, while the coarser ones deposit closer to the wall. This radial distribution by size favours gel chromatography to a certain extent, because, owing to the difference in flow rate between the coarser and the fine material, a flow profile is produced in the column cross-section whose effect is opposed to the wall effect (dealt with in Section 2.1.3 of Part II). At an appropriate flow velocity the two effects may cancel out each other.

To simplify the rather laborious, time-consuming and cumbersome operations described above, extensions (funnel, Mariotte flask) with increasing cross section are connected to the column top (see *Fig. 45*), which can hold the quantity of suspension necessary to fill up the gel bed.

Constant stirring takes care of the even distribution of the grain sizes. Rothstein (1965) recommends stirring the gel suspension in the column during sedimentation, while Altgelt (1965a, b) and Heitz et al. (1969) obtained homogeneous charges by rotation and vibration of the columns.

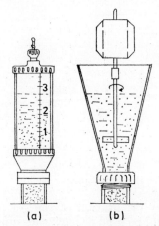

Fig. 45. Extensions connected to the chromatographic column (reservoirs for the packing of gel beds)

(a) a Mariotte flask of the Pharmacia Fine Chemicals AB, Uppsala, Sweden; (b) glass funnel for the LKB Bromma (Sweden) columns with a motor stirrer

Since magnetic stirrers tend to destroy the gel particles, only slowly rotating vane-type stirrers should be used to homogenize suspensions.

In spite of the various solutions and techniques available, the so-called gravity sedimentation of the gel particles should be avoided for two reasons. Owing to the relative slowness of sedimentation the particle size distribution varies along the length of the column; the coarser grains will collect on the bottom, the finer ones in the upper portion of the gel bed. This fact is favourable for chromatography, because the substances to be separated meet the fine particles with higher resolution first. A more important problem of principle is, however, that the structure of gravity-packed columns is loose, the interstices between the particles are large and the flow rate is often unnecessarily high.

Experience gained with compressed columns has indicated that more closely packed gel grains ensure better separation (Edwards and Helft 1970; Fishman and Barford 1970). This seems to be confirmed by Junowicz et al. (1972) who used gel filtration with vibration. This is also supported theoretically (see Section 2.6.1.3 of Part I); the outer volume of the gel bed must be as small as possible in relation to the inner volume. Unlike previous practice, instead of gravity-type sedimentation, Peterson (1970) packed ion exchanger columns under a hydrostatic pressure raised according to a given schedule. His method ensured more stable packings both as regards the flow and mechanical properties. Pressure in this process is increased gradually from approximately 0.1 to 0.7—1.0 atm with increasing thickness of the gel bed. In addition to the Mariotte flask, Peterson recommends the use of a gas (nitrogen) bottle fitted with a reducing valve. The technique described by Widén and Eriksson (1964) (see *Fig. 46*) permits the simultaneous automatic packing of several columns, which makes possible a comparison between packings of similar properties under dissimilar test conditions.

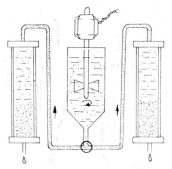

Fig. 46. Set-up for the simultaneous packing of several columns

To achieve satisfactory orientation of the gel particles thin suspensions (1 to 5 g gel/100 ml) should be used to pack the column. If more concentrated suspensions are sedimented the gel grains form a loose structure which will only partially pack under pressure. The subsequent compression of loosely packed columns improves the volumetric distribution of the gel bed

(V_i/V_o) but has no appreciable effect on the steric orientation of the particles. Very fine-grained gels can be sedimented even in more concentrated suspensions.

The type (soft, brittle) of gel, its particle size and the pressure applied in packing are closely interrelated. The optimal pressure is lower (0.1 to 0.3 atm) with softer and finer particles and higher (0.3 to 0.7 atm) with brittle and coarser grains. If the flow-resistance against the packing pressure is excessive, the column should be repacked at a lower pressure. On attaining the requisite column volume, maximum pressure should be maintained until the top of the gel bed adjusts to a constant height. Depending on the gel type, this so-called equilibrium state can be achieved by keeping up the flow across the column for 8 to 24 hours.

To pack organophilic gel particles which have a lower specific gravity than the solvent (for instance, Sephadex LH-20 in chloroform) into a column, a special technique and piston-type columns are required. According to one method, a flow rate sufficiently high to prevent the gel particles from floating is applied. When the required volume has been obtained the packed piston is quickly positioned over the surface of the gel layer wetted by the solvent. To position the piston without producing bubbles needs considerable skill. Another, similar, solution is to place the entire quantity of suspension into the column, to close the column at both ends with plungers and then to start the solvent flow from the bottom upwards, gradually adjusting the final volume of the gel bed with the lower plunger. Organophilic gels may also be packed into columns by allowing the particles to swell first in another solvent of lower specific gravity and then to sediment, thereafter changing the solvent and washing through the finished column. This method requires unrestricted mixing of the two solvents and an approximately identical specific volume of the gel in the two solvents.

2.3.2. REGULATION OF THE FLOW RATE

The velocity of the moving phase is one of the most important parameters of gel chromatography. In accordance with Section 2.6.1.5 of Part I, optimal resolution in column chromatography cannot be achieved unless, under given flow conditions, and with increasing rate of flow, the theoretical plate number generally decreases (Bombaugh and Levangie 1970a, b). This means that the maximum flow rate achievable in practice is not necessarily the optimum from the aspect of the chromatographic properties of the gel bed.

According to their flow properties, rigid and deforming (soft, loose) gel beds can be distinguished. For laminar flow between rigid spheroidal gel particles the Darcy law

$$v = K \frac{p_{H_2O}}{L}$$

holds valid, where v denotes the so-called linear flow velocity (cm³/h · · 1/cm² = cm/h), p_{H_2O} the pressure difference acting upon the gel bed (hydrostatic) in terms of cm of H_2O, L the length of the gel bed in cm;

K the proportionality factor, depending on the gel (grain size, shape, water regain) and the quality of the eluant (viscosity, etc.). $K = K_0 \eta$ and if the viscosity of the eluant $\eta = 1$, then $K = K_0$ where K_0 is the so-called specific permeability of the gel bed.

The specific permeability values of the rigid Sephadex gel particles are shown in *Table 24*. This shows that as the grain sizes become finer, the permeability of the bed decreases (for instance, in Sephadex G-25 and G-50).

According to Darcy's law (assuming constant pressure), the flow rate is inversely proportional to the height of the gel bed, and in any given column this increases proportionally to the pressure difference. With rigid gel particles, the linear flow velocity, to a fair approximation, is independent of the column diameter (the cross-section of the gel bed is included in the linear flow velocity). In the case of diluted aqueous solutions ($\eta \approx 1$ centipoise) the flow velocity of the rigid Sephadex gel beds can be calculated from the data in *Table 24* and Darcy's law. Above 1 cP the flow velocity is inversely proportional to the viscosity of the solution. Through viscosity, temperature influences the flow velocity in the gel bed. For the sake of

Table 24

The specific permeability (K_0) of commercial Sephadex G gels having different particle sizes*

Sephadex type	Particle size (xerogel), μm**			
	Coarse (170)	Medium (100)	Fine (50)	Superfine (30)
	Specific permeability, K_0			
G-10	—	21 (60)	—	—
G-15	—	22 (70)	—	—
G-25	260	94	25	9
G-50	400	145	36	13.5
G-75	—	165	—	16
G-100	—	210	—	19
G-150	—	280	—	26
G-200	—	340	—	31

* Data taken from Pharmacia Fine Chemicals AB, *Separation Bulletin* (1969) and Fischer (1969) (in parentheses).
** The particle size values are harmonic means (in parentheses). The G-10, G-15 in medium size (40—120 μm), the G-75, G-100, -150 and G-200 types in medium (40—120 μm) and in superfine (10—40 μm) sizes are the only sizes available.

orientation, the relative viscosity of a dextran fraction of molecular weight $5 \cdot 10^5$ in a 1% solution is $\eta = 1.5$, that of a fraction with a molecular weight $4 \cdot 10^5$ is $\eta = 1.2$, and that of serum albumin (M = 69 000) is $\eta = 1.065$.

For soft deforming gel particles (Sephadex G-200, Sepharose 2B, 4B, 6B, Bio-Gel P-200, P-300) the Darcy law does not hold valid. The specific permeability of the gel bed depends also on the hydrodynamic pressure and the cross-section of the column. As indicated in *Figures 47, 48, 49* and *50*, as against the nearly linear behaviour of the rigid particles (e.g. Sephadex G-10, G-15, G-50) the irreversible alterations in the structure of the soft gel beds reduce the flow velocity above a defined pressure.

Figure 48 shows the characteristic hysteresis curves of the flow in Sephadex G-200 gel. This proves that the flow rate intimately depends also on the particle size of the gel. The finer the gel particles the smaller the deviation of the ascending and descending legs of the hysteresis curve. The structure of columns consisting of coarser particles is looser and prone to pack to a greater density under pressure. *Figures 49, 50* and *51* show the flow curves for gels with 2, 4, and 6% agarose content, respectively, in columns of different sizes. It is evident that in the softer gels (lower agarose content) and in thinner and longer columns the maximum character of the flow curves is increasing. *Figure 52* shows the changes of permeability in Sepha-

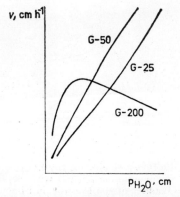

Fig. 47. The dependence of the flow velocity on the hydrostatic pressure (p_{H_2O}) in Sephadex G-25, G-50 and G-200 gel columns

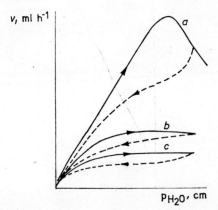

Fig. 48. The flow hysteresis curves of a Sephadex G-200 chromatographic column using *a* 90 μm, *b* 70 μm and *c* 35 μm (average) gel particles (redrawn from Porath, 1972)

rose gels, in a commonly applied column size of 2.5 by 40 cm, which depend on hydrostatic pressure. Gel-bed permeability decreases with decreasing agarose concentration and increasing pressure.

The companies marketing gels generally specify the optimal flow parameters of their soft gels. *Table 25* shows the maximum flow velocities and the related hydrostatic pressures for the deforming Sephadex gels (G-75, G-200) for the three most frequently applied column sizes: 1.5, 2.5 and 5 cm diameter, 100 cm length). *Table 26* indicates the same data on Bio-Gel P

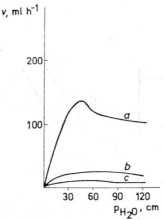

Fig. 49. The dependence of the flow velocity on the hydrostatic pressure in *a* 5×85 cm, *b* 2.5×85 cm, and *c* 1.5×85 cm columns packed with Sepharose 2B agarose (2%) gel. From the Sepharose booklet of the Pharmacia Fine Chemicals AB, 1971

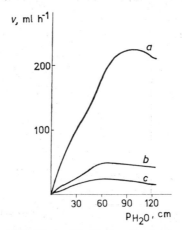

Fig. 50. The dependence of the flow velocity on the hydrostatic pressure in *a* 5×85 cm, *b* 2.5×85 cm, and *c* 1.5×85 cm columns packed with Sepharose 4B agarose (4%) gel. From the Sepharose booklet of the Pharmacia Fine Chemicals AB, 1971

(polyacrylamide) and Bio-Gel A (agarose) gels. In view of the wide range of applicable pressures, *Tables 25* and *26* should be used only for orientation in the determination of the test conditions.

Fischer's (1969) graphic method enables the determination of both the maximum flow velocity and the optimal operating pressures for Sephadex gels, on the basis of the particle size and the dimensions of the column. According to the Kozeny–Carman equation (Carman 1956), good correlations were found with Sephadex gels between the specific permeability (K_0) of the gel beds and the square of the particle diameters (d^2).

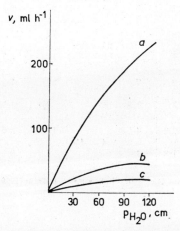

Fig. 51. The dependence of the flow velocity on the hydrostatic pressure in a 5×85 cm, b 2.5×85 cm, and c 1.5×85 cm columns packed with Sepharose 6B agarose (6%) gel. From the Sepharose booklet of the Pharmacia Fine Chemicals AB, 1971

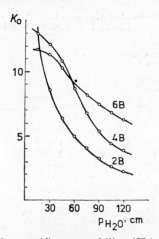

Fig. 52. The dependence of the specific permeability (K_0) of Sepharose (agarose) gels on the hydrostatic pressure in a 2.5×40 cm column. From the Sepharose booklet of the Pharmacia Fine Chemicals AB, 1971

Table 25

Operating hydrostatic pressures (p_{H_2O}) and flow rates for the softer Sephadex G gels*

Column size	1.5×100 cm			2.5×100 cm			5×100 cm		
Sephadex type (particle size), μm	p_{H_2O}	Flow rate		p_{H_2O}	Flow rate		p_{H_2O}	Flow rate	
		ml h^{-1}	cm h^{-1}		ml h^{-1}	cm h^{-1}		ml h^{-1}	cm h^{-1}
G-75 (60)	50—200	44	25	40—160	114	23	38—150	354	18
G-75 (30)		11	6		27	5.5		90	4.5
G-100 (60)	25—100	28	16	24—96	72	15	19—77	234	12
G-100 (30)		7	4		18	3.7		60	3
G-150 (60)	10—40	13	7	9—36	34	7	8—32	108	5.5
G-150 (30)		3	1.8		8.4	1.7		27	1.4
G-200 (60)	5—20	7.2	4	4—16	18	3.6	3—12	60	3
G-200 (30)		1.8	1		4.2	0.9		15.6	0.8

* Data taken from Pharmacia Fine Chemicals AB, Separation Bulletin (1969).

Table 26

Optimum flow rates for the Bio-Gel P (polyacrylamide) and Bio-Gel A (agarose) gels*

Gel type	Linear flow rate, cm h^{-1}			Gel type	Linear flow rate, cm h^{-1}		
	Particle size mesh				Particle size mesh		
	50—100	100—200	200—400		50—100	100—200	200—400
Bio-Gel				Bio-Gel			
P-2	250	150	40	P-200	30	15	—
P-4	225	125	40	P-300	20	8	—
P-6	200	110	40	A-0.5 m	110	35	15
P-10	200	100	35	A-1.5 m	90	30	10
P-30	150	90	—	A-5 m	70	20	9
P-60	125	40	—	A-15 m	50	15	6
P-100	90	40	—	A-50 m	30	10	—
P-150	60	35	—	A-150 m	15	4	—

* Data taken from Bio-Rad Laboratories, Catalogue U/V (1970).
Column size: Bio-Gel P, 1.3×13 cm; Bio-Gel A, 2×95 cm.
Operating pressure: p_{H_2O}, 50 cm.

Figure 53 illustrates the proportionality of K_0 to d^2, while *Figure 54* shows the graphs used in the determination of the flow velocity and the hydrostatic pressure. According to the graphic method, for a given type of gel and particle size, the specific permeability can be read from *Figure 53* (or *Table 24*). *Figure 54* shows the value corresponding to the column height referred to the curve for the gel type and extrapolated to the Y-axis. Joining the point of intersection on the Y-axis with the corresponding value of K_0 the point of intersection P is marked on the auxiliary straight line A. Joining P and the column diameter yields the maximum linear flow velocity

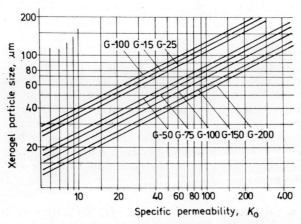

Fig. 53. The correlation between the specific permeability (K_0) of Sephadex G dextran gels and the xerogel particle size (Fischer 1969)

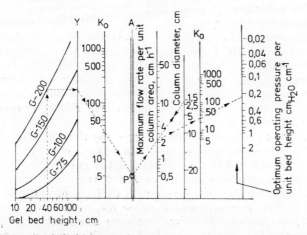

Fig. 54. A graph used for the determination of the optimum operating pressure and optimum flow velocity in columns of known dimensions (height, diameter) packed with different deforming Sephadex G dextran gels (Fischer 1969)

(cm/h). Connecting the flow rate so obtained and the K_0 value, the optimum hydrostatic pressure difference per unit of column length can be read in terms of cm of H_2O. For instance, in the case illustrated in *Figure 54*, to achieve an approximately 2.5 cm/h (4.4 ml/h) flow velocity on a 1.5 by 40 cm column packed with Sephadex G-200 superfine gel, a pressure of 8 to 10 cm H_2O ($0.2 \times 40 = 8$ cm) is necessary.

To reduce the flow resistance of soft gels various additives are used. To Sephadex G-200 gel Craven et al. (1965) added cellulose powder. Sachs and Painter (1972) mixed different deforming gel particles (Sephadex G-100, G-150, G-200, Sepharose 2B, 4B) to silicon-coated glass beads. The beads of optimal size (6 mm diameter) form a rigid network in the column and prevent deformation of the gel particles. Not only does this method increase the column size considerably (by approximately 100%) but (with a larger outer volume but unchanged (?) resolution power according to the authors), at a 5 to 10 times higher pressure than in the columns of identical size and conventionally packed, a ten times higher flow velocity can be achieved. To establish whether the very high flow velocities affect the separation efficiency, further examinations are required. The examination of compressed columns (see Section 2.3.1 of Part II) has shown that optimal resolution of the gel columns is achieved at higher pressures, for any given compression of the gel bed, with a simultaneous decrease in the flow velocity — and not at the maximum flow rate or corresponding pressure difference.

2.3.3. CHECKING THE HOMOGENEITY OF THE GEL BED

The first step in gel chromatography is to check the homogeneity of the gel bed. The evenness of packing and the presence of sedimentation zones, air bubbles, foreign bodies — particularly in softer and more transparent gels — can be revealed by visual inspection or by putting a light source behind the column. A thorough examination of the homogeneity of the gel bed is usually performed simultaneously with the measurement of the volumetric parameters of the column. To determine the outer volume of the bed (V_o) (see Section 2.3.5 of Part II) usually the Blue Dextran 2 000 macromolecular dextran derivative is used. The progress of this blue compound can be followed in the column by naked eye, and in this way macroscopic distortions in the gel bed can be detected. *Figure 55* shows the most frequent distortions of the stained zone.

The quality of packing can be characterized by the degree of zone broadening of various model substances (albumin, cytochrome C, γ-globulin, etc.). According to Flodin (1961) the packing of the gel bed is only then satisfactory if the sample, sure to be excluded from the inner volume and applied in a quantity which corresponds to 10% of the column volume, becomes diluted to a maximum of twice its original volume during the chromatographic process. However, refining the method of packing and the application of the sample, even better results can be achieved. Comparison of zone broadening for the substances excluded from the inner volume, or occupying the complete volume, provides further information on the effects of flow and diffusion phenomena. To optimize the gel chromatographic conditions, preliminary tests must be performed with the same (or similar) substances as are to be used in the process, and each and every factor of the resolution (capacity factor, selectivity, theoretical plate number, etc.) must be determined. These data will numerically characterize the quality of the gel bed and give an indication if the charge or the hydrodynamic parameters need adjustment (compare with Section 2.6 of Part I).

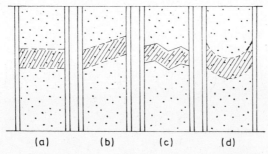

Fig. 55. Checking of the homogeneity of the gel bed using Blue Dextran 2000 macromolecular dextran derivative. The shape of the coloured zone (a) in a correctly packed column, (b) in a column packed in the oblique position, (c) if the surface or the dosage of the sample are uneven, (d) under the influence of the wall effect or when excessively thin gel suspension is used for packing

2.3.4. PREPARATION AND APPLICATION OF THE SAMPLE

The volume, concentration, ionic strength, pH, viscosity and density of the solute to be examined, and the method of application of the sample on the column, are most essential factors of the conditions in which gel chromatography should take place. According to what has been stated in Section 2.3 of Part I, the volume of the sample is determined by the separation volume, the volumetric distribution coefficients and the inner volume of the gel bed. If the difference between the distribution coefficients of the substances is sufficiently great (for instance in desalting), the volume of the sample should be approximately one-third of the inner volume of the gel bed. Under such conditions even a forty-fold increase in the sample volume (!) will not alter the separation (Flodin 1961). However, if substances with similar properties are fractionated, the sample must not be more than 5 to 10% of the column capacity.

The concentration of the sample (e.g. in the case of macromolecules) exerts its effect on the elution curve, primarily through higher viscosity. *Figure 56* shows the elution profiles obtained during the desalting of haemoglobin solutions of different viscosity. A similar case is illustrated in *Figure 57* where dextran mixed with albumin causes distortion of the elution curve. If dextran is dissolved in the eluant, the elimination of the difference between the viscosity of the sample and eluant impedes zone widening for the albumin and restores the original elution profile. Therefore, for the gel chromatography of highly viscous solutions, the viscosity of the eluant should be adjusted to the same value as that of the sample, using neutral additives (e.g. high molecular weight dextran) for the purpose.

In the majority of cases, particularly in the less complicated operations of gel chromatography, the pH and ionic strength of the sample do not play an essential role. For more exact examinations, particularly for the chromatography of low molecular weight compounds, the sample must be

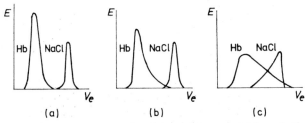

Fig. 56. Elution profiles during the desalting of haemoglobin solutions of different viscosities (Fischer 1969)

(a) dextran = 0, $\eta = 1$;
(b) dextran = 2.5%, $\eta = 4.2$;
(c) dextran = 5%, $\eta = 11.8$

An increase in the relative viscosity of the sample by dextran ($\overline{M}$ = 250 000) causes a gradual distortion of the elution curves

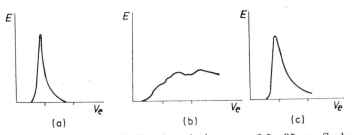

Fig. 57. Gel chromatography of albumin solution on a 2.5×35 cm Sephadex G-25 (fine) column at different viscosity values of the sample and of the eluant (Fischer 1969)

(a) dextran = 0, $\eta = 1$; (b) dextran = 3.5%, $\eta = 4.09$; (c) dextran = 3.5%, $\eta = 1$

In test (a) 5 ml of 0.2% albumin solution was eluted with 0.1 M NaCl; in (b) the sample included also 3.5% of dextran ($\overline{M}$ = 500 000); in (c) the sample was identical with the above. The elimination of the relative viscosity difference prevents distortion of the elution profile

brought into equilibrium either with the buffer solution used as eluant or with the solvent. The buffer change may take place by dialysis or, preferably, by gel chromatography. If more concentrated samples are concerned, diluting with the eluant offers a simple and advantageous solution to the problem.

The application of the sample to the gel bed is an operation which requires very close attention. The advent of modern piston-type columns has simplified this very considerably, since the sample reaches the gel bed evenly through filters which are in direct contact with the gel particles. While the sample is being introduced, the flow velocity in the column must be constant. Care must be taken to prevent air bubbles from getting into the column from T taps and connections. If simpler open columns are used, the sample should be applied over the gel surface manually or dosed via a micropump. During application of the sample, the flow in the column should be interrupted. To prevent mixing, a porous sintered-glass filter plate or a stainless steel sieve corresponding to the column diameter should be inserted. For simplicity and safety, the best technique of sample application is to apply the sample layer at the gel–eluant interface. In this case, however, the specific gravity of the sample must be slightly higher than that of the eluant. The sample is applied with great caution by means of a capillary-tipped pipette in a layer over the gel surface along the column wall, below an eluant layer of a few cm thickness. Then, slowly starting the flow, the sample is made to penetrate into the gel bed. If the sample has been spread over the free gel bed, the eluant should be removed from the gel particles, and only a thin liquid film left behind. Having applied the sample layer and started the flow, penetration should be allowed to continue until the gel surface is just wet. By applying another eluant layer and repeating this operation, the sample should be passed stepwise into the column. Care must be taken not to allow the gel particles to run dry during any one of the operations because this would cause a fast local flow-in of the sample or the washing fluid and render the solute zone uneven.

2.3.5. DETERMINATION OF THE VOLUMETRIC PARAMETERS OF THE GEL BED

The complete volume of the column can be determined by a simple volume measurement (measuring cylinder) before packing. The volume of precision-type columns can be determined by calculation, from the cross-section and the length, after packing. The knowledge of this volume (V_t) is indispensable, not only in the planning of the column dimensions and the quantity of xerogel, but also for calculation of the volumetric distribution coefficients (K_{av}). In this calculation, the sum of the outer and inner volumes of the gel bed must always be less than the overall volume of the column, and the denser the gel structure the greater the difference. Owing to the theoretical and practical difficulties of an exact determination of the inner volume, the elution properties of the solutes are always referred to the overall volume of the gel bed. The inner volume of loose, soft gels may be measured by use of small-molecule substances which fill out the liquid space of the bed completely. Such substances are glucose, salts containing the chloride ion (KCl) or various coloured substances, such as iron(III) thiocyanate, copper(II), chromate, etc. The elution of chloride ions can be simply and quickly followed by a mercury nitrate titration with sym-diphenyl-carbazone indicator (Schales and Schales 1941). With increasing dry-matter content in the gels, the selective retention of various inorganic ions (Zeitler and Stadler 1972) and aliphatic and aromatic adsorption make determination of the actual inner volume increasingly uncertain. Lately, in possession of appropriate instruments, isotopically labelled water has been used for the purpose. Marsden (1971) established that the actual inner volume of gels is indicated by the elution of $H_2^{18}O$ since, unlike tritium, this does not enter into an exchange reaction with the oxygen atoms of the dextran chains.

In principle, any macromolecule excluded from the inner liquid space of the gel can be used for the determination of the outer gel bed volume. Depending on the type of gel, a variety of proteins with different molecular weights (haemoglobin, cytochrome C, etc., enzymes, nucleic acids, polysaccharides or fine-grained colloidal particles — e.g. China ink) may be used. The easily, specifically and sensitively detectable less adsorbent substances are preferable. To reduce adsorption, an eluant of moderate ionic strength (0.02 M), NaCl or buffer solution should be chosen. The elution of proteins (enzymes) and nucleic acids can be traced by their ultraviolet light absorption. To detect polysaccharides, a well proven and simple method is to use the anthrone reaction (Scott and Melvin 1953). For the determination of the homogeneity and outer volume of the gel bed the Blue Dextran 2000 dextran derivative of Pharmacia Fine Chemicals AB, with an average molecular weight of 2 million, stained with Cibacronblau F3G-A (colour index No. 61211) (Blue Dextran 2000, Pharmacia Fine Chemicals AB, Uppsala, 1972) is most frequently applied. *Figure 58* shows the chemical structure of Blue Dextran 2000. Corresponding to the properties of polycyclic chromophorous compounds, Blue Dextran 2000 can be sensitively measured at 280 and 625 nm, its adsorption in dilute (0.02 M) saline or phosphate buffer (pH $= 7.0$; $\mu = 0.15$) being minimal. This product, marketed in lyophilized form, is used in a solution of $0.01-0.02\%$ in a sample which corresponds to $1-2\%$ of the column volume. It should be noted that Blue Dextran 2000 forms an ionic complex with numerous proteins (albumin, lysozyme, haemoglobin, etc.), enzymes (pyruvate kinase, glutathione reductase, acetic acid-succinyl-coenzyme A transferase, etc.) and blood-clotting agents. Its residues can be removed from the column by chromatographing a small amount of protein (e.g. albumin) to prevent the front of the eluting proteins showing erroneously high UV light absorp-

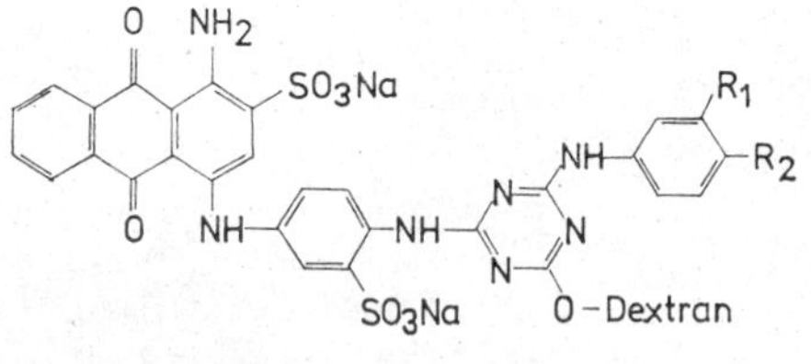

$R_1 = H$ or SO_3Na

$R_2 = SO_3Na$ or H

Fig. 58. The chemical structure of Blue Dextran 2000 (Pharmacia Fine Chemicals AB, Uppsala, 1972)

tion during subsequent chromatography. Blue Dextran 2000 yields unequivocal results, especially when used for the determination of the outer volume of gel beds consisting of smaller- and medium-pored gels. *Figure 59* shows that, on gels with a loose structure and large pores (primarily agarose), the elution profile is asymmetric and fractionation of Blue Dextran 2000 yields an elongated wide zone, indicating the presence of several components.

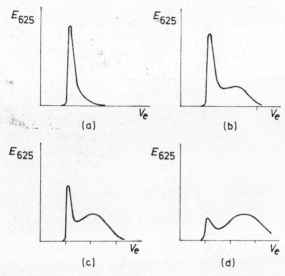

Fig. 59. The variation of the elution profile of Blue Dextran 2000 on gels of different pore sizes
(a) Sephadex G-200; (b) Sepharose 6B; (c) Sepharose 4B; and (d) Sepharose 2B

In such cases the first peak of adsorption (the fraction of highest molecular weight) gives the outer volume. In macroreticular gels, if particles (particulates, viruses, cells, etc.) are excluded with certainty from the inner volume more reliable results will be obtained (Tangen *et al.* 1971).

2.4. GENERAL PROCESSES OF GEL CHROMATOGRAPHY

The general (automated, programmed) operations of gel chromatography include (a) the fractionated collection of the column eluate, (b) the detection, recording and determination of the substances in the fractions, and (c) the documentation and appraisal of the results. Since these operations are common to every chromatographic method, for more detail the reader is referred to existing methodology manuals (Lederer and Lederer 1957; Vámos 1959; Keil and Sormová 1968; Peterson 1970; Mikes 1970). Here, we shall restrict ourselves to a discussion of the problems specific to gel chromatography.

2.4.1. FRACTION COLLECTING

Substances separated by gel chromatography can be separated by fractionated collection of the column eluate. The optimal number and optimal size of the fractions depend on factors which vary from test to test. It is a rule of thumb that the volume of the individual fractions should not exceed 1 to 2% of the total elution volume. Theoretically, the volume of the fractions depends on the resolution which, in turn, depends on the separation volume [the properties of the separated substances and the gel, $V_{sep} = V_{e2} - V_{e1} = (K_{d2} - K_{d1})V_i$] and, further, on the zone broadening of the solutes ($R_S = V_{e2} - V_{e1})/4\sigma$ (see Section 2.6.1.1 of Part I).

Under optimal conditions (at a resolution of 4σ, $R_S = 1$) there is a 4σ distance (volume) between the elution maxima of two substances. To prevent the substances from contaminating one another during fraction collecting, the fraction volume must be less than σ. In practice the volume of the fractions is considerably less than this. The aim is that the substances or transitional zones of most interest for the separation should appear in at least 3 to 5 fractions. From this it follows that the poorer the resolution of the gel bed (low separation volume or large zone broadening) the smaller are the fractions which should be collected. By increasing the column dimensions the volume of the fractions may be increased. The number of fractions to be collected is limited by practical considerations. With decreasing volume not only the amount of the test substances present in the fractions will decrease, but also the number of samples to be analysed will increase. Over certain portions of the elution curve the combination of fractions is justified anyway. There is always a certain error in fraction collecting and the elution volumes can be more precisely determined if volumes not dependent on the separation (for instance V_o) are measured without fractionation.

In simpler operations or during preliminary experiments it is preferable to follow the separation process by manual fractionation, but in major tests, for which several hundred fractions are required, automatic fraction collectors are indispensable. According to the movement of the test tubes, circular (spiral) and linear fraction collectors are known. *Figure 60* shows two characteristic fraction collectors. These must satisfy the following requirements: the tubes should be interchangeable according to a decimal number of pulses, they should be able to operate with different types of tube, they should permit a clear survey of the tubes and their positions, the tubes should be detachable and movable, and the fraction collector should be movable even in reverse.

Fractions may be collected by volume (in syphon systems operating on the balance principle by weight), by drop counting or by time. In analytical gel chromatography (e.g. gel permeation chromatography) time measurement (retention time) is used, almost exclusively, as the elution parameter of the solutes. Each method is liable to certain inherent errors. In syphoning systems the risk exists of intermixing the fractions; in drop counting that of alterations in surface tension must be looked for (Wolkoff and Larose 1976). The correlation between volume and time can be made more reliable by using circulating pumps. If toxic, infectious or radioactive compounds are chromatographed the amount of substance passing between the tubes during fraction change must be reduced to a minimum. In such cases fractionation should take place by drop counting because tube change is then effected merely by the drops themselves. The latest automatic fraction collectors

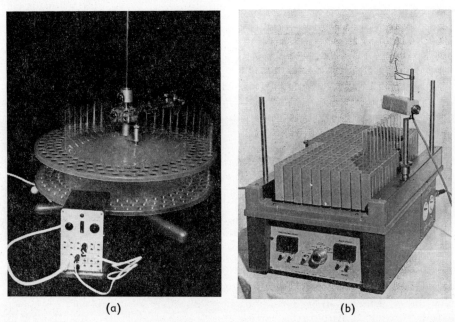

Fig. 60. Modern automatic fraction collectors: (a) SF 62 (Mikrotechna, Prague) (b) Ultrorac 7000 (LKB Produkter, Bromma)

(e.g. the LKB Recychrom apparatus) are fitted with a set of valves which stop the flow at the instant of change. Fraction collecting by drop counting is preferable also if hydrostatic pressure (Mariotte flask) is applied.

2.4.2. DETECTION AND DETERMINATION OF THE SEPARATED SUBSTANCES

The separation process of gel chromatography is characterized by the detection of the substances present in the fractions. The methods of detection may be non-specific or specific. While non-specific methods of detection yield usually qualitative results only, they enable a fast and easy check to be made upon the experimental conditions. The ideal processes are sensitive and quick, consuming little material and retaining the detected substance in its original state. From this aspect the most frequently used non-specific detection methods include UV light absorption, fluorescence, and refractive index and conductance (electric and thermal) measurements. Detection based on radioactivity is a specific process for labelled compounds, but non-specific as regards the method. The labelled atom is usually ^{14}C, ^{3}H or ^{18}O, or a heavy metal (Cu, Cr), a halogen (I) or some other elements (e.g. P, S, etc.). Making use of part of the material, the number of applicable methods can be further increased, for instance by dichromate oxidation or flame ionization detection of organic substances, the determination of the content protein on the basis of the Lowry–Folin biuret-ninhydrin colour reaction or the determination of carbohydrates by the anthrone reaction.

For continuous automatic assessment of the composition of the fractions, relatively compact spectrophotometers fitted with a through-flow cuvette and set to constant wavelength (UV) are used predominantly (LKB Uvicord, Gilson Medical Electronics, VV-254- -2651F, -2801F; Vitatron UFD), which produce monochromatic light by a prism, an interference filter, or a mercury vapour lamp. In some cases (e.g. in the LKB Uvicord apparatus) the 254-nm wavelength of the mercury vapour lamp is converted to the more advantageous 280 nm by fluorescence. These photometers are simple in lay-out and, considering the longer duration of the tests, they aim first and foremost at electric (baseline) stability. A characteristic problem involved with their use is distortion of the linear measuring scale at higher concentrations (below 50% transmission).

The theoretical and practical difficulties of detection have been examined by Morris and Morris (1962) and Mikes (1970) who have published detailed surveys. In certain working processes (e.g. desalting, the following of ionic strength or pH gradient) the measurement of conductivity (with, for example, the LKB Conductolyzer) can be applied with advantage. For organophilic gels and in gel permeation chromatography, variations of the refractive index, which can be measured to a high degree of precision, are used extensively (E-C Apparatus Corp., EC 211; Nester-Faust Manufacturing Corp., 404; Waters Associates, R4). In modern gel chromatography, facilities analogous to gas chromatography, flame ionization (Barber-Colman Co. 5400; Carlo Erba) and argon ionization detectors (W. G. Pye and Co. Ltd.) are used with increasing frequency. In exceptional cases, potentiometry, fluorimetry and dielectrometry may be resorted to. The detectors may be connected to single or multichannel recorders which enable the continuous recording of the operations (LKB Recorder 6510) to be carried out. The more expensive and more sophisticated equipment (Beckman Spinco, Spectromonitor Model 135 A, Hitachi Ltd. Model 034) enable the simultaneous recording of several parameters (optical, pH, conductivity) or the comparison of the results of measurements performed at dissimilar wavelengths to be carried out.

The application of specific detection methods (measurement methods) proceeds usually in parallel with a thorough examination of the chromatographed substances, and calls for an appropriate preliminary study. The specific methods are more sensitive than the unspecific ones (compare, for example, UV adsorption with enzyme reactions) and are suitable for detecting 1% (or less) of sample activity in the fractions. The specific identification methods can be grouped according to the determination of

(a) biological activity (enzyme, hormonal, pharmacological reactions, etc.),
(b) chemical composition (elemental, functional-group and end-group analysis, decomposition and condensation reactions, etc.), and (c) physico-chemical properties (melting and boiling points, molecular weight, etc.).

The high-sensitivity thin layer chromatographic methods and the fast developing automatic analyzers (Technicon, Autoanalyzer) offer assistance for the application of both the specific and non-specific identification

Fig. 61. The LKB Recy Chrom automatic chromatograph in operation

methods. The programming of the various operations forms part of the automation process. Modern chromatographs are available with built-in programmes for the basic operations (Beckman, Spectrochrom; Hitachi ILC-2A). Into the LKB Recychrom system, or into its 4930 A programmer, six different chromatographic functions can be fed. The LKB Ultrograd 11300 and the LKB Ultrograd Level Sensor have been designed specifically for the programming of gradient elution. *Figure 61* shows the LKB Recychrom automatic chromatograph.

For the testing of organophilic substances the Waters Associates, Gel Permeation Chromatograph Model 200 (or its larger variant, the Ana-Prep) are now used widely. For the automatic determination of the molecular weight of polymers, at high pressures and high flow rates up to a temperature of 140 °C, this apparatus uses four Styragel columns connected in series. Chromatography is assessed by a difference-refractometer, fraction volumes being measured by a syphon system. *Figure 62* illustrates the the structural lay-out of this gel permeation chromatograph.

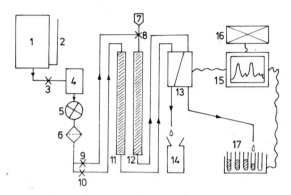

Fig. 62. Schematic set-up of the Gel Permeation Chromatograph of the Waters Associates

1. solvent reservoir; 2. solvent level indicator; 3. 8. 9. 10. valves; 4. deaerator; 5. pump; 6. filter; 7. sample batcher; 11. reference column; 12. sample column; 13. difference refractometer; 14. waster collector; 15. recorder; 16. integrator; 17. fraction collector

2.4.3. DOCUMENTATION AND APPRAISAL OF THE RESULTS

The results of gel chromatography can be illustrated by the elution curves, which show the physicochemical properties of the fractions as functions of the absolute (V_e, V_N, V_{ret}) or relative (V_e/V_o, V_e/V_t) variables of chromatography. If automatic gel chromatographic apparatus is used the (usually non-specific) properties of the eluate (UV absorption, refractive index, conductivity, etc.) can be read from the recorder. This provides a basis for the further (specific) analysis or rechromatography of the fractions, etc. While making use of the theoretical relationships, the parameters of the substances being tested are graphically determined from the elution curves, the details for the determination of the volumetric parameters of the column should also be quoted together with the results of preliminary tests performed with some appropriate model substance. The example below shows the procedure for a test carried out on a Sephadex G-100 column, and its results. The elution curves for the gel chromatographic process are shown in *Figure 63*.

Sephadex G-100 dextran gel particles, in sizes between 40 and 120 μm, were allowed to swell in phosphate buffer (0.01 M, pH 7.4) (Pharmacia Fine Chemicals AB, Uppsala: Lot. No. 7120), packed into a Whatman-type precision glass column of size 1.5 × 45 cm and brought into the equilibrium state by passing a buffer of five times the column volume through it. The gel bed had a volume $V_t = 75$ ml and a height L of 42.5 cm. According to *Table 25* the flow rate in the column was 20 ml/h. For the gel chromatography process an LKB 4912A peristaltic pump, a 8300 Uvicord II sensor (278 nm), a recorder and an Ultrorac 7000 automatic fraction collector were used. Along certain portions of the elution curve (e.g. in the determination of KCl) 0.5-ml fractions were collected by drop counting; the KCl content was measured by titration (Schales and Schales 1941).

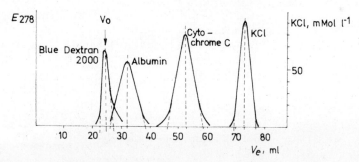

Fig. 63. Gel chromatography of Blue Dextran 2000, bovine serum albumin, cytochrome C and KCl on a Sephadex G-100 column. Particle size, 40 to 120 μm, column size, 1.5×25 cm; V_t, 75 ml; flow rate, 20 ml h^{-1}; phosphate buffer 0.01 M, pH 7.4

1. *Determination of the volumetric parameters of the gel bed*

(a) The outer volume of the gel bed was measured by chromatography of Blue Dextran 2000 (Pharmacia) in 0.2% solution (in phosphate buffer). The sample volume was 1 ml. From the elution curve shown in *Figure 63* the elution volume of Blue Dextran 2000 was $V_e = V_o = 24$ ml, the zone broadening b being 3.9 ml. In line with the theoretical statements, the outer volume of the gel bed amounted to approximately one-third of the total column volume: $V_o = 0.32\ V_t$. From the elution of Blue Dextran 2000 the theoretical plate number for the outer volume of the gel bed can be calculated from the following formula

$$N = 16 \left(\frac{V_e}{b}\right)^2 = 16 \left(\frac{24}{3.0}\right)^2 = 605$$

and the theoretical plate height from

$$\text{HETP} = \frac{L}{N} = \frac{42.5}{605} = 0.07\ \text{cm}$$

(b) To determine the total (available) liquid volume of the gel bed, 1 ml of 0.5 M KCl solution (in phosphate buffer) was chromatographed. From the elution curve $V_e = 73$ ml and the zone broadening of KCl $b = 8.2$ ml. From the elution of KCl the theoretical plate number for the total column volume (N), to a good approximation, is 1267 and the HETP is 0.043 cm.

2. Gel chromatography of serum albumin and cytochrome C

Solutions (1%) of defatted bovine serum albumin (Bovine serum albumin, Fraction V, Armour Pharm. Co.) and cytochrome C (from horse heart) (Koch—Light) were applied to the column in 1-ml samples. In these tests the phosphate buffer contained also 0.1 M of KCl.

(a) Examination of bovine serum albumin. Elution volume V_e, 32 ml, zone broadening b, 13 ml. The theoretical plate number (N) determined with the albumin monomer was 97, and HETP 0.438 cm. From the elution volume and the column parameters the volumetric distribution coefficient of albumin can be computed in the following way

$$K_{av} = \frac{V_e - V_o}{V_t - V_o} = \frac{32 - 24}{75 - 24} = 0.157$$

and, from the Determann equation, the molecular weight of the albumin monomer (for Sephadex G-100)

$$\lg M = 5.941 - 0.847 \cdot \frac{V_e}{V_o}$$

$$\lg M = 5.941 - 0.847 \cdot \frac{32}{24}$$

$$M = 64\,820$$

The molecular weight of the albumin monomer so obtained shows good agreement with the values quoted in the literature. *Figure 63* shows the elution curve of the albumin monomer only.

(b) The examination of cytochrome C. Elution volume V_e 52.5 ml, zone broadening b 12 ml. The theoretical plate number (N) for cytochrome C was 3 310 and HETP 0.137 cm. The volumetric distribution coefficient of cytochrome C, K_{av} was 0.559, and its molecular weight measured by gel chromatography (M) was 12 250 — in fair agreement with the literature.

It is evident from the above that a precisely determined value of the outer volume is one of the most important factors in the computation. It should be noted that the value of 24 ml remained unchanged even during the long series of tests, and throughout a large number of measurements. Nevertheless it is advisable to check on the outer volume from time to time. Similarly important is a precise determination of the elution volumes. Here is an example to prove this: with the column described, a 1-ml deviation in the elution volume would cause an error of approximately 10% in the calculation of the molecular weight.

3. SPECIAL TECHNIQUES

In the course of the development of gel chromatography the relatively simple technique of column chromatography has been enriched by several new techniques, theoretical as well as practical. Such techniques are repetitive or recycling chromatography of substances on one or several interconnected columns, gradient elution based on the properties of the solutes, zone precipitation, and the production of complexes or the application of gel grains by adsorption or polar-apolar phase partition chromatography.

Apart from column chromatography the best known applications of swollen gel particles are in preparative or batch processes and gel phase thin layer chromatography.

3.1. RECYCLING GEL CHROMATOGRAPHY

The best and more effective way to increase the resolution of gel chromatography, among other factors, is to increase the length of the column. However, it is difficult to reproduce the even packing of long columns. Also the maintenance of the flow velocity and the deformation of the gel grains, etc., cause technical difficulties which combine to limit the dimensions of the gel bed. The resolution can be increased also by the connection of several shorter columns in series. This solution, however, consumes a larger quantity of gel and requires not only columns of fairly equal quality (packing) but increases also the mixing volumes at the connections.

To increase the effective length of the gel bed Porath and Bennich (1962) elaborated the method of so-called recycling gel chromatography. Its schematic illustration can be seen in *Figure 64*. The eluate emerging from the column is continuously recirculated to the gel bed in a closed system and chromatographed repeatedly. All the time the UV spectrophotometer fitted with a through-flow cuvette and the automatic recorder register the results of separation per cycle. *Figure 65* shows the separation of thyramine and dopamine by recycling gel chromatography on a Bio-Gel P-2 column (Kalász 1973; Kalász et al. 1975).

Although the number of cycles is theoretically unlimited, practice has proved that the zone of solutes becomes broader in each cycle (giving an increasingly flatter elution curve) and that after a certain number of cycles the faster moving substance catches up with the slower fraction and mix with them *(Fig. 65)*. Therefore, at an optimal number of cycles, which can be computed from the elution volume of the components, chromatography must be discontinued.

For the optimal number of cycles in gel chromatography of a pair of substances having elution volumes V_{e1} and V_{e2}, respectively, Kalász (1973) and Kalász et al. (1976) derived a theoretical relationship. When this optimal

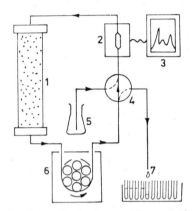

Fig. 64. A simple method of recycling gel chromatography

1. gel column; 2. ultraviolet spectrophotometer with throughflow cell cuvette; 3. recorder; 4. valve set; 5. sample or buffer reservoir; 6. pump; 7. automatic fraction collector

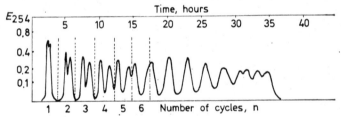

Fig. 65. Separation of thyramine and dopamine (50—50 mg each) by recycling gel chromatography on 3.5×50 cm Bio-Gel P-2 columns at a temperature of +2 °C; eluant 0.2 N CaCl$_2$; LKB Recy Chrom system, Uvicord II, 254 nm (Kalász 1973)

number is attained, the slower component is equidistant from the nth and $(n + 1)$th peak of the faster substance, and

$$n(V_{e2} - V_{e1}) = (n+1)V_{e1} - nV_{e2}$$

whence

$$n = \frac{V_{e1}}{2(V_{e2} - V_{e1})}$$

This optimal cycle number is rarely an integer. Since within n cycles the $(n + 1)$th period of the faster moving substances does not appear, the value of n is rounded upward. Increasing number of cycles enhances the resolution of gel chromatography which, after the nth cycle, will attain

$$R_{Sn} = R_{S0} \cdot n^{1/2}$$

where R_{S0} denotes the original resolution of the gel column.

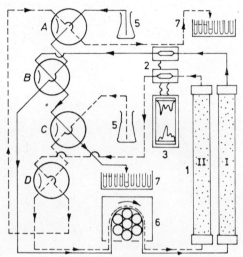

Fig. 66. The flow chart of the recycling gel chromatographic system with two columns. The numbers denote the same parts as in *Fig. 64*. By the correct setting of the valves *A, B, C,* and *D,* different operations can be performed (connecting the columns in series, recycling mode of operation, etc.)

Bombaugh *et al.* (1969b; Bombaugh and Levangie 1970a, b) studied the theory of recycling gel chromatography, mainly for gel permeation chromatography.

If even the optimal number of cycles is insufficient for the separation of some substances then either the column length must be increased or several columns must be connected in series. *Figure 66* shows the interconnection of two gel columns. This technique makes it possible, for instance, for groups of small- and large-molecule substances of various heterogeneous biological systems to be separated on columns packed with gels of different pore sizes. It enables also the fractionation of compound groups by recycling on a suitable column to be achieved.

Recycling gel chromatography may take place also in an intermittent process, by the repeated recycling of parts of the eluate (Morávek 1971). In automatic apparatus (e.g. the LKB Recychrom system), by programming of the valves, the overlapping parts of the cycles can be eliminated and pure solvent can be introduced simultaneously (Welling 1968). In this way (with a loss of material) the cycle number can be further increased and the components separated to the maxima of elution.

3.2. GRADIENT ELUTION IN GEL CHROMATOGRAPHY

The special methods of gel chromatography may call for the application of ionic strength or pH gradients. Gradient elution is used primarily in solubility (salting-in, salting-out), gel chromatography, zone precipitation, complex formation and dissociation; less frequently in adsorption. Continuous (linear, convex, concave) and intermittent (stepwise) gradient formation and the associated theory have been dealt with in detail in the theory of ion exchange chromatography (Novotny 1971; Boross 1968; Peterson 1970; Katz 1963). This book deals only with the quickest and simplest computations and methods in gel chromatographic practice.

The simplest method for gradient formation is to interconnect reservoirs holding solutions of suitable initial and final concentrations (pH), according to the diagram shown in *Figure 67*. The character of the gradient so formed depends also on the shape of the reservoirs. A linear concentration (pH, specific gravity) gradient is obtained if two cylindrical vessels of equal cross-section are used. For such a system Eichenberger (1969) proposed the following general equation

$$C = C_2 - (C_2 - C_1)\left(1 - \frac{v}{V_t}\right)^{Q_1/Q_2}$$

where C denotes the concentration of the liquid in the mixing vessel after discharge of the actual volume v, C_1 and C_2 are the starting and final concentrations, V_t the total liquid volume of the system, and Q_1 and Q_2 the cross-sections of the cylindrical vessels (or vessels having the same geometry).

It will be evident from this equation that, with cylinders of equal cross-section ($Q_1 = Q_2$), the relationship will be linear. The combination of vessels of different cross-sections forms convex ($Q_1 > Q_2$) or concave ($Q_1 < Q_2$) gradients. A similar method of gradient formation is to immerse solid objects of different geometry into the cylindrical vessels of *Figure 67* (cylinder, cone), which alters the cross-section in the same way as do vessels of different form.

Weeke et al. (1972) designed and constructed the very simple and ingenious apparatus illustrated in *Figure 68* to form continuous complex gradients. The partition (1), with sealed edges and adjustable position, divides the plastic reservoir (2) into volumes of different (constant and varying) cross-sections. This set-up can implement every type of gradient formation dealt with so far.

A convex gradient is obtained if a mixing vessel (closed) of constant volume, as shown in *Figure 69* is used. The parameters of this gradient are given by the following equation (Cherkin et al. 1953; Reiner and Reiner 1962; Eichenberger 1969)

$$C = C_2 - (C_2 - C_1)e^{\frac{v}{Vke}}$$

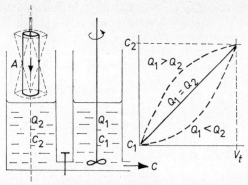

Fig. 67. Formation of linear ($Q_1 = Q_2$), convex ($Q_1 > Q_2$) and concave ($Q_1 < Q_2$) concentration (pH) gradients by the interconnection of cylindrical vessels of equal and dissimilar cross-sections. The immersion of the pieces A (cylinder, cone) reduces the cross section-Q or causes it to change continuously

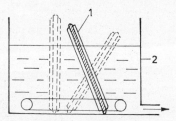

Fig. 68. The set-up designed by Weeke *et al.* (1972) for the production of continuous complex gradients

(1) sealed partition; (2) plexiglass reservoir

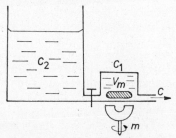

Fig. 69. The formation of a concentration (pH) gradient in a mixing vessel of constant volume (closed). C_1 and C_2 denote the starting and the final concentration, V_{ke} the volume of the mixing vessel, and m the magnetic stirrer

where C denotes the concentration of the solution emerging at the actual volume v, C_1 and C_2 are the starting and final concentrations, and V_{ke} denotes the volume of the mixing vessel.

While this solution is independent of the geometry of the reservoirs, care must be taken to remove air from the mixing vessel.

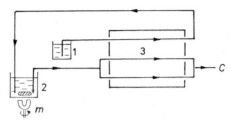

Fig. 70. The method of Ayad et al. (1967) for the formation of linear gradients with a three-way constant-rate peristaltic pump. A linear gradient is produced if the cross-section and the length of the tubes are equal and if the rate of outflow from mixing tank 2 is two times faster than the outflow from 1 (m denotes the magnetic stirrer, 3 the peristaltic pump)

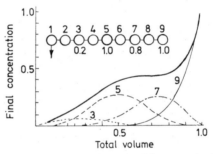

Fig. 71. The formation of a multi-component (or multi-concentration) complex gradient in the nine-chamber Varigrad apparatus (Peterson and Sover 1959; Peterson and Rowland 1961). Curves 3, 5, 7, and 9 show the variation of the concentration of the components passed to the appropriate mixing chamber. Mixing chambers 2, 4, 6, and 8 contain pure solvent. Curve A is the resultant of the components

Ayad et al. (1967) suggested another simple method for linear gradient formation. Its essential features are illustrated in *Figure 70*. Apart from a three-way peristaltic pump, the method requires only the usual laboratory accessories (beaker, tubing, T-connection).

To form multicomponent gradients several open or closed mixing vessels are connected in series (5 to 9 pieces). These hold water or solutions of the required components. When the liquid flow is started different gradients of the components will form in the mixing vessels and pass continuously into the column. Their resultant will form the required mixed gradient. *Figure 71* shows the formation of a complex gradient consisting of four different components in a nine-chamber Varigrad apparatus (Peterson and

Sober 1959; Peterson and Rowland 1961). Similar facilities are marketed by Buchler Instruments Inc. (Fort Lee, New Yersey), Phoenix Precision Instruments Co. (Philadelphia, Pennsylvania), Technicon Chromatography Co. (Chauncey, New York), and Metaloglass Inc. (Boston, Massachusetts). Optional gradients may be formed in the LKB Ultrograd apparatus, which scans the outline of the pre-drawn and cut-out gradient by means of a photoelectric cell, and then sucks the solutions of initial and final concentration in the required ratio, under the control of a valve, first into the magnetic mixing chamber and then to the column. The apparatus of the Beckman Co., the Varipump of the Phoenix Co., and the Dialagrad gradient mixer of the Instrumentation Spec. Co. (Lincoln, Nebraska) operate on similar principles.

3.3. GEL CHROMATOGRAPHIC METHODS BASED ON SOLUBILITY

The structure of the gel offers a specific medium for those separating mechanisms in which, together with solubility, the behaviour of the gel phase to molecules of different sizes also has a decisive role. Owing to their many common features, before going into a discussion of zone precipitation, it is necessary to survey briefly the relatively new process of solubility chromatography.

3.3.1. SOLUBILITY CHROMATOGRAPHY

The processes in which separation is based on the dissimilar solubility of the substances — depending on the pH, the ionic strength or the polarity of the medium — are known by the collective name of solubility chromatographic process (Hoffmann 1969). Preparative separation based on solubility, in the form of fractionated salting-out, has been used for a long time in protein chemistry. The Baker Williams type fractionation of plastics and polymers is one of the solubility chromatographic methods known longest (Smith 1970). The essential feature of solubility chromatography consists of the precipitation of a mixture of solutes, e.g. proteins, on a suitable carrier and subsequent elution by gradual (continuous or stepwise) alteration of the properties of the medium (pH, salt concentration, polarity). Inert column packings (Celite 545, porous glass beads, DEAE-cellulose, etc.) are used generally as the carrier. Solubilization is achieved sometimes by varying the pH of the eluant (Hoffmann 1969; Bjorklund 1971), or more frequently by varying the salt concentration of the solutions.

Depending on the properties of the substances to be separated two chromatographic methods are used: salting-out and salting-in.

During the process of salting-out chromatography, the substances precipitate in a concentrated solution of some suitable compound (e.g. ammonium sulphate, sodium sulphate, etc.), and dissolve at a lower concentration (Hoffmann and McGivern 1969; Behal et al. 1969; Mayhew and Howell 1971; King 1972; Fudano and Konishi 1972). Beling (1961, 1963) in his publications has described most illustratively the principle of salting-out chromatography. Under the effect of the relatively high salt content, the urine oestrogens elicit a strong bond to Sephadex G-25 gel particles brought into equilibrium with distilled water, and give rise to the phenomenon of ion exclusion: during elution with distilled water the salt content of urine migrates faster than the oestrogens. With decreasing salt concentration the oestrogens elute and appear in two fractions — in a zone which is very much sharper than the original sample volume. This process is a typical case of concentration and focussing arising from the effect of the salt gradient. Van Tilburg and Muller (1970) used the same method to refine the separation of conjugated oestrogens from the other chromogenous substances in urine.

In the case of salting-in chromatography the substances precipitate at a low ionic strength and dissolve with increasing salt concentration. A good example of the latter was given in a publication by Epstein and Tan (1961) which, to a certain extent, is a transition towards the technique of zone precipitation. Epstein and Tan chromatographed serum proteins dissolved in 1 M NaCl on a Sephadex G-25 column, brought into equilibrium at a low ionic strength (0.02), to separate pseudoglobulins and euglobulins. Pseudoglobulins, which form the greater part of serum proteins and dissolve readily even at low ionic strength, when leaving the NaCl zone, will elute in the outer volume of the column, while euglobulins, passing into the medium of low ionic strength, will precipitate and emerge only from the NaCl zone (i.e. at approximately the total volume of the column). Although at a rather limited resolution, this method incorporates the basic elements of both solubility and gel chromatography.

The prerequisites of the above methods are that (a) the dissolution and precipitation of the substances should be fast and reversible, (b) that precipitation should not cause the denaturation of the substance, and (c) that at the instant of precipitation the interaction (adsorption) between the carrier and the chromatographed compounds should be loose enough and adequate from the molecular (steric) point of view. The same criteria hold, without any modifications, also for zone precipitation.

3.3.2. ZONE PRECIPITATION

The method of zone precipitation combines the principles of gel and solubility chromatography. Its essential features were described by Porath in 1962. Porath (1962b) introduced a linearly decreasing concentration gradient of ammonium sulphate into a column packed with Sephadex G-100 gel and then applied a solution of the serum proteins to the column. In line with the principle of gel chromatography, the protein molecules (partially) excluded from the inner volume of the gel particles move faster than the concentration gradient of ammonium sulphate.

Figure 72 shows the mechanism of zone precipitation. It is based essentially on the differences of solubility, on gradient elution and on the phenomenon of gel chromatography. These factors, combined, enhance the resolution of both gel and solubility chromatography considerably. As in solubility chromatography, decreasing or increasing concentration may be used also in zone precipitation. Accordingly, one may distinguish between salting-out or salting-in methods of zone precipitation. Although more than ten years have elapsed since the first description of this method, publications on zone precipitation are still few and far between (Sargent and Graham 1964; Swart and Hemker 1970; Little and Behal 1971; Fehrenbach 1971).

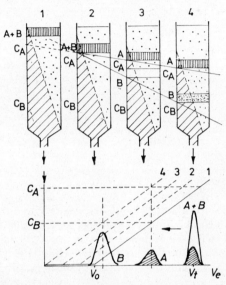

Fig. 72. The zone precipitation method. The locations of the gradient and of the substances to be separated in the gel columns during the different phases of elution are shown

(1) following the linearly decreasing concentration gradient of ammonium sulphate, the sample containing the components *A* and *B* is introduced to the column; (2) the components of the sample travel at a faster rate than the concentration gradient; (3) component *A* attains the concentration of C_A and precipitates while *B* proceeds at an unchanged rate (A is dissolved and carried by the zone of lower concentration); also the *B* component attains the concentration of C_B which is required for precipitation and continues to elute at the rate of the gradient

A characteristic case for the application of salting-in zone precipitation is the separation of low-density human serum lipoproteins, in the form of their polyanion-type macromolecular complexes (Kremmer et al. 1972). The measurement and isolation of very low-density (VLDL) and low-density (LDL) lipoprotein fractions in the examination of various disturbances of fatty metabolism are important from the points of view of physiology and analytical chemistry alike.

A simple method for the isolation of serum lipoproteins is precipitation in the form of polyanion complexes (Burstein et al. 1970). Heparin, dextran sulphate (molecular weight $5 \cdot 10^5 - 2 \cdot 10^6$), sulphonated amylopectin, etc. may be used for the purpose. To form polyanion–lipoprotein bonds, an appropriate pH, bivalent cation (Ca, Mg, Mn) and low ionic strength are necessary. Owing to the ionic character of the bonds the complexes can be dissolved at higher salt concentrations and reversibly precipitated by dilution. The examination of pure lipoprotein fractions isolated by ultracentrifugation has shown that the VLDL and LDL dextran sulphate complexes dissociate at different values of ionic strength, close to the isoelectric point of the proteins (pH 5.4). This enables a separation to be carried out

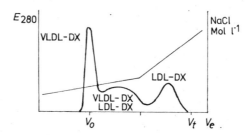

Fig. 73. The dextran sulphate (average molecular weight 2 million) separation of human serum low-density lipoproteins (VLDL and LDL) by salting-in zone precipitation (Kremmer *et al.* 1972)

Column: 2.5×30 cm Sephadex G-200 (medium); V_t, 150 ml of acetate buffer (0.15 M, pH = 5.4; 0.033 M $CaCl_2$); sample volume, 5 ml; NaCl gradient, 0.1 to 0.9 M; VLDL—DX and LDL—DX are the dextran sulphate complexes of the very low and low density lipoproteins, respectively

by salting-in zone precipitation. *Figure 73* shows the elution curve for the zone precipitation of lipoprotein complexes on a Sephadex G-200 gel column. In agreement with Porath's observations, the use of different gel types (Sephadex G-25, G-75, G-100) verified that the pore size of the gel matrix is an important, even determinant, factor in the separation mechanism.

3.4. COMPLEX FORMATION IN GEL CHROMATOGRAPHY

A combination of the examination of gel chromatography and molecular interactions has helped solve numerous theoretical and practical problems. The methods elaborated serve both for the analysis of association–dissociation equilibria, the bonds in the complexes protein–protein and protein–small-molecule compounds (ligands), and isolation of the components.

For the conditions of zonal separation, frontal analysis and gel chromatography, the analytical processes follow the so-called Hummel—Dreyer technique. Wood and Cooper (1970) have given a detailed survey of the chromatographic analysis of the interactions of proteins and small molecules according to the above division. In the course of zonal separation a relatively small amount of a mixture of the protein and the ligand is chromatographed on a column brought into equilibrium with the solvent (buffer). The two maxima of the ligand indicate the compounds bound to protein and the free compounds (e.g. the bond of phenol red to albumin — Lee and Debro 1963). If irreversible complexes and covalent bonds are forming, the distribution of the ligand and the ratio protein-to-ligand permit a quantitative assessment of the complex formed. Slowly or quickly forming reversible equilibria or the occurrence of intermediate substances, however, alter the ratios and the results can then be only qualitatively assessed.

Nichol et al. derived theoretically that, in analogy to ultracentrifuging and electrophoresis, gel chromatography also permits the examination of quickly developing reversible and irreversible equilibria by frontal analysis

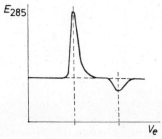

Fig. 74. The method elaborated by Hummel and Dreyer (1962) for the examination of the process of complex formation on pancreas ribonuclease and 2'-cytidylic acid on a Sephadex G-25 column. For more details, see the text

of the elution zone of solutes (Nichol and Winzor 1964; Nichol et al. 1967a, b). Several publications have dealt with the interactions of proteins from the aspects of the mechanism of the separation in gel chromatography (Ackers and Thompson 1965; Ackers 1967; Zimmermann et al. 1971; Warshaw and Ackers 1971).

The method elaborated by Hummel and Dreyer (1962), the essential features of which are shown in *Figure 74*, is a characteristic example of the gel chromatography of complex formation. Hummel and Dreyer examined the equilibrium for forming ribonuclease and 2'-cytidylic acid complex on a Sephadex G-25 column, in which the cytidylic acid content of the buffer was equal to that of the sample. During elution first the peak of the enzyme (and the cytidylic acid bound to it) emerged, followed by a drop to the cytidylic acid concentration (negative peak). The amount of this drop (the area of the negative peak) corresponds to the quantity of ribonuclease-bound cytidylic acid applied to the column, under equilibrium conditions characterized by the concentration of the free ligand (cytidylic acid). By a relative increase in the concentration of the ligands the bonding sites of the enzyme can be saturated. This makes possible, for instance, the determination of the stoichiometry of the enzyme's interactions with various substrates, inhibitors, effectors, etc. In such cases the concentration of the ligand in the sample and in the eluant is identical and there is no negative peak on the chromatogram. If, on the other hand, the concentration of the ligand in the sample is, from the outset, higher than that in the column, then the protein peak will be followed by a positive peak in the ligand. The association constants of the complexes can be determined for ligand concentrations below saturation (Bryan and Frieden 1967; Colman 1972). A precondition of the Hummel—Dreyer method is that the complex and its components should progress in the column at dissimilar velocities. Another requirement is that the complex and the protein should not separate during chromatography. This requirement can be met in two ways — either by making the equilibrium between the protein and the complex to occur so fast that the attainment of dynamic equilibrium should precede separation, or by eluting the protein and its complex at an identical rate. The latter criterion can be easily met by the choice of an appropriate type of gel, with which, for example, both the protein and its complex elute with the outer volume of the gel. In addition, to attain equilibrium, complex formation must take place at an appropriate rate.

Fairclough and Fruton (1966) examined the competition of acetyl-L-tryptophane and acetyl-L-tryptophanamide with albumin. They observed negative peaks in the concentration of both compounds. Their ratio indicated the degree of competition.

The basic idea in the preparative application of complex formation is that, on a properly chosen type of gel, the size and the elution properties of the complex molecule differ greatly from the properties of one or both components. Porath (1960) has proposed the use of complex formation to enhance the effectiveness of gel chromatography. Before this Lindner *et al.* (1959) separated peptide hormones, oxytocin and vasopressin from the protein extract of hypophysis by means of a special protein bond, the so-called Van-Dyke protein. Later Frankland *et al.* (1966) refined this method. A characteristic example for preparative processes is the separation of the B_{12} vitamin content of sea water making use of the strong vitamin binding capacity of a special peptide, known as the "intrinsic factor" of the gastric mucosa (Daisley 1961). The earliest preparative applications of complex formation include the isolation of serum haptoglobin in the form of a haemoglobin complex (Lionetti *et al.* 1963; Ratcliff and Hardwicke 1964; Hodgson and Sewell 1965). The haptoglobin–haemoglobin complex elutes earlier than other serum proteins and then haemoglobin is added in excess, making possible the quantitative measurement of the haptoglobin content. A similar technique was used in the determination of the insulin binding capacity of serum (Manipol and Spitzy 1962; Rivera *et al.* 1965; Toro-Goyco *et al.* 1966) and the binding of thyroid hormones (triiodothyronine, thyroxine) to serum proteins. Gel chromatography permits the easy separation of the inorganic from the protein-bound iodine content of blood plasma (Spitzy *et al.* 1961), and thyroid hormones (Lissitzky *et al.* 1962, 1963; Jacobson and Widström 1962; Lissitzky and Bismuth 1963; Makowetz *et al.* 1966; Müller 1967). The analysis of the protein–hormone bonds led to the feasibility of measuring the thyroid-hormone capacity of the plasma (Shapiro and Rabinowitz 1962; Pearson Murphy and Pattee 1964).

Auricchio and Bruni (1966) verified with gel chromatography the bonding of a proteolytic enzyme to the substrate protein. In their tests they chromatographed trypsin on a gel column (Sephadex G-100) whose inner volume was made available to the enzyme. Owing to the slight difference between their molecular weights, trypsin cannot be separated from the ribonuclease present by the usual method of gel chromatography, but if casein is added to the eluant, trypsin separates easily from ribonuclease. The trypsin–casein complex elutes faster from the column than either free trypsin or ribonuclease.

The most recent trend in preparative gel chromatography based on complex formation is the application of the macromolecular dextran derivative Blue Dextran 2000, dealt with in Section 2.3.5 of Part II. Haeckel *et al.* (1968) extracted the pyruvate-kinase content of a yeast extract by dissolving 1% Blue Dextran 2000 in it and chromatographing on a Sephadex G-200 column. Pyruvate-kinase with Blue Dextran 2000 forms a complex of high molecular weight which elutes with the outer volume of the column, ahead of all contaminants. The complex dissociates in an ammonium sulphate solution of higher concentration and can be sharply separated from Blue Dextran 2000 by a similar chromatographic operation. A similar method was used to purify the glutathione-reductase (Staal *et al.* 1969) and pyruvate-kinase (Staal *et al.* 1971) content of erythrocytes. Blume *et al.* (1971) proved by further investigations that the complex of erythrocyte pyruvate-kinase and Blue Dextran 2000, the allosteric effector of the enzyme, readily dissociates in the presence of fructose-1,6-diphosphate. This shows that for complex formation the conformation of the protein is of decisive importance. As regards the scope of this method, similar complex-forming properties of numerous proteins have been described in the literature (White and Jencks 1970). Swart and Hemker (1970) separated blood coagulation factor in the form of their Blue Dextran 2000 complexes, using the ionic strength gradient.

3.5. APPLICATION OF GEL PARTICLES FOR ADSORPTION AND LIQUID-LIQUID PARTITION CHROMATOGRAPHY

The use of gel particles for the purposes of adsorption chromatography does not fall closely into the domain of gel chromatography and also its practical significance is less than that of other well-known adsorbants. Nevertheless, in gel chromatographic practice the effect of adsorption must be dealt with, because, in certain groups of compounds, primarily small-molecule aromatic compounds and the denser gel types, adsorption constitutes an important factor in the mechanism of separation (Gelotte 1960; Porath 1960; Glazer and Wellner 1962; Marsden 1965). For the sake of completeness it should be added to what has been stated on adsorption in Section 2.7.3 of Part I that the three types of interactions between the crosslinkage of the gels and the adsorbent molecules — namely (1) the adsorption of aromatic compounds, (2) the hydrogen bridge, and (3) the ionic bonds (the latter to a negligible extent only) — can be distinguished by proper choice of the solvents used for elution and for the chromatographic separation. While polar eluants (e.g. chloroform, benzene on Sephadex LH-20 gel) repress the aromatic adsorption of the solutes (Wilk *et al.* 1966), compounds containing hydroxyl and/or carboxyl groups (e.g. peptides) tend to remain on the column (Tentori *et al.* 1967). The eluants forming hydrogen bridges (water, acetic acid), on the other hand, affect the hydrogen bridge bonds between the gel matrix and the solutes. Streuli (1971b), having chromatographed small-molecule polar organic substances in different solvents on a Sephadex LH-20 column, established that in dimethylformamide it is primarily the molecular sizes (gel chromatography), in tetrahydrofuran it is the hydrogen bridge bonds, and in methanol and acetonitrile (to varying degrees) it is both the aromatic adsorption and the hydrogen bridges, on which the success of separation depends. Westernack (1972a) and Williams (1972) performed similar examinations. On the basis of the above, solvent mixturesar e used for adsorption chromatography which provide a wide transition between the two effects (e.g. chloroform–ethanol mixtures (Wilk *et al.* 1966) on Sephadex LH-20). By the correct selection of the solvent, on the denser types of gel, the aromatic and hydrogen bridge bonds can be repressed and the phenomena of gel chromatography given predominance. Carnegie (1965b), on a column of Sephadex G-10, separated aromatic peptides with amino acid content according to their molecular weights with a mixture of phenol–water and acetic acid. Also urea and potassium thiocyanate solutions reduce aromatic adsorption. In aqueous solutions, on the other hand, increasing the ionic strength by the removal (partial) of the hydrate shell of the gels will enhance the adsorption–aromatic compounds (Janson 1967).

The methods of sorption chromatography of gels are the same as those applied in the general process of column chromatography (see Section 2 of Part II). Adsorption is indicated by volumetric distribution coefficient values (K_d) higher than unity. In sorption chromatography usually Sephadex G-10 (G-15, G-25), Sephadex LH-20 and Bio-Gel P-2 (P-4, P-6) are used. Although their properties are slightly different from those of the dextran gels (see Section 2.7.3 of Part I) polyacrylamide gel grains are used in a similar manner for the separation of basic proteins, nucleosides, nucleotides and nucleic acid bases (Bonilla 1969; John 1970).

Gel particles allowed to swell in polar or apolar solvents are applied also in partition chromatography. In such cases it is solvent molecules bound to the gel matrix which constitute the stationary phase, as against the moving phase flowing between the grains. It is a characteristic property of the gel particles that they retain sharp phase limits even in solvents which completely mix with one another, although the composition of the solvent content of the moving phase and of the gel particles may vary, depending on the hydrophilic or hydrophobic properties of the matrix. This latter property of the gel phase permits a sensitive separation based on the differences in solubility. For partition chromatography Sephadex G-25 and Sephadex LH-20 gels are used predominantly. In practical work, when preparing for column chromatography, first a solvent mixture is made which separates into polar and apolar (lower and upper) phases (e.g. butanol–benzene–pyridine–acetic acid). In one phase (generally the polar phase) the gel grains are made to swell and packed into the column, then the substances are chromatographed and eluted with the other phase.

Use of the partially apolar Sephadex LH-20 has proved advantageous in the partition chromatography of hydrophobic compounds and has, in fact, been widely accepted in the examinations of lipids (Nyström and Sjövall 1965a; Therriault and Poe 1965; Siakotos and Rouser 1965; Wuthier

1966; Williams and Merrilees 1970; Calderon and Baumann 1970; Terner et al. 1970; Winkler et al. 1972). A characteristic example of the use of partition chromatography is the purification of lipid extracts from other impurities (soluble in water and methanol) (Wells and Dittmer 1963; Maxwell and Williams 1967).

Murphy (1967, 1971) separated steroid hormones on a Sephadex LH-20 column with different solvent mixtures. The advantages offered by Sephadex LH-20 are speediness, good recovery, the feasibility of washing and re-use, and high capacity. Unlike paper and thin layer chromatography, in Sephadex LH-20 the presence of other substances does not interfere with the chromatography of steroids. This fact renders the results easily reproducible.

3.6. THE PREPARATIVE (BATCH) PROCEDURES OF GEL CHROMATOGRAPHY

The most frequently used preparative processes of gel chromatography — concentration, desalting and buffer change — follow basically the principles of group separation and are comparable with lyophilization, exhaustive dialysis or membrane filtration (Flodin et al. 1960). These processes are eminently suited for the separation of large-molecule compounds (salts and macromolecules) which, in the case of concentration, holds specifically for the solvent molecules.

In the preparative methods, apart from the technique of column chromatography (see Section 2.1 of Part II), the so-called batch process is applied. The essence of this process lies in the mixing of dry or swollen gel particles in a suitable vessel (beaker, centrifuge tube, sintered glass filter) with a solution of the substances tested and, when equilibrium is attained, in the centrifuging or syphoning off of the liquid volume from between the gel particles. A set-up for the purpose is illustrated in *Figure 75*. Both in its implementation and efficiency, this process is less sensitive than column chromatography, and to achieve satisfactory separation it must be repeated in most cases. However, the equipment is simpler, the process is faster and the amounts of the material that can be handled are larger. If large volumes are concerned, basket centrifuges (e.g. MSE 3000) lined with filter cloth or a polyamide insert are used for separation of the gel particles (Petterson et al. 1963; Gelotte and Emnéus 1966; Emnéus 1968). For the washing of the particles pure solvent or, if concentrated, part of the original solution is used (Painter and McVicar 1963).

3.6.1. CONCENTRATION OF SOLUTIONS OF MACROMOLECULES USING XEROGELS

The capacity of gels to swell, i.e. the strong water (solvent) absorptivity of hydrophilic (or organophilic) gel matrices, can be used to advantage for the concentration of solutions of macromolecules. Dry xerogel grains, while swelling in macromolecular solutions, adsorb the solvent. If a gel of suitable pore size is used, the macromolecules cannot penetrate into the inner volume of the gel and become concentrated in the outer volume, between the particles. In a manner similar to solvent molecules, salts (and also other smaller molecules depending on the pore size of the gel) may ingress into the gel structure without altering the ionic strength and pH of the macromolecular solution. This effect, in addition to the mildness factor, supports the use of xerogels rather than lyophilization or evaporation, in the course of which, in parallel with the concentration of macromolecules, the salt concentration (or that of other substances) would also increase. According to Flodin et al. 1960) to concentrate protein solutions such hydrophilic gel-forming substances should be used which (a) are prone to swell quickly, (b) can easily be separated completely from the concentrated solution, (c) are all but insoluble and not susceptible to contamination by the solutions to be concentrated and (d) do not adsorb the proteins. Since dextran gels (Sephadex G-25, G-50) meet all these requirements, they have found wide acceptance. When selecting the best pore size of a gel for concentration, an important criterion is to ensure that the protein has a molecular weight which ensures that it will be excluded from the inner gel volume, and as a result the water absorptivity of the xerogel (W_r) will be maximal. Thus, to concentrate, for example, enzymes, proteins and macromolecules having molecular weights above 30,000, Sephadex G-50 should be used instead of Sephadex G-26, not only because its water adsorptivity is nearly double that of G-25 but also because owing to its looser gel structure, there is less risk of adsorption. A further advantage offered by xerogels is that they can be regenerated by washing and drying (see Section 1.5 of Part II). The usual method of concentration is the preparative process. According to practice, under continuous mixing, approximately one-third part of the weight of small pored coarse-grained Sephadex G-25 or G-50, Molselect G-25 or G-50, or Bio-Gel P-2, P-4, P-6 or Acrilex P-2, P-4, P-6 xerogel is added to the macromolecular solution. The swollen grains are then, after 10 to 15 minutes stirring, separated by centrifuging or by passing through a Büchner funnel (sintered glass filter) by suction. To recover the concentrated solution entrapped between the grains as inclusions, the gel cake is washed with a small amount of solvent (buffer) or with part of the original solution. Thereafter the washing fluid is mixed with the concentrated solution. This brings about a two- to threefold concentration in a single step and the operation with the filtrate can be repeated at will. Hydrophobic macromolecules dissolving in organic solvents can be concentrated in the same manner as the organophilic gels mentioned in Part I. Unlike the requirements of column chromatography there is no need for complete swelling of the xerogel. If heat-sensitive substances are handled, mixing may be performed in a refrigerator. In the course of concentration the material-loss amounts to 5 to 10% per step, on an average. The effect of adsorption can be examined in a model test, by the ratio of the parallel concentration of the macromolecule and a non-adsorbent indicator with a high molecular weight (e.g. Blue Dextran 2000, Ficoerythrin, etc.). Similar control examinations should be performed to make sure that the composition of the solvent mixture remains unchanged after interaction with the gel (Painter and McVicar 1963).

Concentration by the use of xerogels was first described by Gelotte and Krantz (1959) for pepsin. Later Flodin *et al.* (1960) published a precise description of this preparative technique. Petterson *et al.* (1963) concentrated the cellulase content of Polyporus versicolor culture medium from large volumes. Deutsch *et al.* (1963) concentrated haemoglobin, Kibukamusoke and Wilks (1965) urine proteins. McGrath (1966) concentrated proteins with Sephadex xerogels. In recent years this method has become routine in laboratory practice.

3.6.2. DESALTING AND BUFFER CHANGE

The desalting of solutions of biological macromolecules and proteins, the exchange of salts with another low molecular weight compound, i.e. buffer, and the change of the buffer are frequent operations in the laboratory. For this purpose exhaustive dialysis, semipermeable films, bags or microporous membrane filters (Millipore, Diaflo, Diapor) are usually used. The operation takes several hours even under optimal conditions and if larger amounts are handled it might take days to accomplish. Gel chromatography offers an incomparably faster and better solution to the problem. In line with the principle described in the previous section, the better part of the salts and small molecules passing from the macromolecular solution into the inner volume of the gel remains in the gel phase while the excluded macromolecules may be removed with the outer volume. If the preparative process is applied for desalting or for buffer change then, obviously, part of the salts remain in the solution of the macromolecule and, to extract it, the operation must be repeated. The preparative technique of column chromatography permits a more perfect separation. For this purpose coarse- or medium-grained Sephadex (Molselect) G-25, G-50 or Bio-Gel (Acrilex) P-2, P-4, P-6 gel particles swollen in water or buffer solution and equilibrated are packed into a column, at 3 to 4 times greater capacity than the volume of the solution to be desalted. Calculations relating to desalting have been dealt with in Section 2.1.2 of Part II. Short columns with larger diameters and a simple lay-out are also well suited for desalting and the flow velocity may be quite considerable (10 to 12 cm/h). Having applied the sample, elution is continued with the equilibrated buffer solution using distilled water or weak saline solution (if desalting is concerned). The salts or the buffer to be changed will stay in the top portion of the column while the macromolecules, having passed through the outer volume, emerge in the eluant. A detailed description of the method of desalting was published by Flodin (1961). Initially, distilled water was used for elution but since most polymers, biological macromolecules and proteins are prone to precipitate at low ionic strength and adsorb on the gels, it proved preferable to perform desalting by a buffer change. Complete desalting is seldom required and, owing to the Donnan equilibrium of charged macromolecules, in principle it must be implemented in another way (by the lyophilization of volatile buffers). A job frequently encountered is the removal of high salt concentrations (ammonium sulphate) or other small molecules (phenol, fluorescent dyes, excess isotopes, intermediates, initial reagents, etc.) so as to recover the macromolecule in a buffer solution of defined composition and defined concentration. The capacity of the new buffer solution must be large enough to maintain the requisite pH of the solution, in spite of the eluted protein.

The monographs by Flodin (1962), Determann (1969) and Fischer (1969) provide surveys of the applications of desalting and buffer change. A specific field of application for this method is the cautious and mild variation of the pH of various substances and enzymes sensitive to pH differences. Basically, solutions of the same buffer with different pH are used in the process, which offers the advantage that, unlike the inevitable local overacidification or overalkalinization caused by the batching of alkali or acid, the pH of the solution in the column varies continuously.

The superiority of preparative gel chromatography over dialysis has been confirmed by many facts, over and above speediness and mildness. Not only does the recovery of both macro and (if necessary) small molecules approximate to 100% but the service life of the gel columns, compared to that of dialysing films, is practically unlimited. A drawback of gel chromatography is that, depending on the volume of the column, the sample is liable to become diluted. The separation processes of the gel particles and of dialysis are both based on the equilibria of partition. During dialysis the substances diffuse across the membrane and the minutest defect of the membrane puts the success of the test in jeopardy. The speed of the operations depends on the surface of the membrane. The membrane surface can be increased to a limited extent only and the solvent volumes required for dialysis are substantial. The gel particles, on the other hand, have a large surface, equilibrium can be attained in a minimum of time and the volume of the solvent which is required is negligible. Nor do slight irregularities of packing affect the results. Adsorption is mostly slight and suppressible. The exclusion limits of dialysing or microporous membranes related to molecular sizes are inaccurate and their statistical deviations are considerable. If gel particles are used the range of scatter is less. Finally, while dialysis does not permit further fractioning of the solutes, this is quite natural in gel chromatography.

3.7. THIN LAYER GEL CHROMATOGRAPHY

The use of gel particles in thin layer gel chromatography is a process new even among the special methods of gel chromatography. The first publications on the techniques appeared as long ago as the early nineteen-sixties (Determann 1962; Johansson and Rymo 1962, 1964; Andrews 1964; Morris 1964) but a systematic and organized production of the base materials and the apparatus began only in the second half of the nineteen-sixties (Determann and Michel 1965; Radola 1968a, b; Jaworek 1970a, b; Thin-Layer Gel Filtration with the Pharmacia TLG-Apparatus, 1971; *Separation News*, March 1972, Pharmacia Fine Chemicals AB, Uppsala).

A comparison of the use of gels in applications with known adsorbants (silica gel, aluminium oxide, cellulose, etc.) reveals many similar and dissimilar features. As in the classical method of thin layer chromatography, the gel particles are also spread over a glass, plastic or metal plate, in a uniform thin layer. The method and the techniques, the development and assessment of the chromatograms — in general the principle of "open column chromatography" — correspond to the conventional process. The first essential difference in the use of acrylamide, agarose, etc. gels is that as opposed to aerogels, chromatography takes place on swollen particles with continuous fluid flow taking place between them. Under these conditions, no so-called solvent front arises, and, owing to the limited adhesion of the gel particles, only the downward flow technique of chromatography can be practiced. For similar reasons the gel bed cannot be used in the vertical position and the slope of the plates in relation to the horizontal must not be more than 20°. Gels spread over plates and dried cannot be used because not only do some gels (e.g. agarose-types) undergo irreversible changes while drying but the drying out of the sample causes the denaturation of numerous biological substances (e.g. proteins). Layers formed of reversibly swellable gels are liable to crack when dried and become uneven. The xerogel grains pick up solvent when they swell, therefore the flow velocity in the solvent front is much lower than over the moist areas. What happens is that the material applied will migrate with the solvent front and fail to separate.

There are considerable differences in the application of gels with respect to the molecular weight of the substances to be separated and the mechanism of separation. Contrary to the adsorption of small-molecule compounds, thin layer gel chromatography is used for the separation of sensitive substances of high molecular weight. The indisputable advantages of the method are that it is much simpler and quicker than column chromatography, that its resolution is superior to that of columns of the same length, that it allows the simultaneous and parallel examination of several samples (standards, markers) on the gel bed and that the examinations need very small amounts of the substances (maximum of 10 μg). Also the solvents and buffer solutions applied to the gel layers are the same as in column chromatography. A slight constraint is that, owing to the large surface involved, volatile solvents or buffers (ammonia) must be used with caution. Thanks to the advantageous properties outlined, thin layer gel chromatography is used to determine the optimal test conditions of (gel) column chromatography or as an indicator system of column chromatography, for the analysis of the fractions. The shortcomings of the method are that it makes the quantitative assessment of the chromatograms rather cumbersome and that its preparative possibilities are limited.

3.7.1. THE MATERIALS AND TECHNIQUES USED IN THIN LAYER GEL CHROMATOGRAPHY

The first and most important criterion for the preparation of gel beds is to choose the proper particle size. For thin layer chromatography superfine xerogel particles of 10 to 40 μm diameter are required. It will be evident from *Tables 2* and *5* that these granular sizes are available in most commercial qualities. Most of the experience refers to Sephadex dextran gels whose different types, G-50, G-75, G-100, G-150 and G-200, cover the entire fractionation range of the peptides and globular proteins between the molecular weights 1500 and 250 000. If superfine gel particles are used, their proper swelling and their purification by sedimentation and appropriate concentration of the gel suspension are of particular significance. *Table 27* shows the optimal composition of suspensions with Sephadex gels. If the suspension

Table 27

Optimal concentration of superfine Sephadex gel suspensions for thin layer gel chromatography*

Sephadex gel type (superfine)	Concentration of suspension g xerogel per 100 ml eluant
G-50	10.5
G-75	7.5
G-100	6.5
G-150	4.8
G-200	4.5

* Data taken from the publication Thin-Layer Gel Filtration with the Pharmacia TLG-Apparatus, Pharmacia Fine Chemicals AB, Uppsala, 1971.

is too thin, the gel will flow off the plate; if it has a higher than optimal concentration, the particles will not adhere to the plate. While slight differences can be compensated by adjusting the state of equilibrium to ensure the reproducibility of the results, it is advisable to apply a suspension of the same composition to each plate. As regards the handling and storage of gel suspensions, see Section 1 of Part II.

To spread the gel bed, any one of the applicators (Desaga, Camag, Shandon) known from thin layer gel chromatography may be used. According to the literature, the thickness of the bed may vary between 0.25 and 1 mm — usually between 0.4 and 0.6 mm. *Table 28* indicates the quantity of Sephadex gel suspension required for the most frequently applied plate sizes and layer thicknesses.

Table 28

Amount of superfine Sephadex suspension required for the preparation of gel layers*

Thickness mm	Gel suspension required, ml		
	Plate size, cm		
	10 × 20	20 × 20	20 × 40
0.4	12	25	50
0.6	17	35	70
0.8	22	45	90
1.0	27	55	110

* From the publication Thin-Layer Gel Filtration with the Pharmacia TLG-Apparatus, Pharmacia Fine Chemicals AB, Uppsala, 1971.

Depending on the dimensions of the space available or the apparatus, the gels are applied to 10×20, 20×20 and 20×40 cm glass plates, in an even layer. Most factory-made applicators provide uniform beds. *Figure 75* shows a simple laboratory set-up for the preparation of the gel bed. The technique used in the TLG apparatus of Pharmacia Fine Chemicals AB is

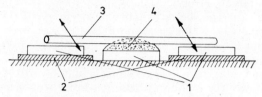

Fig. 75. A simple laboratory set-up for the preparation of the gel layer
1. chromatographic glass plates; 2. backing plates (filter paper, thin glass plate); 3. glass rod, 4. gel suspension.
The arrows indicate the direction of application

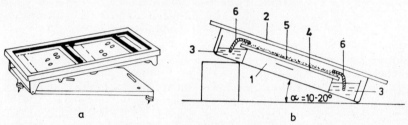

Fig. 76a. The TLG apparatus of Pharmacia Fine Chemicals AB, Uppsala, constructed for thin layer gel chromatography
Fig. 76b. A simple laboratory set-up for thin layer gel chromatography
1. chamber; 2. cover plate; 3. solvent reservoirs; 4. chromatographic glass plate; 5. gel bed; 6. filter paper strips

similar to this. For thin layer gel chromatography the glass plates must be carefully cleaned and degreased. Manufacturers' catalogues recommend their storage in a concentrated soda solution.

Figures 76a and *b* illustrate the lay-out of the Pharmacia apparatus and the schematic chart of a simple process of thin layer gel chromatography. As seen in *Figure 76b*, fluid flow in the gel bed (5) applied to the glass plate (4) is ensured by thick filter paper strips (6) (Whatman 3MM, MN 214) immersed in the solvent reservoirs (3), in contact with the gel bed. The 1—2-cm wide filter paper strips saturated by solvent (buffer) are fitted on the gel bed with slight pressure, taking care to ensure uniformity of pressure. Chromatography takes place in the equilibrium steam space of the chamber (1) filled with the solvent phase. Fluid flow starts with adjustment of the chamber to a suitable angle.

3.7.2. SAMPLE APPLICATION AND INSPECTION OF THE GEL LAYER

Before performing thin layer gel chromatography the gel particles and the moving phase must be brought into a state of equilibrium. To this end buffer solution (eluant) must be circulated across the gel layer in the chamber for 10 to 15 hours (overnight). This operation reveals defects in the experimental conditions of thin layer gel chromatography. If the gel suspensions are too thin (or the flow velocity is too high) the bed becomes thinner and uneven, indicating a reduction of the flowing phase or displacement of the grains. Thickening of the layer, on the other hand, shows that the suspension is too concentrated. Having achieved equilibrium, 5 to 10 μl of the sample is applied to the appropriate points of the gel layer with a micropipette or a microsyringe (Hamilton, Eppendorf) in a line or in dots. Samples should be applied to plates in the horizontal position, to form 4 to 10 mm diameter patches on the gel, depending on the thickness of the gel bed. On a plate of 20-cm width 10 to 15 samples can be accommodated. The position and the sharpness of the separation of the fractions depend on the concentration (viscosity) of the samples. Highly concentrated and viscous substances will migrate at a relatively slower rate and produce tailed, hard-to-separate, fractions. Therefore, protein solutions at a concentration higher than 2% should not be used.

To check on the quality of the gel layer and make sure that the test conditions are correct, standard or marker substances — mainly coloured compounds, chromates, copper(II) ions, haemoglobin, cytochrome C, Blue Dextran 2000, etc. — are chromatographed parallel with, or before the samples. By letting the coloured markers overrun, the gel layer can be re-used without disturbing the system. On a uniform gel layer, at a suitable rate of flow, the coloured compounds yield homogeneous, easy-to-localize patches with diameters not larger than twice the diameter at the point of application.

3.7.3. DETERMINATION OF THE FLOW RATE

The rate of flow of the moving phase (eluant, buffer) is an essential parameter of thin layer gel chromatography. In addition to the difference in the levels of the solvent chambers, the flow rate is determined by the type of gel and the thickness of the bed. Experience has shown that in superfine Sephadex gels, with a 0.6-mm thick layer and 10 to 20° slant, the optimal flow rate of the substances completely excluded from the gel phase is approximately 3 cm/h. Accordingly, on a 20-cm long plate, chromatography will take 4 to 6 hours. Using softer gels (Sephadex G-100, G-150, G-200) in a similarly positioned chamber, the flow rate is slower than with Sephadex G-50 and G-75, which include a larger number of crosslinks. To determine the flow rate, non-absorbent coloured macromolecules, haemoglobin, myoglobin, cytochrome C or fluorescein isothiocyanate conjugated proteins are used, of the kinds which are sure to be excluded from the inner volume of the gel.

The behaviour of substances tested during thin layer gel chromatography is characterized by the so-called relative migration, or its reciprocal value (for cytochrome C: R_{cyt}; for albumin: R_{alb}, etc.). The basis for the relationship — as in the formulation of the R_f value (retention factor) — is the distance between the (fastest) macromolecules excluded from the gel phase or those used for the standard, and the point of application. The reciprocal

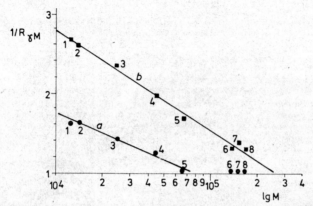

Fig. 77. Thin layer chromatography of proteins on *a* Sephadex G-75 and *b* Sephadex G-200 (superfine) gel beds. The correlation between the reciprocal value of the relative distance of migration ($1/R_{\gamma M}$) and the logarithm of the molecular weight. For particulars of the proteins used in the determination of the base points see *Table 29*. From Separation News, March 1972, of Pharmacia Fine Chemicals AB, Uppsala

value of the relative migration is proportional to the logarithm of the molecular weight (or the Stokes radius) of the compounds tested (Wasyl et al. 1971; for more detail see Section 2.6.3.4 of Part I).

Figure 77 shows the relationship between the reciprocal value of the relative migration on Sephadex gel layers and the molecular weight of a number of standard proteins.

Table 29 supplies information on the properties of the standard proteins used in thin layer gel chromatography. To plot the calibration point of *Figure 77*, and to perform similar tests, the booklet of Pharmacia AB (*Separation News*, March 1972) recommends the following process. Allow 7.5 g of Sephadex G-75 superfine xerogel to swell in 100 ml of 0.05 M phosphate buffer containing 0.1 M NaCl (pH 7.6) over a hot water bath for five hours. Treat 4.5 g of Sephadex G-200 superfine xerogel in a similar way. Apply 0.6-mm thick layers of the suspensions to 20×20 cm chromatographic plates. Bring the layers into equilibrium in a chamber at a 10° slant during one night.

From the samples 5 μl is applied to the layer by a microsyringe. The concentration of the sample is 10 to 20 mg/ml. Chromatography is performed at a slant of 15°. The run of the Sephadex G-75 layer takes 2.5 hours, that of

Sephadex G-200 4 hours. The chromatograms are developed by the replica technique, on Whatman 3MM paper. The proteins are stained by a methanol–acetic acid (9 : 1) solution of 0.1% bromophenol blue, the paper being bleached by 5% acetic acid. The chromatogram gives a reading of the distance of migration (d and $d_{\gamma M}$) of the proteins and of the γ_M globulin used as reference, from which the relative migration can be calculated according to

Table 29

Physicochemical and chromatographic parameters of standard proteins on superfine Sephadex G-75 and G-200 gel layers*

No.**	Protein	$R_{\gamma M}$***		Molecular weight	Stoke's radius Å
		G-75	G-200		
1	Cytochrome C	0.61	0.37	12 400	16.4
2	Ribonuclease	0.62	0.38	13 700	19.2
3	Chymotrypsinogen A	0.71	0.43	25 000	22.4
4	Ovalbumin	0.82	0.50	45 000	27.3
5	Bovine albumin (monomer)	0.91	0.58	67 000	35.5
6	Bovine albumin (dimer)	1.00	0.76	134 000	43
7	Aldolase	1.00	0.73	158 000	45
8	γglobulin	1.00	0.76	160 000	53

* Data taken from *Separation News* (March 1972), Pharmacia Fine Chemicals AB, Uppsala.
** Proteins designated according to the calibration points, *Fig. 87*.
*** Relative migration distances referred to γM-globulin.

the relationship $R_{\gamma M} = d/d_{\gamma M}$. With the knowledge of $1/R_{\gamma M}$, the molecular weight of the proteins tested can be read from the calibration line shown in *Figure 77*.

3.7.4. DEVELOPMENT AND EVALUATION OF THE CHROMATOGRAMS

The process of thin layer gel chromatography can be followed from the displacement of the various markers, or coloured compounds (fluorescein, macroglobulins marked with amidoblack or bromophenol blue) of high molecular weight. The substances separated can be detected directly or indirectly.

In simple cases the direct method consists of the determination of the position or concentration of the coloured fluorescent or radioactive proteins and the measurement of the distance of migration (Roberts 1966). Iodine vapour may also be used for the purpose. Williamson and Allison (1967) and Coleman et al. (1968) assessed the chromatograms by direct autoradiography and scanning densitometry, respectively, of the labelled compounds. James et al. (1968) assessed the chromatogram obtained from a gel layer applied to a quartz plate from direct UV absorption by a special spectrophotometer.

Another possibility for direct development techniques is to dry out cautiously the gel layer (at 50 to 60°C for 15 to 30 min) and develop the separated substances by a suitable staining method. To detect proteins ninhydrin dissolved in ethanol and 1% amidoblack 10B dissolved in methanol–water–acetic acid (in the ratio of 75 : 15 : 15) proved most effective. Also the staining processes applied in paper and thin layer gel chromatography (electrophoresis) are used.

In the indirect methods (replica or reprint techniques — Radola 1968) a thick sheet of chromatographic filter paper (MN 2140) is carefully placed upon the moist gel layer with a slight pressure in the direction of flow (Whatman 3MM). One or two minutes later, when the greater part of the chromatographed substance has been absorbed by the filter paper, the latter is removed, dried out and stained according to the method best suited to the substance in hand. In another interesting variant of the indirect development technique, the enzymes are detected by means of filter paper impregnated with coloured (or discolouring) substrate solution (Johansson and Rymo 1964).

As in the classical methods of paper and thin layer chromatography, chromatograms after development are assessed by the accurate determination of the position of the spots and their distance from the point of application. In the computation of the relative distances of displacement, the shape and tailing of the spots play an important role. In practice the distance of migration is measured from the front or from the centre of the spot. The precise measurement of the distance is particularly important in molecular weight determinations (James et al. 1968; Jaworek 1970a, b; Heinz and Prosch 1971; Klaus et al. 1972; Waldmann-Meyer 1972). The assessment of the chromatograms depends also on the aim and purpose of the application. Gilbert (1966) examined the dissociation of proteins on the gel layer. Most clinical applications consist of the fractionation of serum proteins and the detection of pathological macroglobulins (Bergström 1966; Harrison and Northam 1966; deGoldman et al. 1970; Carter and Hobbs 1971). The resolution of thin layer gel chromatography can be increased by a combination of the technique of electrophoresis or immunodiffusion (Hanson et al. 1966; Agostini et al. 1967; Kohn 1968; Dizik and Knapp 1970; Clinton et al. 1972). More recently hydrophobic gel layers (Sephadex LH-20) have been used in organic solvents (Klimisch and Stadler 1972).

PART III
APPLICATIONS OF GEL CHROMATOGRAPHY

1. APPLICATIONS IN PROTEIN CHEMISTRY

1.1. SEPARATION OF PROTEINS FROM SALTS AND SMALL-MOLECULE SUBSTANCES

The fast acceptance and spread of gel chromatography may be attributed to the fact that it simplifies the routine operations of protein chemistry, e.g. dialysis, to a great extent. Before its introduction, inorganic salts and small-molecule substances were usually separated from proteins by using semipermeable films, in processes which took one or two full days to accomplish. Using gel columns the same operation can be completed in a few hours. Almost the first publications on gel chromatography directed attention to this feature (Porath and Flodin 1959; Flodin 1961, 1962).

When removing inorganic salts, the aim is only the elimination of some salts or ions and not complete desalting, with the protein being recovered in a defined buffer. The protein solution is applied to a gel column washed with this buffer. Electrolytes cannot be completely eliminated in columns washed with distilled water even though the protein concerned may dissolve in it and does not precipitate on the column during the gel chromatographic process. This is because most gels contain a small number of acid groups which through ion exclusion, prevent the ions with the same charge from diffusing into the gel and their separation from large molecule substances (Gelotte 1960; Flodin 1962). Furthermore, proteins may adsorb on the gels from distilled water or from solutions of low ionic strength. Therefore, for the complete desalting of proteins the column must first be washed with a buffer solution containing volatile salts, for instance ammonium formate, then, after completion of gel chromatography, the protein solution must be lyophilized (Flodin 1962; Male 1967).

For the removal of inorganic salts, and for the buffer change of proteins, small-pored gels are used (e.g. Sephadex G-25, Bio-Gel P-10) which completely exclude the proteins (see Sections 1.2.2.1 and 1.2.3.1 of Part I).

According to Flodin (1962), if protein solutions are desalted for preparative purposes, the amount of solution that may be applied to the gel column is approximately 30% of the column volume. If desalting is performed to improve the analytical process, or if electrolyte change takes place, the volume of the protein solution must not exceed 15% of the column volume. If the HETP is known, the dimensions of the gel column required for the separation of a defined amount of substance can be precisely computed (Flodin 1962; compare Section 2.6.1.4 of Part I).

Organic small-molecule substances can be separated from proteins by gel chromatography, in the same way as inorganic salts. This facility is of great help in the isolation of individual proteins and in the removal of excess of the reagents used in their treatment. The latter operations include the marking of antibodies by radioactive or fluorescent substances (Killander et al. 1961).

If the small-molecule substance is capable of producing a complex with the protein, separation will only then be successful if the dissociation constant of the complex is high enough, i.e. if the complex is unstable and if the process of dissociation (or association) is fast, compared to the gel chromatographic process. If this requirement is not satisfied, separation will be only partial. Such complex forming is characterized by concentration of the small-molecule substances which will never be zero between the protein peak and the small-molecular peak. Furthermore, if the protein fraction is rechromatographed on the same gel column, partial separation similar to the first chromatogram will be obtained. Such small-molecular substances, forming relatively stable complexes, can only then be separated from proteins by gel chromatography if their dissociation can be enhanced by the addition of certain substances (e.g. detergents, inhibitors), or by acidification, or by increasing their ionic strength. The gel chromatography of some specific protein complexes will be considered later.

Griffith (1975) has shown that dodecyl sulphate and other small-molecular weight compounds can be separated from denatured proteins with Sephadex LH-20 gel and 70% (v/v) formic acid as eluant. This method seems to be useful for removing the unbound dyes from proteins after a dyeing process. The formic acid solvent can be removed after gel chromatography by dialysis against water and, following this, a 0.5% solution of ammonium hydrogen carbonate.

Franek and Hruska (1976) used a gel-filtration-centrifugation method for the separation of free and protein-bound ligands in radioimmune assays.

According to Flodin's experiments (1961), if the solution is highly viscous the separation of proteins from salts or from small-molecular substances will never be perfect. According to this author, the zones of protein and dextran will not separate from a concentrated dextran solution and both zones will become asymmetric.

In the course of desalting or buffer change by gel chromatography, the electrolyte composition of the protein solution emerging from the column will never be equal to that of the buffer used for the pre-washing of the column, and it will in addition have a lower concentration. This effect can be attributed to Donnan equilibrium and it will be more definite the higher the concentration of the protein solution.

1.2. PROTEIN FRACTIONATION ON GEL COLUMNS

Proteins can be fractionated on gel columns either by differences in their molecular weights or differences in their solubility.

For the fractionation of organic extracts, tissue extracts, serum proteins or urine proteins, etc., by molecular size, large-pored gels may be applied into which proteins can diffuse, such as Sephadex G-100 and G-200, dextran gels of analogous properties, Bio-Gel P-100, P-200 and P-300 or polyacrylamide and agarose gels with corresponding characteristics.

Figure 78 illustrates the separation of serum proteins on three different dextran gels. In solutions which contain several protein components, gel chromatography can achieve only partial separation. For the gel chromatographic fractionation of solutions containing a mixture of several proteins,

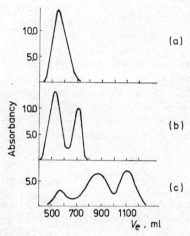

Fig. 78. Fractionation of serum proteins on (a) Sephadex G-75, (b) G-100 and (c) G-200 dextran gel columns (Flodin 1962)

the elution volumes and the column dimensions can be calculated on the basis of the characteristic parameters of the gels and the molecular weight of the proteins to be separated (cf. Section 2.6.3 of Part I).

In the absence of protein-protein or protein-gel interactions, the chromatographic zone of the proteins yields a nearly symmetric bell curve. That the K_d value of the proteins depends on the concentration to a small degree only has been observed during the gel chromatography of concentrated solutions (see below). This fact may cause slight asymmetry of the elution curve.

The fractionation of basic proteins is hampered by their bonding on the gels, based on cation exchange. This phenomenon can be eliminated by using buffer solutions of concentrations between 0.1 and 1.0 M. According to Craft (1961) and Deibler *et al.* (1970), strongly basic proteins should be fractionated in acidic medium, e.g. in 0.01 M hydrochloric acid. Deibler *et al.* (1970) established that if gel chromatography takes place in such acid solutions, the elution volume will be dependent not only on the molecular weight but also on the quantity of the average ionic charge per amino acid residue.

The very basic proteins, histones, can be fractionated from 0.01 M HCl solution on a Bio-Gel P-2 column (Candido and Dixon 1972; Louie *et al.* 1973; Kecskés *et al.* 1976). Roark *et al.* (1976) have shown that histone fractions have anomalously high molecular weights on crosslinked dextran gels in 0.05 M sodium acetate (pH 5), and also on polyacrylamide gel (Bio-Gel P-30), if 0.5% dextran is present in the solvent. The specificity of this puzzling effect of the immobilized and soluble dextran appears to be peculiar to histones.

Proteins may be fractionated on gel columns also by their dissimilar solubility, using the zone precipitation (solubilization, salting-in) method (cf. Section 3.3 of Part II). Epstein and Tan (1961) washed the gel column before application of the protein solution with a dilute electrolyte solution (0.02 M phosphate buffer) in which only one of the protein components could be dissolved. The proteins applied in a solution with higher salt concentration passed into the dilute buffer solution in the top of the column, causing those components to precipitate which do not dissolve at lower salt concentrations. The authors used this method in the analysis of human serum, primarily for the separation of euglobulins from pseudoglobulins. The euglobulins not dissolving in the dilute buffer and precipitating in the top of the gel column were dissolved again during chromatography and moved together with the solution of higher salt concentration while the pseudoglobulins passed through the column without precipitation and emerged in the effluents earlier than the euglobulins.

Porath (1962b), who introduced the term "zone precipitation", prepared an ammonium sulphate concentration gradient (up to a saturation degree of 85—20%) along the column, before application of the protein solution and applied the human serum fraction partially separated previously to a Sephadex G-200 column. He obtained three protein fractions in the effluent of which the first contained albumin, ferritine and a small amount of α_1 globulin, the second α_1, α_2, β and "fast" globulins, and a small amount of albumin, the third two globulins which could be distinguished from other globulins by their electrophoretic migration.

Porath (1962b) used the zone precipitation method also to purify α-hydroxysteroid-dehydrogenase. According to this author, in zone precipitation it is advantageous if the proteins precipitate within the gel. He therefore proposes the use of gels into the pores of which part of the proteins to be separated can penetrate. He also calls attention to the fact that, when the change of concentration of ammonium sulphate takes place, the volume of the gel column may also change and reduce the rate of through-flow. Porath showed that the use of large-pored Sephadex G-100 and G-200 gels is better suited to this purpose than G-25, because the volume of the latter increased by 20 to 25% (Porath 1962b).

Hoffmann (1969) also applied the zone precipitation process to fractionate serum proteins. Unlike Porath he caused the salt concentration gradient to form during the elution of the protein applied in a dilute sodium acetate buffer and not previously. For column packing he used the relatively small pored Bio-Gel P-10 polyacrylamide gel, but in this case the maximum salt concentration was likewise moderate, i.e. 0.3 M. The protein fraction which eluted under the effect of the salt solution was purified 40 to 150-fold, in contrast to the fraction which first emerged from the column and contained most of the proteins, and showed only a 1.2—1.3-fold purification.

For the solubilization chromatography of serum proteins, Hoffmann and McGivern (1969) used a Sephadex G-50 column washed with a 2.7 M ammonium sulphate solution. Having applied the proteins, they were eluted with a non linear decreasing ammonium sulphate concentration gradient by directing the flow from the bottom upward, opposite to the application. They found that Sephadex G-50 dextran gel was better suited for this method than the Bio-Gel P-10 polyacrylamide gel because the latter exhibited a greater change of volume with changes in salt concentration. They stress that chromatography should take place at a slow rate since dissolution and precipitation are relatively slow processes and because above a rate of 2.5 cm/h the zones are liable to trail.

Rapp and Lehmann (1971) applied zone precipitation for the isolation of α_1 fetoprotein. Little and Behal (1971) also performed this method on porous glass with a Bio-Glass 200 column for the preparation of arylamidase of human liver. They washed the column first with a 2.8 M ammonium sulphate solution and then applied the liver extract and eluted the precipitated proteins in several steps, decreasing the salt concentration. They succeeded in removing approximately 90% of the other proteins from the enzyme, whereafter they were able to purify the enzyme further by ion exchange chromatography. They emphasize that porous glass is preferable to gels because its volume is independent of changes in the ionic strength and thus applicable in a wide range of ammonium sulphate concentrations.

Kremmer *et al.* (1972) fractionated serum lipoproteins by zone precipitation (see Section 3.3 of Part II).

The potentialities inherent in the processes of zone precipitation are still far from being exhausted, and important results may be foreseen in this field. The use of porous glass for column packing holds out particularly good promise. Little and Behal (1971) call attention to the fact that fractionation may be used also to vary the pH between wide limits.

Zone precipitation fractionation can be applied not only for the purification of proteins but also for the determination of the salt concentration necessary for the crystallization of various proteins (Porath 1962a, b).

1.3. CONCENTRATION OF PROTEIN SOLUTIONS

Gelotte and Krantz (1959) were the first to establish that protein solutions can be concentrated by Sephadex products. Dry gel added to the protein solution absorbs water and swells, and, if it is sufficiently small pored, the proteins are excluded from the gel particles and become concentrated.

According to Flodin *et al.* (1960), for such a type of concentration, water-absorbent substances should be used (1) which swell rapidly in water, (2) which can easily and quantitatively be removed from the concentrated solution, (3) which are insoluble and do not contaminate the protein solution to be concentrated, and (4) which do not bind the protein to be concentrated.

The dextran and polyacrylamide xerogels meet all these criteria to a fair degree. However, in analytical work it should be borne in mind that the base material of the gel matrix will dissolve even from these substances, but only to a negligible extent — negligible from the point of view of contamination.

An advantage in concentration with xerogels which makes it superior to lyophilization and evaporation is that the solution will concentrate only with proteins (i.e. substances of large molecular weight), leaving the concentration of inorganic salts and small-molecule substances unchanged. Another favourable property of this process is its rapid performance (see Section 3.6.1 of Part II).

In the ideal case, when the gel is swollen approximately two-thirds of the water quantity is inside the gel particles and one-third in the interstices, and a nearly threefold concentration can be achieved. After removal of the swollen gel particles, the process can be repeated and, in principle, any optional degree of concentration can be obtained.

Flodin *et al.* (1960) concentrated a filtered medium of the Polyporus versicolor culture on Sephadex G-25 (coarse) gel in three consecutive steps from 7500 to 410 ml. The concentrate contained 87% of the activity of the cellulase enzyme of the initial solution.

Deutsch *et al.* (1963) used Sephadex G-25 for concentration of dilute haemoglobin solutions, adding 1 g of dry gel to 5—6 ml of solution. They mixed this for ten minutes, then passed the mixture through a sintered glass filter. If the gel stayed red they washed it with 1—2 ml of the original solution. In this way, with three consecutive treatments, from the 120 mg% solution they obtained a 1.8 g% one which contained 85—90% of the haemoglobin. The authors observed no denaturation of the protein.

Kibukamusoke and Wilks (1965) concentrated urine proteins on Sephadex G-25 xerogel.

Painter and McVicar (1963) proved by experiment that concentration with xerogel can be applied with good results even on proteins dissolved in glycerin. These authors concentrated the solution of fibrinolysin using glycerin as the stabilizer. Mixing protein-free aqueous solvents containing 40, 50 and 60% of glycerin on Sephadex G-25 dextran gel, they proved that after treatment, the glycerin concentration of the interstitial solution increased slightly (by 1—2%). In the tests concerned this rise did not affect the applicability of the method and they were able to achieve a nearly threefold concentration of solutions containing fibrinolysin without an appreciable loss of protein. An even greater concentration (approximately 4.8-fold) could be obtained when they did not wash the residual interstitial protein solution from the gel, however, in this case there was an approximately 20% loss of protein.

In *Table 30* some data published in the literature are presented for concentration and protein recovery. As can be seen, each xerogel treatment brings about a 2.5-fold concentration and causes 5 to 10% protein loss.

1.4. CHARACTERIZATION OF ISOLATED PROTEINS

Gel chromatography can be used not only in preparative operations but also in the analysis of purified proteins. The simplest analytical processes include checking on the homogeneity of a protein preparation and the deter-

Table 30

Concentration of protein solutions by Sephadex G-25 xerogel particles

Proteins	Degree of concentration	Loss of proteins, %	References
Cellulase	2.53	5	Flodin et al. (1960)
	2.58		
	2.53	19	
	1.83	13	
Haemoglobin	2.32	5	Determann (1969)
	2.43	14	
Lactic dehydrogenase	1.91	9	Determann (1969)
Pepsin	1.36	19	Determann (1969)
Fibrinolysin	4.7	19	Painter and McVicar (1963)
	2.62	3	

mination of the molecular size of the isolated protein. In the former, a preparation having a single elution peak is not necessarily homogeneous, it may contain impurities whose K_d value is equal to that of the protein. However, since such uncertainties are concomitant to every method of protein analysis, to verify the homogeneity of the preparation it is advisable to perform two separations, based on different principles. From this point of view gel chromatography, as a method which separates by molecular size, offers a favourable complement to the fractionating methods based on the differences of electric charge (ion exchange chromatography, electrophoresis). For examination of homogeneity by gel chromatography, the same conditions are required as for the determination of the molecular weight.

Numerous authors have used gel chromatography for the determination of the molecular weight of proteins: Wieland et al. 1963; Whitaker 1963; Ackers 1964; Andrews 1964; Squire 1964; Auricchio and Bruni 1964; Davison 1968; Determann and Mätter 1969; Jaworek 1970a; Fish et al. 1969; Deibler et al. 1970; Heinz and Prosch 1971; Eipper and Mains 1975; Miller et al. 1975; Carter et al. 1976, etc. The various calculation methods of the authors were detailed in Part I.

It is always the migration rate of the protein in the gel which must be determined, and the column (or thin layer) must be calibrated with standard proteins. As regards the proteins used for calibration, see *Table 15* (p. 85).

In the case of a thin layer gel chromatographic molecular weight determination, the use of prestained proteins is advantageous; the dyeing procedure does not affect significantly the migrations of the various proteins (Miller et al. 1975; Epton et al. 1975).

The precision of the determination of molecular weight by gel chromatography depends primarily on the accuracy of measurement of the migration rate and on the reproducibility of the measurements. In optimal cases, i.e.

when a gel of suitable pore size and suitable standard proteins are used, the molecular weight can be determined to an accuracy of a few per cent. The molecular weights established by gel chromatography, and by other methods, do not generally show greater deviations than the numerical values quoted by various authors for the molecular weights of the same proteins. It should, however, be taken into consideration that the rate of migration during gel chromatography is a function of the size and shape of the molecule and that the molecular weight can be more accurately determined if the gel chromatographic process is complemented by sedimentation and diffusion tests (Ackers 1964; Rogers et al. 1965; Siegel and Monty 1966; Ackers 1970).

Attention must be paid also to the fact that the rate at which proteins migrate depends not only on the molecular size and shape of the molecules but also on other factors — namely, the concentration of the protein solution and, in some cases, the sorption of the protein on the gel particles.

Winzor et al. have shown that, for proteins, the distribution coefficient may depend on the concentration (Winzor and Scheraga 1963; Winzor and Nichol 1965). In this case the isotherm of distribution assumes a concave shape and can be expressed in the following formula

$$Q = k_1 c_0 + k_2 C_0^2$$

where Q denotes the quantity of protein in the gel phase, c_0 the protein concentration of the solution in the interstices of the gel grain, and k_1, k_2 are protein- and gel-type-dependent constants, respectively.

According to Winzor and Nichol (1965) ovalbumin chromatographed on Sephadex G-100 gives a k_1 value of 0.32 and a k_2 value of 0.0046.

The dependence of the distribution coefficient of proteins on concentration may be interpreted by the fact (Brumbaugh and Ackers 1968) that distribution as an equilibrium state is a function of activities and not of concentrations, and that the activity coefficients of the solution and gel phases do not change in the same way with changes in concentration. If the ratio of the activity coefficients is a linear function of concentration, that is, if

$$\frac{g_0}{g_1} = 1 + k' c_0$$

where g_0 is the activity coefficient in the solution, g_1 that in the gel phase, and k' a constant, then the concentration dependence of the distribution coefficient is

$$K_d = K_{d0}(1 + k' c_0)$$

where K_{d0} denotes the distribution coefficient of an infinitely dilute protein solution.

Owing to the dependence of the distribution coefficient on concentration, the elution volume increases with increasing protein concentration of the protein applied to the column and the chromatographic zone will become asymmetric. The results of the tests carried out by Winzor and Nichol (1965) with ovalbumin are shown in *Figures 79* and *80*. This type of concentration-dependence of the distribution coefficient must not be mistaken for concentration-dependence arising from association of the protein molecules, which will be dealt with later.

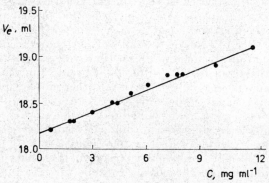

Fig. 79. The concentration dependence of the elution volume in gel chromatography of ovalbumin on Sephadex G-100 (Winzor and Nichol 1965)

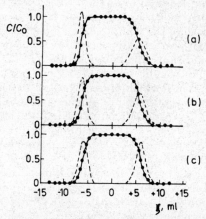

Fig. 80. Effects of concentration dependence of elution volume upon the protein distribution in gel chromatography (Winzor and Nichol 1965). Positive values of ξ refer to advancing sides. The solid lines represent calculated patterns, the circles are experimental points and the broken lines are derivative curves obtained using 0.5 ml increments in volume. Concentration of the 12 ml ovalbumin samples applied:
(a) 12 mg ml^{-1}, (b) 7 mg ml^{-1}, (c) 1.6 mg ml^{-1}

As seen from *Figures 79* and *80*, the dependence of the distribution coefficient on concentration is relatively small, the value of k' in the formula being between 10^{-2} and 10^{-3}, but for the determination of the molecular size of proteins such deviations are significant.

According to Brumbaugh and Ackers (1968), by determining the dependence of the distribution coefficient on the concentration, and the value of g_0, that of g_1 and its concentration-dependence can be established.

Winzor and Nichol (1965) proved also that the Johnston–Ogson effect known from sedimentation analysis can be observed also in gel chromatography if the zones of the different proteins are only partially separated from one another. In such cases the concentration of the slowly migrating component is different in the separating zone and in that part of the zone which contains the mixture.

The molecular size of proteins cannot only be characterized with isolated proteins. Auricchio and Bruni (1964), experimenting with lactic acid dehydrogenase preparations purified to different degrees, found that the elution volume of the enzyme is independent of the other proteins present. Downey and Andrews (1965) came to the same conclusion.

However, this statement holds valid only for such unpurified protein solutions whose components do not form complexes with one another. Enzymes, for instance, since they do form reversible complexes with those substrates of high molecular weight occasionally present in the solutions, have elution volumes which depend on the concentration. Auricchio and Bruni (1966) showed that with constant casein concentration and increasing trypsin concentration, the elution volume of the enzyme decreases. They observed the same effects in the gel chromatography of ribonuclease and ribonucleic acid. The complexes of enzymes with their specific inhibitors of high molecular weight, as the complex of trypsin and soybean trypsin inhibitor, elute earlier than the components alone (Krieger et al. 1976). Such type of interactions with substrates or inhibitors of high molecular weight provide favourable conditions for the separation of enzymes from other proteins of similar molecular weight *(Fig. 81)*. This fractionation process may be extended also to enzymes which transform substrates of small molecules, provided that the substrates or their analogues are bound to a carrier of high molecular weight by stable covalent bonds. Phosphofructokinase and pyruvate kinase (Haeckel et al. 1968; Blume et al. 1971) form a complex with Blue Dextran 2000 because the latter's chromophor portion contains a substrate-analogue structural element.

Downey and Andrews proved that milk-tributyrinase, if associated to casein micellae, is eluted earlier than the free enzyme. They established furthermore that pancrease-lipase also forms a complex with casein, unlike other enzymes, for instance wheat germ esterase, xanthine oxidase, milk-alkali phosphatase.

Not only association to other proteins may alter the gel chromatographic migration of the enzymes; specific interactions with the gel substances also bring about this effect. This takes place when the basic substance of the gel is similar to the enzyme substrate. Dextran gels, for instance in the case of α-amylase and lysozyme, behaved as analogues of the substrate, and may slow down migration in the gel (Gelotte 1964a, b; Whitaker 1963). Such interactions may depend on the pore size of the gel, for instance amylase on small-pored dextran gels elutes according to its molecular weight.

From distilled water or solutions of low ionic strength proteins may associate to gel substances by ionic bonds. Miranda *et al.* (1962) have described the ionic bonding of scorpion toxin, ribonuclease and lysozyme on dextran gels. Glazer and Wellner (1962), studying the ion exchange properties of dextran gels, established that the protein binding capacity of an 8.5×1.2 cm Sephadex G-50 gel column based on ion exchange is 7.2 mg for lysozyme,

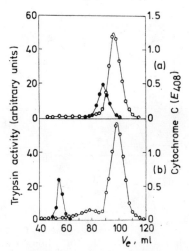

Fig. 81. Gel chromatogram of trypsin and cytochrome C chromatographed on Sephadex G-100 (a) in the absence of casein and (b) in the presence of casein. (Auricchio and Bruni 1966). (●—●) the trypsin activity; (o—o) extinction measured at 408 nm

2.8 mg for ribonuclease and 2.4 mg for bovine serum albumin. Ionic interaction with proteins may take place even in agar gel containing agaropectin, which contains sulphate groups. In desulphurated agar gels, prepared according to Porath (1971), and in agarose gels, no ionic interactions are encountered.

From solutions of low ionic strength proteins may be sorbed also on acrylamide gel. This was proved by Bonilla (1970) on the basis of his experiments with Bio-Gel P-2. Egg white lysozyme, pseudomonas cytochrome C and the snake venom neurotoxin readily bind on the gel-column from distilled water. The proteins so bound can be eluted by salt solutions of higher concentration.

Since gels generally adsorb aromatic compounds, theoretically a similar sorption can be expected in proteins containing aromatic amino acids. This is of small importance in practice because the aromatic side-chains are found mostly in the inner regions of the protein molecules. Sorption based on hydrophobic interactions may, however, be essential if lipoproteins are concerned; these may bind to gels even from solutions of higher ionic concentration.

The adsorption of protein molecules on controlled pore glass may be more pronounced, especially in the case of small protein molecules, with a high aromatic amino acid content; this side-effect can be eliminated by the use of 6 M urea plus 0.5% sodium dodecylsulphate as solvent (Frenkel and Blagrove 1975). Mizutani and Mizutani (1975) have shown that the adsorption of serum albumin on glass can be prevented with amino acid buffers. The zwitterion character of these compounds must play an important role in preventing adsorption on the —SiO— repeating structure of the glass.

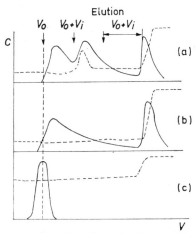

Fig. 82. Graphs on the virtual fractionation of a homogeneous protein susceptible to adsorption. The solid lines represent the protein concentration, the dotted lines the salt concentration. The protein emerging at V_0 corresponds to the non-adsorbed and therefore normally eluting portion. (a) The chromatogram obtained if a protein sample with a high salt content is applied to a column equilibrated with a dilute salt solution, and if first the dilute then subsequently a more concentrated salt solution is used for elution; (b) the same as (a) with a protein sample of low salt content; (c) the usual elution curve obtained when a protein sample of the same salt concentration as used for equilibration of column is applied

The interactions between proteins and gels may distort and falsify the findings both quantitatively and qualitatively: homogeneous proteins may show up as heterogeneous, and conversely, heterogeneous proteins may appear to be homogeneous. The former case holds if the sorption of the protein depends strongly on the ionic concentration and if the column is washed with an electrolyte or buffer solution of a lower concentration than the concentration of the sample in which the protein was applied. In such cases the portion of adsorbed protein is eluted by the slowly migrating and more highly concentrated salt solution *(Fig. 82)*, while the homogeneous protein is eluted producing a zone with two maxima. For the reverse case, Stevanson's (1968) experiment provides a good example. He proved that, during the first application on a Sephadex G-150 column, human immunoglobulin appears to be homogeneous but on repeated applications an aggregate of higher molecular weight also emerged. Further, on the first use of the column after alkaline regeneration, again only one single elution peak was obtained. His explanation of this phenomenon is that during the first chromatographic process, the faster migrating aggregate becomes adsorbed, saturates the protein binding sites and allows only the component of lower molecular weight to emerge.

The elution volume of proteins in gel chromatography may depend not only on the ionic concentration but also on the concentration of some other components. According to Hellsing (1968), on Sephadex G-200 gel the K_{av} value of serum albumin, using buffer solutions with 0 to 4% dextran content, rises from 0.42 to 0.72 with increasing dextran concentration.

It will be evident from the above that, for proper characterization of protein, the gel chromatographic elution volumes and the distribution coefficients must be measured under carefully selected conditions. Even so, only proteins having related properties should be compared. For instance, the elution volumes of fibrous proteins or glycoproteins should not be compared with the elution volumes of simple globular proteins (Andrews 1965, 1967a, b). Notwithstanding these constraints, gel chromatography has proved an invaluable tool for the characterization of protein molecules.

1.5. GEL CHROMATOGRAPHY IN THE EXAMINATION OF PROTEIN COMPLEXES

In the gel chromatography of complexes of protein-protein molecules or of proteins and compounds of low molecular weight, the rate of the association or dissociation of the complex in question, compared to the speed of chromatography, is an important feature. If migration in the column is much faster than the restoration of the association equilibrium upset by fractionation, then the individual complexes and components will emerge in separate zones at the end of the column, and their separation will be the sharper the slower the rate at which equilibrium is restored. The extreme case is represented by the aggregates which form practically irreversibly. These separate sharply (e.g. the monomer and dimer forms of serum albumin). If protein complexes which form reversibly but slowly are gel chromatographed, the shape of the chromatogram will depend on the concentration of the initial solution. The experiments performed by Kakiuchi *et al.* (1964) with bacterial α-amylase provide a good example of this case. If concentrated enzyme solution is applied, the associated form of higher molecular weight compound will be in excess, while, if a dilute enzyme solution is chromatographed, the dissociated form will predominate *(Fig. 83)*. That the association equilibrium of this enzyme is slow to evolve has been confirmed also by kinetic measurements of the heat denaturation (Piukovich and Boross 1972).

In the cases referred above, the gel chromatography of the complexes enables the determination of their molecular size and, thereby, their composition. Basically the same holds true for complexes of proteins with small molecules. If the complex dissociates at a very slow rate, gel chromatography helps establish the stoichiometry of the components, and it is only the composition of the protein fraction and the quantity of the small-molecule substances moving together with the protein which must be determined.

If the association equilibrium of the protein complexes develops at a relatively fast rate, the chromatogram will show an anomalous course, different from the gaussian bell shape and the elution volume will decrease with increasing concentration of the protein solution. For the gel chromatographic analysis of protein-protein interactions of this nature, the method of "broad zone" or "frontal analysis" is applied; here the volume of the sample of the

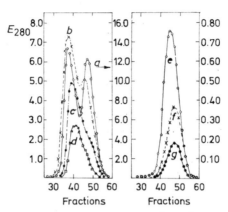

Fig. 83. Gel chromatographic patterns at varying the concentration for *Bacillus subtilis* amylase using Sephadex G-100 column (540×25 mm) (Kakiuchi et al. 1964)

Concentration of the 2 ml protein samples applied to the column: a: 11.4%; b: 5.68%; c: 2.85%; d: 1.05%; e: 0.40%; f: 0.25%; g: 0.14%

The chromatography was made in 0.1 M sodium chloride and 0.005 M calcium acetate at pH 7.5. Fraction volume 3.2 ml. Temperature 15 to 20 °C; flow rate 65 to 45 ml h^{-1}

protein solution applied to the column is sufficient to form a plateau of constant composition in the chromatogram, in which the overall protein concentration is equal to that of the solution applied to the column. If simple proteins are concerned, the elution volume can be determined from the so-called centroid point of the sigmoid curve connected to the leading or trailing end of the plateau. The centroid point is known also by the term "equivalent sharp boundary" because it is at this point where the protein solution would be eluted with a sharp jump in concentration if there were no dispersion of concentration along the column during chromatography.

The simplest case of protein-protein complex formation is when the molecules of a given protein are capable of associating reversibly with one another. Winzor et al. (1967; Winzor and Nichol 1965; Winzor and Scheraga 1963, 1964) proved that, in the frontal analysis of such proteins, the concentration gradient at the leading end of the plateau is steeper, i.e. the frontal portion of the zone becomes "sharper". At the trailing end of the chromatographic zone, the situation is the reverse and the concentration gradient is flatter. The shape of the concentration gradient forming here depends on the type of association: the function dc/dv (which is the differential quotient (slope) of the chromatographic curve) has one maximum in monomer-dimer systems of association and two maxima and one minimum if more than two protein molecules are able to associate reversibly with one another (Figs 84 and 85).

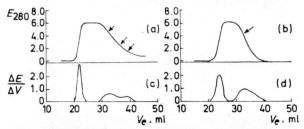

Fig. 84. Elution profiles obtained in the gel chromatography of (a) α chymotrypsin (3.8 mg ml⁻¹) and (b) DIP-chymotrypsin (3.2 mg ml⁻¹) on Sephadex G-100 (1.25 × 32 cm) (Winzor and Scheraga 1963). The curve (a) represents the elution pattern if two or more associated forms are forming; the curve (b) the same pattern with only one associated form; (c) and (d) are the respective first derivative curves, 0.50 ml having been used as the increment in volume. The arrows indicate inflexion points

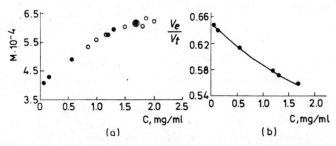

Fig. 85. The concentration dependence of the molecular weight (a) and the gel chromatographic migration rate (b) of bovine thrombin on Sephadex G-100 column (Winzor and Scheraga 1964). The points represent the values obtained by gel chromatography, the open circles indicate the molecular weights determined by ultracentrifugation

In such systems the size of the "unity" of the protein molecule can be calculated from the elution volume of the chromatographic process performed at a very low concentration (Winzor and Scheraga 1964; Andrews 1964). The determination of the numerical values of the association constants by the analysis of the chromatographic curve is a rather cumbersome and time-consuming operation in which the best-fitting theoretical curve must be found by computer. Several authors have worked on the theoretical definition of the chromatography and with the analysis of the experimental curves (Gilbert 1963, 1967; Cann and Goad 1965; Ackers and Thompson 1965; Winzor 1966; Winzor et al. 1967; Chiancone et al. 1968; Chun et al. 1969; Henn and Ackers 1969). Ackers (1967), in a summary, has presented a detailed survey of the subject.

The complex bond in the protein complexes (e.g. in enzymes built up of several polypeptide chains, or in proteins bound to natural membranes, etc.) is sometimes so strong that it cannot be made to dissociate to the individual components by dilution. In these cases it is advisable to treat the protein complex with concentrated urea, guanidine hydrochloride solution or ionic,

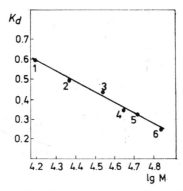

Fig. 86. A calibration line for the determination of the molecular weight of protein subunits by gel chromatography in 6 M guanidine-hydrochloride solution (pH = 6.5) containing 0.1 M 2-mercaptoethanol (Giorgio et al. 1971)

1. haemoglobin; 2. trypsin; 3. pepsin; 4. ovalbumin; 5. glutamate dehydrogenase; 6. serum albumin

or neutral detergents, and perform the gel chromatography in similar solvents. For the examination of the quaternary structure of enzymes and for the determination of the molecular weight of the subunits on an agarose column from 6 M guanidine hydrochloride solution, good examples were quoted by Fish et al. (1969), Klasu et al. (1972) and Giorgio et al. (1971). The correlation between K_d and the logarithm of the molecular weight under the conditions used by the latter authors is shown in *Figure 86*. For gel chromatography in urea solution, Patterson and Lennarz (1970), Takagi and Iwanage (1970), and Senior and McLennan (1970) have presented examples. Stracher (1969) applied gel chromatography in 4 M LiCl solution to separate the light and heavy chains of myosine.

In some cases, to cause the dissociation of protein complexes, acidification of the solution is sufficient; in others, variation of the ionic strength is necessary. If there are also disulphide bonds among the polypeptide chains of the protein, these must be first split by adding cystein, 2-mercaptoethanol or some other reagent, or by electroreduction. To prevent reformation of disulphide bonds, gel chromatography should be performed in a nitrogen atmosphere or in the presence of a reducing agent (Pákh et al. 1972; Fish et al. 1969). If the disulphide bridges are split by sulphitolysis, this is unnecessary (Porter 1971). It is customary also to modify the liberated sulphhydryl groups by alkylation after the reduction, before gel chromatography is carried out (Clenn 1971; Franck Nezlin 1963; Dorrington et al. 1967).

The association constants of the complexes of proteins formed with small molecule compounds can be easily determined by gel chromatography.

Hummel and Dreyer (1962) were the first to describe a method for the gel chromatographic analysis of these complexes. While studying the complex of ribonuclease and 2'-cytidylic acid, they washed the column first with a buffer containing 2'-cytidylic acid and then applied the ribonuclease dissolved in the same solution. The enzyme forms a complex with cytidylic acid whose quantity remains unchanged while migrating in the gel column, because it moves in a solution of the same cytidylic acid concentration. Accordingly, when the protein peak appears, the concentration of the cytidylic acid in the effluent increases by the quantity bound to the protein. Behind the protein zone the cytidylic acid concentration will reset to the constant level and subsequently, at the point corresponding to the rate of migration of the small molecule substances, a "hole" will appear on the chromatogram (cf. Fig. 74). The shortage of cytidylic acid here is equal to the quantity bound to and eluted together with the protein.

Fairclough and Fruton (1966) used this method in the examination of the complexes of serum albumin, with tryptophan derivatives and tryptophan peptides. They determined the protein-bound quantities of the small molecules and the $\bar{v}$ values, i.e. the average quantity bound to a single protein molecule, by gel chromatography performed with different C concentrations of the small-molecular substances. Then, plotting the quotient $\bar{v}/c$ as a function of v (the Scatchard process, 1949) they obtained a straight line, the slope of which, K', denotes the negative value of the apparent association constant of the complex.

Pfleiderer (1964) examined by gel chromatography the binding of reduced nicotinamide-adenine-dinucleotide on lactate dehydrogenase, Alvsaker (1965) and Sheikh and Moller (1968) studied the binding of uric acid to plasma proteins. Cooper and Wood (1968) and Keresztes et al. (1972) used frontal analysis with gel chromatography to study the bond of small-molecule substances to proteins. According to their method, a large volume of the solution containing the protein and the substances tested are applied to a gel column, washed previously with buffer, in a quantity sufficient to cause a well-defined plateau in the eluate. In this plateau (*Fig. 87*, β region) the quantity of the small-molecule material, partly bound, partly free, corresponds to the association-dissociation equilibrium. In the subsequent region of the chromatogram (γ) the amount of the small-molecule substance corresponds to the concentration of the free molecules in the equilibrium state. This GELFAC method (gel-frontal analysis chromatography) is applicable if the complex and the free protein move at identical rates in the gel column, but the substance bound to the protein proceeds at a considerably lower rate. The migration of the protein complex must also be independent of the concentration of both the protein and the binding substance (Cooper and Wood 1968).

Gel chromatography is used frequently for the qualitative or quantitative characterization of the binding of various substances to proteins. The determination of the dissociation constant is only then difficult if the protein has more than one binding site, some of which form complexes of dissimilar strength, and if the dissociation constant of the complex has a low value and, therefore, practically no dissociation takes place during chromatography. Although, in principle gel chromatography can be used even in this latter case, with very diluted enzyme solutions in which the concentrations can no longer be accurately determined.

For enzymes with several coenzyme binding sites, a good example is provided by D-glyceraldehyde-3-phosphate-dehydrogenase, in whose four active centres the binding of NAD can be characterized by different dissociation

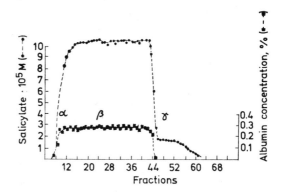

Fig. 87. Study of the complex formation of serum albumin with salicylic acid by gel frontal analysis chromatography (Keresztes *et al.* 1972)

constants. On a gel column only part of the enzyme-bound coenzyme is liberated (Boross 1965). Therefore, gel chromatography is suitable for the establishment of the deterioration of the coenzyme binding centre (e.g. structural changes after specific modifications of a side-chain of the enzyme);

1.6. GEL CHROMATOGRAPHY OF PEPTIDES AND AMINO ACIDS

Although the long-chain polypeptides yielded by the partial hydrolysis of proteins migrate in gel columns in the same way as proteins, adsorptive interactions are likely to take place in them. After proteolytic or selective chemical splits, the liberated apolar and aromatic side chains will more readily bind to the gel matrix. This, however, is not unequivocally detrimental since the different strengths of adsorption may separate polypeptides of equal molecular size, provided that their aromatic amino acid content is different (Eaker and Porath 1967; Tentori *et al.* 1967). Since sorption can be prevented by the use of acetic acid, urea or mixtures of organic solvents, two consecutive chromatographic runs will achieve very effective fractionation. Langley and Smith (1971), for instance, fractionated the bromocyanide splitting products of glutamate dehydrogenase first from 30% acetic acid then from 4 M guanidine-hydrochloride solution.

The use of organic solvent mixtures may be expeditious also because of the poor water-solubility of some peptides. Suitable chemical modifications may, however, enhance the solubility of the peptides. Sajgó (1969), for instance, maleylated the peptides obtained by the bromocyanide splitting of aldolase, then chromatographed them in an ammonium hydrogen carbonate buffer, instead of Lay's (1968) fractionating process, using a pyridine–acetic acid–water solvent mixture.

For the gel chromatography of natural peptides, the same rules are valid as for those obtained by the hydrolysis of proteins. An excellent example for the fractionation of peptides containing different aromatic amino acids is the gel chromatographic separation of thyrocydin A, B and C from a 10% acetic acid (Mach and Tatum 1964; Ruttenberg 1967).

The molecular weight of various peptides can be determined by gel chromatography from a phenol-acetic acid–water mixture in which aromatic amino acids do not adsorb (Synge and Youngson 1961; Carnegie 1965a).

The use of eluants containing guanidine hydrochloride (Eipper and Mains, 1975), urea and dodecyl sulphate (Frenkel and Blagrove 1975), formic acid (Jamaluddin et al. 1976) or acetic acid (Bogardt et al. 1976; Marshall et al. 1974, etc.) may be also suitable. Like proteins, peptides move in the gel column according to their molecular size rather than according to their molecular weight. This may explain why the more space-demanding gramicydin elutes faster than its moleculer weight would justify.

Small-size peptides and amino acids can be chromatographed on small-pored gels. The rate of their migration can be influenced by aromatic adsorption or by the charge of the molecule (Gelotte 1960; Eaker and Porath 1967; Janson 1967). In acid medium, around the basic amino acids, a thick solvate shell and ionic double-layer are formed and these are, therefore, excluded from the small pores of the gels. For similar reasons acidic amino acids are eluted rapidly in alkaline solution. The presence of salts, e.g. sodium chloride in high concentration (2 M), reduces the thickness of the solvate shell and increases the elution volume. From a neutral solution, tyrosine is adsorbed on the gel to a greater extent than phenylalanine, but in an alkaline solution, owing to the dissociation of its phenolic hydroxyl group, its elution is faster than that of the latter.

The variation of the pH and the electrolyte concentration enables the separation of any two amino acids, except the pair isoleucin–leucine.

The aromatic adsorption also changes when the aromatic amino acid forms a complex with small molecules, e.g. sodium dodecyl sulphate, present in the eluant (Nandi 1976). A complex having a reduced or no effective charge would be partially retained. The concentration of the small-molecule reactant shows a minimum in the depleted front which permits the calculation of the association constant for complex formation.

Ziska (1970) found that, from among the tyrosine- and tryptophan-containing glycine dipeptides, those which contain glycine in the N terminal position show a stronger adsorption to dextran gels from 0.2 M acetic acid or from acetic acid solutions containing sodium chloride. In an alkaline medium tyrosine-peptides are excluded from the gel, owing to dissociation of the phenolic hydroxyl group. The addition of salt enhances aromatic adsorption by reducing the thickness of the hydrate shell of the compound.

Amino acids and peptides can be separated on dextran gel columns also by distribution chromatography. Wieland and Bende (1965) separated

L-alanyl-L-tyrosine and D-alanyl-L-tyrosine isomers on a Sephadex G-50 column by chromatography from a 1 : 1 mixture of pyridine and water. For the chromatography of oxytocin, glucagon and secretin, Yamashiro *et al.* (1966) used the butanol–benzene–pyridine–acetic acid–water solvent mixture.

Veatch and Blut (1976) separated gramicydin C and its dansylated product on Sephadex LH-20 gel from methanol. This fractionation seems to be based on aromatic adsorption rather than molecular sieving, since the dansylated compound is eluted later.

Lakshmi and Nandi (1976) have found that hydrophobic interaction increases if the eluant contains sugars. The retention of the esters of N-acetyl-aromatic amino acids increases with increasing sugar concentration. The "affinity number" $A = (V_e - V_t)/g$, where V_e denotes the elution volume, V_t the total volume, and g the weight of the gel, increased more in sucrose solutions than in those of glucose or fructose. The enhanced retention may be due to a stronger hydrophobic interaction with the gel matrix or to the altered solubility of the aromatic esters in the mobile and stationary phases.

A further possibility for the separation of oligopeptides, polypeptides, or aromatic amino acids by gel chromatography is offered if these are capable of binding to large-molecule substances. For instance, tryptophan bound to serum albumin is eluted together with it (Fairclough and Fruton 1966). Oxytocin and vasopressin gel chromatographed in a pyridine–acetic acid buffer will migrate bound to large-molecule proteins, while if chromatographed from 1 M formic acid they elute at the points corresponding to their molecular weights (Lindner *et al.* 1959). In such case, two consecutive gel chromatographic runs from different solvents ensure a high degree of purification.

2. GEL CHROMATOGRAPHY IN THE CHEMISTRY OF NUCLEIC ACIDS

2.1. PURIFICATION AND CHARACTERIZATION OF NUCLEIC ACIDS

Nucleic acids occurring in nature are linear polymer macromolecules in sizes which may vary between wide limits. The steric structure of the molecules is not irregular. Nucleic acid molecules with appropriate nucleotide sequences (complementary) are capable of forming double helices. Such double helices may be formed even between parts of a single nucleic acid molecule. The present knowledge about the steric structure of the natural nucleic acids is relatively poor but it is beyond doubt that their space requirement is widely different from that of globular proteins and that in their native state they cannot be regarded as simply elongated filament like formations. Since the gel chromatographic behaviour depends on the shape and size of the molecules, the rate at which nucleic acids migrate in the gel column is not equal to that of the migration of proteins or other polymers of the same molecular weight.

To a first approximation, this holds true also for nucleic acids, those having higher molecular weights migrate faster in gel columns. Gelotte (1961) proved that, on a Sephadex G-25 column, nucleic acids separate from nucleotides and from other substances of low molecular weight and that, on a Sephadex G-75 column, the polynucleotides elute according to their molecular size. Zachau (1965) on the basis of his experiments, demonstrated that the problem is not simple, because the migration of oligonucleotides and nucleic acids depends on the molecular weight in a different way. Specifically, on the one hand, oligonucleotides are not helical and therefore require a greater space and, on the other, they are prone to aggregate. This explains why, in Zachau's tests, the transfer ribonucleic acids and the products of their partial hydrolysis showed hardly any separation.

To separate nucleic acids from small-molecule substances and inorganic salts by gel chromatography is as easy as it is in the case of proteins. Therefore phenol, which is generally added to the solution to cause the proteins in the extracts to precipitate, may also be removed from nucleic acids. For such purification of the nucleic acids, and for buffer change, small-pored gels (Sephadex G-25, Bio-Gel P-2, etc.) have proved well suited. Levinson et al. (1976) used Sephadex G-25 gel for the separation of an acriflavine-labelled DNA from the unbound reagent.

The fractionation of nucleic acids on large-pored gels is not as simple. The molecular size of the natural deoxyribonucleic acids (DNS) differs greatly from that of the ribonucleic acids (RNS) and they can readily be separated *(Fig. 88)*. Following denaturation, the separation will only be partial. Further fractionation of DNS and RNS by gel chromatography is a problem still unsolved. The elaboration of separation processes is hampered by the fact that nucleic acids cannot be detected according to their biological activities as easily as in the case of enzymes. From the UV

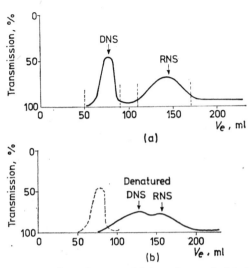

Fig. 88. Gel chromatogram of a mixture of (a) native and (b) denatured DNS and RNS (Bartoli and Rossi 1967)

light absorption of the effluent leaving the column, it cannot be unequivocally established whether or not the partial separation of the individual nucleic acids actually took place in the large-molecule zone. If the difference of molecular weights is considerable, it is the division of the zones which indicates fractionation. The situation of transfer ribonucleic acids of nearly identical size is exceptional because they can be more easily distinguished by biochemical measurements and accurately detected in the eluate. This naturally facilitates the establishment of best conditions for separation on a gel column. The problem of their fractionation will be dealt with in more detail in the next section.

In spite of the difficulties outlined above, gel chromatography has given assistance in the study of many problems of nucleic acid chemistry. It has proved a useful auxiliary method in the examination of the steric structure of nucleic acids. With suitable enzymes, it is possible to split the so-called single stranded parts of the nucleic acid and, after enzymatic hydrolysis, the size of the products from gel chromatography permit conclusions to be drawn as to the ratio of single-stranded and double-stranded parts of the molecules, their intramolecular distribution and the resistance of the molecules to enzymatic decomposition.

According to Schutton (1971), by splitting single-stranded DNS by exonuclease, the acid-soluble products include large-molecule fragments and 40 to 150 nucleotides containing shorter chains which in 6 M urea solution are separated readily on a Sephadex G-200 column from the small-molecule substances. Spenzer and Walker (1971) used a similar method in the examination of the structure of 50 S ribosome subunits. After a specific enzymatic splitting they were able to separate two fractions on a Sephadex G-100 column, one consisting of approximately 50 nucleotide units with a molecular weight of 15 700, and another fraction of low molecular weight consisting of free nucleotides and oligonucleotides. This confirms the authors' assumption that the chain of the native macromolecule forms loops, these arranging into double-strands at certain parts, while at other parts — in the loop bends — it is single-stranded, and it is only at these sites that it can be split by the enzyme.

Wagner and Ingram (1966) studied the hydrolysis of soluble ribonucleic acids with ribonuclease by gel chromatographic analysis. Working with a Sephadex G-25 column they obtained better separation from 0.1 M ammonium acetate solution than from a more concentrated, 0.5 M, solution. Formaldehyde or heat treatment of the hydrolysis product caused no alteration in the shape of the chromatogram, from which the authors concluded that the products of hydrolysis have no secondary structure which would be altered by the treatment or, if there was one, it had no appreciable effect on the gel chromatographic migration.

Lift and Ingram (1964), Pierce and Zubhoff (1964), Armstrong et al. (1965) and Thang et al. (1971) analysed various products split by ribonucleases from different transfer ribonucleic acids by gel chromatography and drew conclusions from their findings regarding the mechanism of enzymatic splitting, and the factors which render the structure resistant to splitting.

Maruyama and Mizuno (1965) examined the enzymatic hydrolysis of ribosomal ribonucleic acids, separating the oligonucleotides obtained from the intact nucleic acids on a Sephadex G-50 column and chromatographing them on DEAE cellulose.

Yogo et al. (1971) drew conclusions from their gel chromatographic analyses about the dissociation of 50 S ribosome subunits, brought about from the effect of inorganic phosphate ion. They separated three major fractions on a Sephadex G-100 column *(Fig. 89)*; the first was a mixture of 23 S and 16 S nucleic acids, the second contained small-molecule 5 S nucleic acid, while the third contained 4 S nucleic acid. From the findings they established that, if 30 S subunits are also present in the initial solution, no 5 S nucleic acid will form in the dissociation process.

Jones et al. (1964) fractionated synthetic polynucleotides partially on Sephadex G-75 and G-200 columns and studied, from the fractions so separated, the effect of the chain length on the template behaviour of the polyribonucleotides in the polypeptide synthesis.

Konig et al. (1971) applied gel chromatography to the investigation of nucleic acid metabolism and studied the incorporation of inorganic phosphate labelled with ^{32}P isotope in rabbit brain, and the decomposition of labelled nucleic acids. On a Sephadex G-200 column, they separated nucleotides, inorganic phosphate, phenol and other small-molecule substances from nucleic acids and separated the DNA and a RNA fraction of high molecular weight from the transfer ribonucleic acids. The high molecular weight RNA fraction was subsequently separated from the DNS's by a second gel chromatographic process on an agarose column (Bio-Rad A-50 or A-150).

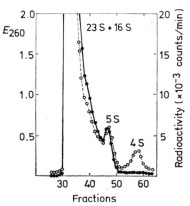

Fig. 89. Gel chromatography of the ribonucleic acids of the 50 S ribosome subunit (Yogo et al. 1971)

Gel chromatography was used in model experiments on the enzymic binding of t-RNS on ribosomes by Ringer et al. (1976).

Schell (1971) used gel chromatography for the analysis of the cell uptake of polynucleotides and stated that isotope-labelled double-stranded synthetic poly(A)poly(U) and triple-stranded poly(A)2poly(U) nucleotides penetrated in their original form into the cells of Ehrlich ascites. The author detected these in the phenolic cell extract by gel chromatography, the synthetic polymers eluted from a Sephadex G-150 column immediately after the ribosomal RNAs.

Bosch et al. (1961) attempted the separation of soluble ribonucleic acids from the similarly behaving microsomal RNA fraction on Sephadex G-75, but without success because the pore size of the gel was too small to let the nucleic acids penetrate. On Sephadex G-50 they succeeded in separating the soluble RNSs from the mono- and oligonucleotides.

The use of agarose gels in the purification of DNA from mammalian tissue was published by Loeb and Chauveau (1969), that from plant materials by Kado and Yin (1971) and recently by Lurquin et al. (1975).

Rogers et al. (1972) applied gel chromatography for the examination of the molecular weight of DNAs secreted by lymphocytes. They performed the fractionation on an agarose (Sepharose 2B) column in a tris-hydrochloride —EDTA—sodium dodecylsulphate solution. For reference (marker) substance ^{14}C labelled HeLa cell 32 S, 28 S and 18 S ribonucleic acids were used. According to the gel chromatographic analysis, the molecular weight of the secreted DNA was between 3 and 12 million dalton. To determine the dead volume of the column the authors used E. coli cells marked with tritium.

Hallick *et al.* (1976) used Sephadex B gel column for the isolation of transcriptionally active chromosome from *Euglena gracilis*.

The "native", extremely long, DNS molecules can be easily broken into smaller macromolecular fragments by physical or chemical treatment. Experiments of Tung *et al.* (1974) showed that gel chromatography is a useful tool in the characterization of DNA molecules after various physical and chemical treatments.

Ribonucleic acids from ribosomes were fractionated on agar gel, on the basis of adsorption of r-RNA on the gel from a solution of high ionic strength, and eluted with dilute salt solution (Popovic 1975).

2.2. FRACTIONATION OF TRANSFER RIBONUCLEIC ACIDS ON GEL COLUMNS

The transfer ribonucleic acids specific to individual amino acids (t-RNA) are easy to distinguish from one another. This property enabled the determination of the optimum conditions for their separation on gel columns to be carried out. However, only some of these methods may be regarded as purely gel chromatographic process, because, in many cases, fractionation was performed by distribution chromatography on gel columns, using organic solvent mixtures.

Schleich and Goldstein (1964) studied the gel chromatographic behaviour of t-RNAs fractionated with countercurrent distribution. The fractions which emerged first during the countercurrent distribution, when chromatographed on a Sephadex G-100 column from a solution of 1 M NaCl, showed three peaks. The third, with the lowest molecular weight, corresponded to the ribonucleic acid with transfer activity. The first two, after heat treatment in urea at 70 °C and rechromatography on the gel column, gave three peaks again, i.e. the third, active, component has also emerged. This indicates the susceptibility of t-RNSs to aggregate, on the one hand, and the efficiency of the gel chromatographic process for the separation of aggregates, on the other. The authors succeeded in separating by gel chromatography also a fourth inactive small molecule substance among the fractions which emerged later during the countercurrent distribution, which was supposed to contain decomposition products.

Röschenthaler and Fromageot's experiments (1965) showed that t-RNSs specific to individual amino acids do not separate when gel chromatographed in a solution of 1 M NaCl, but they can be partially fractionated by elution in a suitably altered ion concentration. Using decreasing NaCl–acetic acid concentration gradients, the authors obtained three peaks of which the second and third contained active t-RNAs. One contained glutaminic acid-, leucine-, tyrosine-, glycine- and serine-specific, the other phenylalanine-, lysine-, proline-, aspartic acid-, alanine-, cysteine-, and valine-specific t-RNAs. The t-RNA specific to histidine produced two peaks and yielded maxima at the points of both the second and the third nucleic acid fraction. Also the proline-specific t-RNA proved to be heterogeneous; it had two maxima but each within the third nucleic acid fraction. The authors attribute the fractionation effect to the fact that the conformation of t-RNAs specific to differentamino acids depends on the ionic environment to a different degree.

Zachau (1965) found that the serine-acceptor RNA migrates faster than the other t-RNAs, and this rate may be further accelerated if serine is bound to it. The author discovered, furthermore, that the gel chromatogram of cytoplasmic–nucleic acids is different in a 6 M urea and in an ammonium acetate solution, owing either to changes in the secondary structure of the nucleic acids or to adsorption.

Tanaka et. al. (1962) examined nucleic acids on Sephadex G-25 with partition chromatography. They washed the dextran gel with the lower phase of a mixture of butanol–tributylamine–acetic acid–dibutylether–water (100 : 130 : 10 : 2.5 : 2.7). The RNA dissolved in the upper phase was then applied to the column and eluted with the upper phase. This method enabled the partial separation of serine-, tyrosine-, leucine-, threonine-, and valine-specific acceptor RNAs.

Bergquist and his co-workers (Bergquist and Robertson 1965; Bergquist et al. 1965) performed distribution chromatography with a solvent mixture of similar composition on a Sephadex G-25 column, but eluting with a decreasing concentration gradient of dibutyl ether. Not only did they succeed in partially separating several t-RNAs but they verified the heterogeneity of the serine-specific t-RNA.

The rate of migration of RNAs from solvent mixtures on gel columns is determined by minor chemical differences in the molecules rather than by their molecular weights. This has been confirmed by the dependence of the elution volume of the individual t-RNSs on the polarity of the solvent mixture (Tanaka et al. 1962).

2.3. GEL CHROMATOGRAPHY OF OLIGONUCLEOTIDES, MONONUCLEOTIDES, NUCLEOSIDES AND NUCLEIC ACID BASES

As long ago as the early nineteen-sixties, Gelotte (1960, 1961) established that the migration of elementary compounds which build up nucleic acids is a function both of the molecular weight and of the adsorption. On a Sephadex G-25 column, he was able to separate nucleic acids, oligonucleotides, nucleoside di- and triphosphates, adenylic acid and nucleosides. He demonstrated that adsorption of purine bases and purine nucleosides is strongest.

From a solution in distilled water, nucleotides migrate at a rate faster than expected on the basis of their molecular weights. This might be explained by ion exclusion. Exclusion will be absent if chromatography takes place from electrolytic solutions (*Table 31*). From a dilute, 0.01 M ammonium hydroxide solution, ion exclusion is still observable but with a NaCl solution of 0.05 M the adsorption effect is already dominant.

According to Zadrazil et al. (1961), on Sephadex G-25 only up to pentanucleotides can be fractionated because oligonucleotides larger than this blend with the nucleic acid zone and elute with it. The authors confirmed Gelotte's observation (1961) according to which nucleotides can be separated on gel columns from a mixture of nucleosides and bases, because the latter become adsorbed. They showed that some bases do not always separate from their nucleosides. Purine bases and their nucleosides, for instance, yield separate peaks but cytidine and cytosine emerge in the same zone. Pyrimidine bases can be separated from purine bases but neither can be fractionated from each other *(Table 31)*.

Table 31

Volumetric distribution coefficient (K_d) of nucleic acids, nucleosides and nucleotides for Sephadex G-25 gel

Compounds	Mol. weight	A	B		C			
		0.01 M $(NH_4)_2CO_3$	0.005% $(NH_4)_2CO_3$ pH=7.6	0.02 M phosphate, pH=7.6	0.01 M NH_4OH pH=10.6	H_2O	0.05 M NaCl	0.05 M phosphate, pH=7
Nucleic acid bases								
Thymine	126	1.55	—	—	—	—	—	—
Uracil	112	1.54	1.82	1.80	—	1.1	—	1.2
Cytosine	111	1.69	1.80	1.85	1.3	1.6	—	1.4
Guanine	151	3.23	3.40	3.35	—	—	—	—
Adenine	135	3.62	3.38	3.35	1.2	2.2	2.4	—
Nucleosides								
Thymidine	242	1.23	—	—	—	—	—	—
Uridine	244	1.27	1.66	1.70	—	1.0	—	1.0
Cytidine	243	1.42	1.66	1.70	—	1.2	—	1.2
Guanosine	283	2.30	2.60	2.55	—	1.6	—	1.8
Adenosine	267	2.50	2.59	2.50	1.8	1.7	1.8	—
Nucleotides								
Thymidylic acid	322	0.89	—	—	—	—	—	—
Uridylic acid	324	0.69	0.66	0.70	0.1	0.1	0.8	0.7
Cytidylic acid	323	0.73	0.67	0.70	0.1	0.1	0.8	0.7
Guanylic acid	363	1.04	0.91	0.89	0.1	0.4	1.3	0.9
Adenosine-2-phosphate	347	1.18	—	—	—	—	—	—
Adenosine-3'-phosphate (Adenylic acid)	347	1.27	0.85	0.80	0.1	0.1	1.2	—

Notes: A: Hohn and Pollmann (1963); B: Zadrazil *et al.* (1961); C: Gelotte (1960).

As demonstrated by Hohn and Pollmann (1963), the position of oligonucleotides during elution is an exponential function of the chain length *(Fig. 90)*. Their tests threw light on the different behaviour of nucleic acid derivatives, proteins and peptides during gel chromatography: for oligonucleotides (polynucleotides) the limit molecular weight for exclusion on Sephadex G-25 gel is 2000 dalton (instead of 4500) and on Sephadex G-100 dextran gel it is 35 000 dalton (instead of 100 000). *Tables 31* and *32* show the K_d values as determined by Hohn and Pollmann (1963) for oligonucleotides, mononucleotides, nucleosides and bases.

Hayes *et al.* (1964) studied the migration of thymidine-oligonucleotides from triethylammonium carbonate solution on dextran gels of different pore sizes as a function of the molecular size and the number of terminal phosphate groups They demonstrated that, from among the oligonucleotides

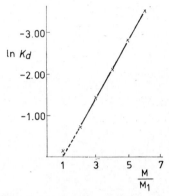

Fig. 90. Correlation between the molecular weight and the distribution coefficient of oligothymidylic acids (Hohn and Pollman 1963). M corresponds to the molecular weight of the oligomer, M_1 is the molecular weight of the thymidylic acid

Table 32

The volumetric distribution coefficient (K_d) for Sephadex G-25, G-50 and G-75 gels

Oligo-thymidylic acid	M	K_d			
		A	B		
		G-25	G-25	G-50	G-75
		0.01 M $(NH_4)_2CO_3$	0.005 M Triethylammonium carbonate		
Monothymydilic acid	322	0.98	0.31	0.50	0.80
Dithymidylic acid	626	0.48	0.13	0.30	0.65
Trithymidylic acid	930	0.24	0.06	0.22	0.53
Tetrathymidylic acid	1 234	0.12	0.02	0.13	0.44
Pentathymidylic acid	1 538	0.06	—	—	—
Hexathymidylic acid	1 842	0.03	—	0.06	0.50
Decathymidylic acid	3 362	—	—	—	0.15

Notes: A: Hohn and Pollmann (1963); B: Hayes *et al.* (1964).

with identical number of nucleotide units, the derivatives free from terminal phosphate groups produced the highest K_d values. These were followed by those which contained one (3' or 5' position) phosphate group, while the lowest values were obtained with those which contained two phosphate groups. In this respect, the differences are considerably greater than calculated from the molecular weight increase; it may be that the strong association of phosphate ester with the adjacent water molecules gives rise to a higher rate of migration. The experiments showed that the K_d values of thymidine-oligonucleotides containing 5'-phosphate, on Sephadex G-75 gel, are different up to approximately decanucleotide (see *Table 32*). The

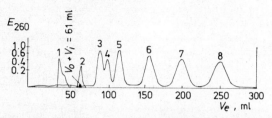

Fig. 91. Fractionation of purine and pyrimidine bases on a Sephadex G-10 column (Sweetman and Nyhan 1968)

1. dextran; 2. acetone; 3. uracyl and cytosine; 4. thymine; 5. hypoxanthine; 6. xanthine; 7. guanine; 8. adenine

Sephadex G-50 gel seems to be suitable for fractionation of derivatives containing no more than 4—5 nucleotide units, even when slower migrating oligonucleotides without terminal phosphate groups are involved.

Schwartz *et al.* (1965) studied the chromatography of nucleotides, nucleosides and bases on polyacrylamide gel. As in dextran gels, here too adsorption takes place and adenine migrates faster than leucine, which has a nearly equal molecular weight. Cytidylic acid and uridylic acid migrate faster than sodium chloride, adenine being considerably slower. Uridylic acid, uridine and adenine can be separated from each other. The zone of adenosine falls between those of uridine and adenine. According to the authors, gel chromatography makes fast group fractionation possible.

Sweetmann and Nyhan (1968) thoroughly investigated the separation of purine bases by adsorption on Sephadex G-10 gel. They found that protonation of the purine derivatives tested reduced adsorption, whereby the elution volume of adenine, under the given test conditions is 198 at pH 5 and 239 ml at pH 7. The elution volumes decrease also in alkaline solutions because the hydroxyl ion, which is also capable of adsorption, competes with the bases. The separation of mixtures from neutral solutions is shown in *Figure 91*.

Koike *et al.* (1971) applied gel chromatography to the purification of adenylic acid-3',5'-nucleosides, adenyl-3',5'-guanosine-2',3'-cyclic phosphate and synthetic oligoadenylic acids. Denamur and Gaye (1971) carried out chromatography on a Sephadex G-25 column, as one of the steps in the isolation of four natural uridyl-diphosphate-trisaccharides.

The separation of 8-aza-adenine metabolites on Sephadex G-10 gel was performed by Melgunow (1975).

Uziel and Cohn (1965) studied the desalting of nucleotides by gel filtration. Their results proved that, on a Bio-Gel P-2 column, the elution volume of oligo- and mononucleotides was smaller than the volume of the column, that chloride ion emerges after one volume, bromide ion and urea emerge somewhat later, and that ammonium sulphate and ammonium hydrogencarbonate emerge before one column volume. Ribonucleosides adsorb strongly, while bases adsorb still more strongly — dividing into two groups, pyrimidine and purine bases. This shows that the strongly adsorbing derivatives can be desalted because the migration of the electrolytes in the column takes place at a faster rate. In the case of mononucleotides the sample volume must be relatively small and never exceed 10% of that of the column. This prevents overlapping with the zone of the salts.

Gel chromatography is useful in the study of complex formation of nucleotides with metal ions or small molecules. Birch and Goulding (1975) have shown the formation of a Li-ADP complex. The binding of ammonium formate to purine nucleotides was found by Bernofsky (1975) in the chromatography of ATP or GTP on Sephadex G-10 gel.

3. GEL CHROMATOGRAPHY OF CARBOHYDRATES

Apart from proteins and nucleic acids, large quantities of carbohydrate also occur in nature. Some of the carbohydrates are polysaccharides of high molecular weight, others are metabolic products or intermediates of low molecular weight. The chromatographic separation of the various carbohydrates is of significance in biochemistry as well as in physiology and pathology.

Numerous tests have shown that carbohydrates can be readily fractionated by gel chromatography. Even in the initial period of gel chromatography it was proved that fractionation takes place according to the molecular weight, or, more precisely, according to the molecular size (Granath and Flodin 1961; Flodin 1962). The larger oligo- or polysaccharides are eluted faster than the smaller ones. Also, gel chromatography of carbohydrates has properties which differ from those of proteins and nucleic acids, for instance the special interactions observed between the gel material and the carbohydrate molecules, depending on the ionic strength or correlations between the molecular weight and the elution volume which are not the same as in the case of globular proteins (see Section 2.6.3 of Part I).

The application of dextran gels to the chromatography of carbohydrates does not involve the risk of the dextran dissolving from the gel and falsifying the results. According to Flodin (1962) if Sephadex G-25, G-50 and G-75 are continuously washed, then 0.002 to 0.003% will dissolve per day; from Sephadex G-200, 0.005% will dissolve per day. The latter is the only gel which produces a positive reaction with anthrone reagent.

In the following, the laws and regularities of the gel chromatography of various polysaccharides and oligosaccharides are surveyed shortly, illustrating with examples the different applications of the method in the chemistry of carbohydrates.

3.1. GEL CHROMATOGRAPHY IN THE CHEMISTRY OF POLYSACCHARIDES

If polysaccharides are analysed, the removal from the solution of inorganic salts and small-molecule compounds may become necessary. These operations can be easily performed by chromatography on small-pored gels. Examples are available in the work of Chaterjee *et al.* (1961) on the removal of iodate ions, or by Östling (1962), and Flodin (1962) on the separation of glucose from dextran or from inuline.

In view of the therapeutic applications of dextran, the determination of the molecular weight, and the distribution by molecular weight of the derivatives obtained by partial hydrolysis of natural dextran, is of great importance. Several authors have studied the gel chromatographic fractionation of dextran. *Figure 92* illustrates the gel chromatography of three dextran derivatives of dissimilar average molecular weights (Hellsing 1968). The correlation of the gel chromatographic mobility with the average molecular weights were studied by Granath and Flodin (1961), their results are shown in *Table 33* and *Figure 93*.

Granath and Kvist (1967) made a detailed study of the applicability of gel chromatography for the determination of the molecular weight distribution of dextran preparations and extended their tests to other carbohydrates also.

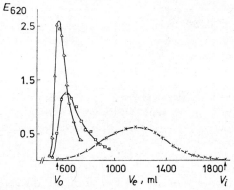

Fig. 92. Gel chromatograms of dextran samples having different average molecular weights (Hellsing 1968)

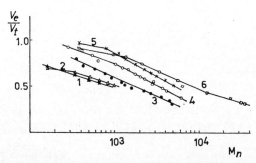

Fig. 93. The dependence of the gel chromatographic migration of dextran- and cellulose-oligosaccharides or polysaccharides on the molecular weight using dextran gels of dissimilar porosity (swelling capacity) (Granath and Flodin 1961). For the experimental conditions see *Table 33*

Table 33

The migration rate of dextran and cellodextrin oligosaccharides and polysaccharides chromatographed on dextran gels of different crosslinkages*

The curves of Fig. 93	Water regain of the gel W_r g g^{-1}	Mole number–molecular weight of the oligo(poly)saccharide $\overline{M}_n$	The equation for the V_e/V_t ratio	Sample
1	1.9	162— 1 000	$1.41-0.31 \lg \overline{M}_n$	Dextran oligosaccharides
2	2.3	162— 1 000	$1.35-0.28 \lg \overline{M}_n$	Cellulose oligosaccharides
3	2.3	< 5 000	$1.81-0.41 \lg \overline{M}_n$	Dextran
4	4.3	< 1 000	$2.19-0.48 \lg \overline{M}_n$	Dextran
		1 000— 1 700	$1.73-0.33 \lg \overline{M}_n$	
5	7.3	< 1 000	$1.35-0.17 \lg \overline{M}_n$	Dextran
		1 000—13 000	$2.07-0.40 \lg \overline{M}_n$	
		13 000—40 000	$1.58-0.28 \lg \overline{M}_n$	
6	8.7	800—(15 000)	$2.36-0.50 \lg \overline{M}_n$	Dextran

* Granath and Flod in (1961).

On *Figure 94* the elution curve of an inulin preparation is shown, together with the molecular weight distribution curve calculated from it. Other authors have studied also the distribution of pullullane by molecular weight. The determination of the molecular weight distribution of clinical dextran was studied recently by Nilsson and Nilsson (1974). Aspinall and Miller (1973) have shown that prior dyeing of dextran samples did not affect their migration in thin layer gel chromatography; the use of previously dyed polysaccharides is advantageous because the migration and separation of the components become visible (Anderson et al. 1969).

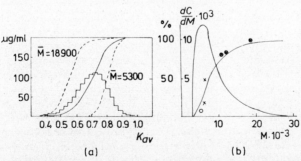

Fig. 94. Gel chromatographic analysis of inulin samples (Granath and Kvist 1967). (a) The chromatogram and the related cumulative curve, and the cumulative curves for two standard samples of 18,900 and 5300 $\overline{M}$; (b) the molecular weight distribution curve calculated from the chromatogram and the integral curve

The gel chromatographic analysis of dextran can be used to advantage in the examination of the permeability of various membranes and holds out good promise also in clinical practice, for instance in the analysis of urine specimens collected at different times following a dextran infusion.

For the production of dextran and ficoll preparations of different molecular sizes, Laurent and Granath (1967) performed various experiments.

For the fractionation of dextran by molecular weight, chromatography on porous glass beads may also be used. The logarithm of the molecular weight and the elution volume show a linear correlation.

In the examination of starch and glycogen, gel chromatography has been introduced only recently, although attempts were made earlier. Nordin, in 1962, fractionated dextrins obtained from starch on a Sephadex G-75 column, to separate amylose from amylopectin. A special enzymatic treatment and subsequent gel chromatography proved applicable also for the analysis of the structure of starch molecules isolated from various plants, i. e. for the determination of the length of the linear chains and the branches (Lee et al. 1968; Oka et al. 1971; Akai et al. 1971b). The same applies to the analysis of the fine structure of glycogen (Akai et al. 1971a; Harada et al. 1972; Mercier and Whelan 1970).

Recently Dintzis and Tobin (1974) used control-pored glass in an analysis of the molecular weight of amylose and dextrin.

Also the dispersity units of natural starch can be analysed by size, using gel chromatography on an agarose column (Oka et al. 1971).

In the field of research into the structure of acid polysaccharides found in different conjunctive tissues, gel chromatography has been recently introduced. A report on the most promising experiments came from Constantopoulos et al. (1969) and Wasteson (1969).

Constantopoulos et al. (1969) studied the chromatographic behaviour of acid mucopolysaccharides of different origin on a Sephadex G-200 column. The correlation they established between the logarithm of the molecular weight of the standard preparations and the elution volume is illustrated in *Figure 95*. Wasteson (1969) made standard chondroitin-4-sulphate preparations by preparative gel chromatography and used them to establish the correlation between the rate of migration and the molecular weight. He elaborated a micromethod to establish the distribution of different acid polysaccharides by their molecular weights. In a later publication (Wasteson 1971) he extended his examination also to an agarose (Sepharose 2B) column and found that the ion concentration of the eluant influences the elution volume, because at a low ion concentration the gyration radius of the acid polysaccharide molecules is greater. The concentration of a 0.2 M NaCl solution is sufficiently high and a further increase of the salt concentration does not cause further alterations in the elution volume. Variation of the polysaccharide concentration caused no significant effect, unless a relatively large quantity of carbohydrate was applied to the column. This is in agreement with the earlier findings of Winzor and Nichol (1965) regarding dextran. Wasteson's (1971) results are illustrative of the difference between the gel chromatographic behaviour of proteins and polysaccharides: the K_{av} value of chondroitin-4-

sulphate, with a molecular weight of 13 000 dalton, is nearly equal to that of serum albumin having 69 000 dalton molecular weight.

Wasteson et al. (1972) used gel chromatography to study the proteoglycanes in different regions of cartilage tissue.

Roberts and Gibbons (1966) analysed the molecular weight of the materials arising from the treatment of epithelial mucopolysaccharides by neutral hypochlorite with gel chromatography, calibrating the agarose gel column with dextran preparations of known molecular weights.

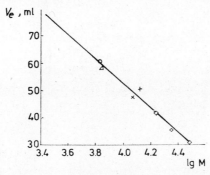

Fig. 95. Relationship between the elution volume and molecular weight for acidic mucopolysaccharides in gel chromatography on Sephadex G-200 (Constantopoulos et al. 1969)

According to Tortelani and Romeguli (1975), chromatography on a Sephadex G-100—cellulose (4 : 1) thin layer is useful for the determination of the molecular weight of acidic mucopolysaccharides; the migration distance varies linearly with the logarithm of the molecular weight.

Taniguchi (1970) analysed acid mucopolysaccharides of urine by thin layer gel chromatography. Tsiganos et al. (1971) examined the polydispersity of cartilage proteoglycane fractions by column chromatography on agarose gel. Schulsatz et al. (1971) carried out fractionations on acrylamide gels while studying the structure of the cornea, whilst Lyon and Singer (1971) examined the structure of the polysaccharides of the nucleus pulposus. Kresse et al. (1971) performed separations on dextran and agarose gel columns in the analysis of the aortic proteoglycanes. Revell and Muir (1972) applied gel chromatography in their examinations of the resorption and secretion of isotopically labelled chondroitin-4-sulphate injected into guinea pigs. Heinegard (1972) and Hardingham et al. (1972) used gel chromatography to study the structure of various proteoglycanes.

The literature provides information also on the gel chromatography of other natural polysaccharides. Nelsestuen and Suttie (1972) used gel chromatography fractionation to examine the carbohydrate portion of bovine prothrombin, Thieme and Ballen (1972) used it to study phosphomannane found in the cell wall of the *Kloeckera brevis* yeast, Lloyd (1972) in his investigation into the *Cladosporium Verneckii* galactomannane, Preston et *al.* (1969a, b) to examine the structure of the extracellular polysaccharide produced by *Penicillium charlesii*. Troy *et al.* (1970) studied the biosynthesis of the polysaccharide coating of *Aerobacter aerogenes*, fractionating the intermediates and the partially decomposed products on dextran gel columns. Örgen and Lindahl (1971) analysed the degradation of heparin in mouse mastocytome tissues by gel chromatography.

Gel chromatographic analysis of the molecular weight, and the distribution by molecular weight, of cellulose is difficult because of the high viscosity of the solutions. As regards their converted products, first on cellulose

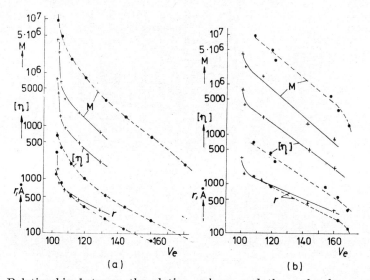

Fig. 96. Relationship between the elution volume and the molecular parameters of cellulose nitrate samples (Meyerhof 1970)
The solid lines (+) represent the values for cellulose nitrate in chromatography (a) on four Porasil columns (F, F, 1500, 1000) and (b) on four Styragel columns (10^7, 10^6, 10^6, 10^5 Å) from acetone solutions; the dotted lines with points represent those for polymethacrylate samples

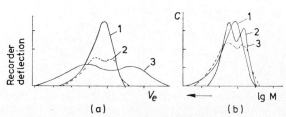

Fig. 97. Chromatographic pattern (a) and the calculated molecular weight distribution curves (b) for a mixture (1 : 1) of two cellulose nitrate fractions of dissimilar average molecular weight ($P\eta = 6700$ and 1200 respectively) (Meyerhof 1970)
1. Porasil columns, acetone solvent; 2. Styragel columns, tetrahydrofuran solvent; 3. Styragel columns, acetone solvent

nitrate derivatives, results are promising (Meyerhof 1965a, b, 1970; Segal 1966, 1968).

According to Meyerhof, the well-known method of gel permeation chromatography, i.e. the use of tetrahydrofuran eluant on a Styragel column, has not proved applicable for cellulose nitrate, owing both to poor solubility conditions and to adsorption. He therefore experimented with various column packings on inorganic bases, e.g. with Porasil and Bio-Glas derivatives, ceramic materials deposited on a silicon aluminium oxide base, with acetone as eluant. In comparative tests, Meyerhof also chromatographed synthetic polymers and found that adsorption on glass beads is considerable. Ethyl cellulose binds to glass beads in a practically irreversible way. Although the pore size and the adsorption properties of the silicon aluminium oxide based column packings would meet requirements, their particle sizes and particle shapes do not because they are produced by crushing, which makes packing very difficult.

Meyerhof, chromatographing on columns packed with Porasil, obtained good results using cellulose nitrate and polymethacrylate. However, under such conditions, cellulose nitrate can only then be fractionated if its polymerization degree is less than 3000 *(Figs 96 and 97)*. He achieved better results with Styragel columns on which separation was feasible up to a polymerization degree of 8000. Adsorption takes place also in this case, especially if columns of high resistance and if packings of pore sizes below 10^5 Å are used.

Meyerhof (1970) used this method to examine also the nitrate of natural cellulose, but the findings with respect to the molecular size distribution were not satisfactory.

In the purification of bacterial cell wall lipopolysaccharides, Romanowska (1970) used Sepharose gels and obtained a successful separation from the ribonucleic acids present in the phenol-water extract.

3.2. GEL CHROMATOGRAPHY OF OLIGOSACCHARIDES

The members of the oligosaccharide homologous series with low molecular weight can be separated on small-pored gels. Flodin and Aspberg (1961) used a Sephadex G-25 column to isolate the members of the cellodextrin series emerging from the acetolysis of cellulose and subsequent deacylation, up to the sixth member. Flodin (1962) used a similar method for the fractionation of the members of the homologous series of isomaltoses.

According to Flodin et al. (Flodin 1962; Flodin et al. 1964) the small molecular weight members of the chondroitin-sulphate oligosaccharides give sharper peaks when fractionated with a highly concentrated salt solution *(Fig. 98)*. The authors fractionated also hyaluronic acid oligosaccharides on a Sephadex G-25 gel column.

Martin et al. (1972) showed that, chromatographing on dextran gel, the elution volume of the various oligosaccharides is greater than that of the corresponding polyethylene glycol oligomers used. This relative retention of the carbohydrates is explained by sorption, i.e. the formation of hydrogen bonds between the carbohydrate molecules and the dextran gel. These authors have characterized the chromatographic migration rate of carbohydrates by the R_g value, i.e. the ratio of their elution volume to that of glucose. The R_g value of large molecules excluded from the gel is 0.56, that of the small molecules which readily diffuse into the gel is 1.27. *Table 34* shows the R_g values of different disaccharides, pentoses and hexoses. The authors

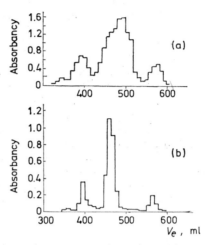

Fig. 98. Elution diagrams of oligosaccharides, derived from chondroitin sulfuric acid, from a column of Sephadex G-25 (200—400 mesh) (Flodin 1962) (a) 0.1 M NaCl solvent, (b) 1.0 M NaCl solvent

Table 34

The relative elution volume (R_g) of some carbohydrates*

Pentose	R_g	Hexose	R_g	Disaccharide	R_g
Ribose	1.065	Psicose	1.040	Saccharose	0.926
Lyxose	1.062	Mannose	1.029	Maltose	0.922
Xylose	1.047	Sorbose	1.026	Cellobiose	0.907
Arabinose	1.024	Fructose	1.018	Trehalose	0.902
		Glucose	1.000	Lactose	0.895
		Galactose	0.990	Melibiose	0.877

* From Martin *et al.* (1972).

assume that, in an aqueous solution, the volume of the carbohydrates is proportional to the molecular weight of their crystalline hydrate. The value of lg R_g and lg (R_g — 0.56), illustrated as a function of these "corrected" molecular weights, yield a straight line.

Martin *et al.* (1972) experimented also with permethylated carbohydrates in which the hydroxyl groups (assumed to cause sorption) are blocked. Such permethylated carbohydrates migrate faster than those with corresponding molecular weights but untreated. This means that methylation increases their migration rate to a greater degree than would be expected on the basis of an increase in the molecular weight alone. This confirms the significance of the formation of hydrogen bonds in the gel chromatography of carbohydrates.

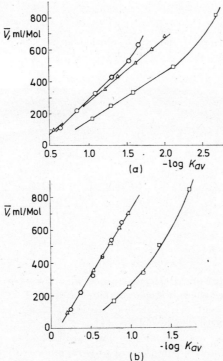

Fig. 99. Relationships between $-\lg K_{av}$ values and the partial molar volume ($\overline{V}$) of cellodextrins (○), xylodextrins (△) and polyethyleneoxide (□) (Brown and Chitumbo 1972)
(a) chromatography on Sephadex G-15 gel; (b) chromatography on Bio-Gel P-2 gel

Brown et al. thoroughly studied the gel chromatographic behaviour of various oligosaccharide series, for example cellodextrines (Brown 1970), mannodextrines (Brown and Andersson 1972) and xylodextrines (Brown and Chitumbo 1975). It was found that there is a linear correlation between the partial molar volume of the molecules and the negative logarithm of the K_{av} values *(Fig. 99)*. The lines relating to the different homologous series will only then coincide if chromatography is performed on Bio-Gel P-2 polyacrylamide gel. Even in this case, the straight line will differ from the line which refers to the homologous series of polyethylene glycol. This accords with the comments of Martin et al. (1972) who attributed the effect to interactions between the carbohydrates and the gels. They proved that these interactions depend on the ionic strength of the solution, namely in opposite direction in the uses of dextran gels or polyacrylamide gels: the increasing salt concentration of the solution weakens the interaction with polyacrylamide gel, and increases it with dextran gel. *Table 35* shows the K_{av} values relating to members of the oligosaccharide series studied by the authors.

Table 35

The volumetric distribution coefficient (K_{av}) of oligosaccharides chromatographed from 0.1 M NaCl solution at 25 °C on Sephadex G-15 and Bio-gel P-2 gels*

Number of monomer units	K_{av}					
	Mannodextrins		Cellodextrins		Xylodextrins	
	P-2	G-15	P-2	G-15	P-2	G-15
1	0.80	0.53	0.78	0.52	0.81	0.57
2	0.68	0.40	0.68	0.42	—	—
3	0.60	0.30	0.59	0.33	—	—
4	0.51	0.21	0.52	0.27	0.58	0.29
5	0.45	0.15	0.46	0.21	0.51	0.24
6	0.40	0.12	0.42	0.18	0.47	0.19
7	0.33	0.08	—	—	0.49	0.15
8	—	—	—	—	0.37	0.13

* Brown and Chitumbo (1972), Brown and Andersson (1972).

Brown and Andersson (1972) examined also the temperature dependence of the interaction between oligosaccharides and gels. At higher temperatures they observed an increase in the elution volumes, i.e. stronger sorption *(Fig. 100)*. (Cf. Section 2.7.3 of Part III.)

Recently Sabbagh and Fagerson (1976) obtained a similar temperature dependence in the gel chromatography of glucose oligomers. The authors

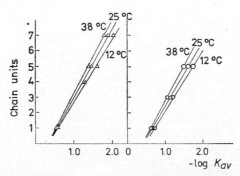

Fig. 100. The effect of temperature on the gel chromatographic migration of xylodextrins and cellodextrins (Brown and Andersson 1972). The designations are the same as in *Fig. 99*

studied the effect of various parameters on the fractionation and resolution of the oligomers, up to 12 glucose units, and the separation of the isomers, e.g. isomaltose from maltose, etc. According to the authors, the separation process should not be interpreted on the basis of a single mechanism and solvent-gel, solute-solvent and solute-gel interactions must all be considered.

Gel chromatography can be used for the fractionation and isolation of the natural oligosaccharides or the hydrolysis products of various polysaccharides. Cunningham and Simkin (1966) studied the glycopeptides extracted from the acid glycoproteins of guinea pig serum by gel chromatography. Holking *et al.* (1972) found oligosaccharide of abnormal molecular weight in cattle suffering from pseudolipases, by gel chromatography on Sephadex G-25 gel. Wadström and Hisatsune (1970) applied fractionation on a Sephadex G-10 column to an investigation of the mechanism of a bacteriolytic enzyme. Raftery (1969), Tsai (1970) and Akabosi *et al.* (1972) fractionated chitin oligosaccharides by gel chromatography.

As will be clear from the earlier discussion, the gel chromatographic behaviour of carbohydrates differs from that of proteins and peptides in several respects. Because of these differences, the gel chromatographic migration of various carbohydrate-containing proteins will differ from that of proteins free from carbohydrates (Andrews 1964, 1965). No quantitative relationship has so far been established between the carbohydrate content and the gel chromatographic behaviour of glycoproteins.

4. GEL CHROMATOGRAPHY OF SYNTHETIC POLYMERS

The success of the gel chromatographic fractionation of natural compounds acted as a powerful stimulant for analyses in synthetic organic chemistry. Even during the early period of the application of gel chromatography Vaughan (1960) experimented with the fractionation of polystyrene of a broad molecular weight distribution, from a benzene solution on styrene-divinylbenzene copolymer resin. He found that separation was satisfactory only in the range of lower molecular weights, while the larger polystyrene gel molecules were impermeable. Almost concurrently, Vaughan (1960) chromatographed low-vulcanized rubber latex from organic solvents and established that polyisobutane can be separated from mineral oil of lower molecular weight in n-heptane, since the latter passes through the column at a slower rate.

In the development of the gel chromatography of synthetic polymers, Moore's (1964) experiments marked a decisive juncture. Setting out from the fact that, in the preparation of dextran gels for the gel chromatography of hydrophilic compounds, the gel matrix is obtained in solution, i.e. diluted by a solvent, he produced copolymers from styrene and divinylbenzene in organic solvents. Into the gels so obtained polymers of considerably greater molecular weight were able to diffuse than into the types prepared by polymerization without the use of a solvent. Thus, for instance, while the copolymer containing 1% divinylbenzene prepared in the traditional way as an intermediate of ion exchange resins proved impermeable for polymers with molecular weights above 3500 dalton, the gels produced in a polymerization mixture consisting of 90 parts of styrene, 11 parts of divinylbenzene and 80 parts of toluene became impermeable only for polymers with a molecular weight above 250 000 dalton.

Moore (1964), altering the ratio of the components to the solvent, prepared gels for use in widely varying ranges of molecular weight. He examined their applicability with standard anionic polystyrene and polypropylene glycol preparations with a relatively narrow molecular weight distribution and different average molecular weights. The author found that, in line with the laws of gel chromatography, first the large molecules emerge in the bottom of the column and that, for every type of gel, there exists a range of molecular weights in which there is a linear correlation between the logarithm of the average molecular weight of the polymer applied and its elution volume.

Moore made the fundamental observation (1964) that the presence of polymers with high molecular weights does not influence the chromatographic migration of polymers with lower molecular weights. Since separation depends on the permeability of the gels to molecules of different sizes, Moore recommended the use of the term "gel permeation chromatography" for the process.

This term is still widely used for the gel chromatography of synthetic polymers. Although this field of application does not differ basically from the separation of natural macromolecules or of materials of low molecular weights on a gel column, in practice it represents a more or less separate field of gel chromatography. The difference lies in the fact that the molecular weights of the molecules being fractionated are very close to one another rather than that organic solvents are used for the chromatography of generally water-insoluble polymers. At the present stage of development, two adjacent members of a polymer series built up of many thousands of monomer molecules cannot be separated by gel chromatography because the difference between the molecular weights is only a fraction of 1% of the total molecular weight. In principle, even adjacent members migrate at different rates during chromatography; the greater the difference between the degree of their polymerization, the greater the difference between their elution volume, i.e. the easier their separation.

Accordingly, during the gel chromatography of a synthetic polymer sample, the composition of the eluate changes continuously; initially the larger, and later on the smaller molecular size polymers will be in excess. This allows an approximate fractionation of the polymers by their chain lengths which, however, will never be perfect since only polymer fractions of a narrow molecular weight distribution can be prepared. On the other hand, the method is well suited for analytical purposes; the analysis of the chromatographic curve permits the establishment of molecular weight distribution of the polymer sample. Such analyses are of considerable importance in the characterization of the differently prepared polymers (Lengyel 1968; Muzsay 1970). Obviously, this trend in gel chromatography developed rapidly and today automatic equipments are readily available.

The gel chromatographic migration of polymer molecules is determined primarily by their molecular size and not by their molecular weight. Even Moore (1964) came to this conclusion from the results obtained during the gel chromatography of low molecular weight organic compounds. However, the determination of the molecular size of synthetic chain polymers is a less simple problem than with, for instance, globular proteins whose steric structure is relatively stable, or varying within narrow limits. No doubt the molecular size of a synthetic polymer depends also on the number of monomer units of which it is built up. If substances are polymerized from the same monomer, logarithm of their elution volume and either the molecular weight or the number of units will yield a linear relationship in the range of applicability of the gel. With standard substances of known molecular weights a calibration curve can be plotted in the linear part of which, knowing the elution volume of a sample of unknown molecular weight, the unit number or the molecular weight can be established. However, the calibration curves of polymers prepared from different monomers do not coincide even under absolutely identical experimental conditions because, as regards their diffusion into the gel pores, the molecules of different polymers having the same molecular weight are dissimilar in size.

Calibrating with polystyrene samples in order to determine the molecular weight of a chemically different polymer on the basis of the elution volume, the value read on the calibration curve must be multiplied by a factor Q which depends on the polymer used for calibration and the material of the sample.

Hendrickson and Moore (1966), in an effort to obtain a uniform calibration curve, introduced the concept of the "effective C atom number" ($\#C$). In organic substances of relatively small molecular weight, from the bond intervals and bond angles, the "effective chain length" values were calculated, the summation of which for each molecular segments, characterizes the size of the complete chain molecule *(Table 36)*.

In the range of separation of a given gel, the logarithm of the effective carbon number is a linear function of the elution volume, therefore, the calibration curve so plotted will coincide if different organic compounds are gel chromatographed under identical conditions.

Table 36

The "effective chain length" values of selected atoms*

Atom	Radius for covalent bond, Å	Van der Waals radius, Å	Group	Angle of band, deg.	Chain length # C number basis	Å
C	0.77	—	C–C–C	109.5	1.00	1.25
O	0.66	—	C–O–R	105	0.67	0.84
N	0.70	—	C–N–R	109.5	0.91	1.14
S	1.04	—	C–S–R	104	1.00	1.25
F	0.60	1.35	C–F		0.27	1.84
Cl	0.99	1.80	C–Cl		1.09	2.60
Br	1.14	1.95	C–Br		1.32	2.88
I	1.36	1.95	C–I		1.54	3.15
H	0.30	1.00	C–H		0.00	1.25
O	0.51	1.60	C=O	120	0.61	2.00

* Hendrickson and Moore (1966).

The effective C number could be used with advantage in the interpretation of the chromatographic migration of organic compounds of lower (a few hundred dalton) molecular weights (see Section 4.5 of Part III). However, if linear polymers with long chains are concerned, the space requirement of the molecule will no longer be an unambiguous function of the chain length because the chains are not extended in the solution but assume the structure of a random coil. Several authors have established that, to plot a universal calibration curve, the hydrodynamic volume of the polymers rather than the molecular length or the molecular weight should be used (Benoit 1968; Benoit et al. 1966; Boni et al. 1968; Cantow and Johnson 1967; Cantow et al. 1967a, b). The hydrodynamic volume is proportional to the product of the numerical average of the molecular weight of the molecules ($\overline{M}_n$) and their intrinsic viscosity. Representing the logarithm of this product as a function of the elution volume, regardless of the structure of the polymer and of the numerous conditions under which chromatography is performed (e.g. temperature, solvent), the calibration curves will coincide *(Fig. 101)*.

If the distribution coefficient (K_d) is used in the calculations instead of the elution (or retention) volume, the calibration curve also becomes practically independent of the column size. However, even this type of calibration will not be completely universal; it will be independent only of the chemical structure of the polymer but will depend on the gel and on the conditions of chromatography, e.g. on the concentration of the sample and the speed of the gel chromatographic process. Neither can it be used if adsorp-

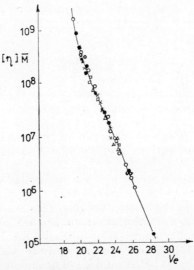

Fig. 101. Relationship between the elution volume and the logarithm of the hydrodynamic volume of the molecules (Benoit 1968). The ordinate represents the product of the weight-average molecular weight of the chromatographed samples and the intrinsic viscosity of the solutions in a logarithmic scale. The designations refer to samples polymerized from different monomers

tion or some other interaction takes place between the gel and the polymer molecules.

Cantow et al. (1967a, b, c) compared the elution volumes of polyisobutene and polystyrene linear polymers at a temperature of 150 °C by gel chromatography from trichloro-benzene on polystyrene gel. They found that the calibration curves plotted on the basis of the chain length of the extended molecule do not coincide, but those obtained on the basis of the distances between the chain ends calculated from the intrinsic viscosity of the solution and the average molecular weight do.

The calculations in the latter case are based on the work of Ptizin and Eizner (1966) and refer to the actual size of the molecules in solution. Cantow et al. (1967a, b, c) examined also the effect of temperature on the elution rate of polyisobutene and established that, with rising temperature, the elution volume tends to decrease. Illustrating the logarithm of the average molecular weight of the samples as a function of the elution volumes measured at temperatures of 35°, 70°, 110°, and 150 °C, they obtained parallel curves, but when they plotted the average molecular weight multiplied by the viscosity values measured at different temperatures on a logarithmic scale, the calibration curves coincided *(Fig. 102)*. This shows that the effect of rising temperature on the decrease of elution volume can be interpreted by the alteration of viscosity, or, more precisely, the alteration of hydrodynamic volume.

Increasing concentration of the samples used for calibration increases the elution volume *(Fig. 103)* (Cantow et al. 1967c; Troth 1968). This effect, in polymers having high molecular weight, is especially strong.

According to Troth (1968), the effect of increasing sample concentration upon the elution volume can be expressed, to a good approximation, by the following equation

$$V_a = V_o + [\eta]C$$

where V_a denotes the elution volume measured at the sample concentration C, V_o the elution volume extrapolated to zero concentration, and $[\eta]$ the intrinsic viscosity.

As shown by *Figure 103b*, the variation of $[\eta]$ with increasing concentration is practically linear but the straight line does not pass through the origin.

Studying the gel chromatography of low molecular weight hydrocarbons, Hendrickson and Moore (1966) observed that the branching of the carbon chain has no influence on the elution volume. This is valid also for synthetic polymers provided that the calibration curves are plotted in accordance with their hydrodynamic volumes (Carmichael 1968; Le Page et al. 1968; Drott and Mendelson 1968). Gel chromatography and viscosity measurements permit the assessment of the number of branchings; namely the weight average (n_S) of the branching points is related to the ratio of the gyration radii of the branched and linear polymer molecules and the latter can be determined by viscosimetry. Drott and Mendelson (1968) published tables for the calculation of n_S and a simple nomogram which, knowing the elution volume and the intrinsic viscosity, helps determine both the molecular weight of the polymer and the average number of branchings *(Fig. 104)*.

Gel permeation chromatography of various natural and synthetic polymers was studied recently by Belenkii *et al.* (1975a, b). In the chromatography of flexible-chain polymers on macroporous swelling sorbents, the authors observed a deviation from the Benoit principle of universal calibration. It was suggested that these are caused by different degrees of thermodynamic compatibility of the eluted polymers with the sorbent matrix.

If the relationship between the molecular weight and the elution volume is known, it is possible to examine the molecular weight distribution of synthetic polymers. This subject has been dealt with in the section on the gel chromatography of dextran preparates. The average molecular weight of the polymers in the various fractions can be established from the calibration

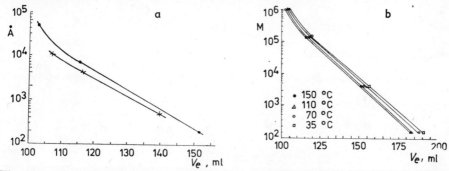

Fig. 102a. Correlation between the elution volume and the projected extended chain length (in Å) of polyisobutene (●) and polystyrene (×) samples chromatographed on Waters polystyrene gel at 150 °C from trichlorobenzene (Cantow *et al.* 1967a)

Fig. 102b. Variation of the correlation between the molecular weight and the elution volume of polyisobutene samples with temperature (Cantow *et al.* 1967a)

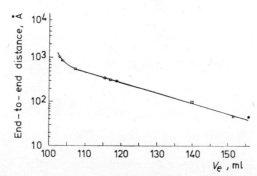

Fig. 102c. The calibration curve for synthetic polymer samples independent of the temperature and the sample material (Cantow *et al.* 1967a). (●) refers to polyisobutene samples chromatographed at 35 °C, (○) to the same samples at 150 °C; (□) to polystyrene samples chromatographed at 150 °C

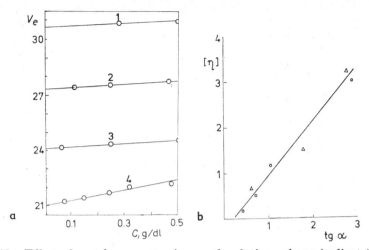

Fig. 103a. Effect of sample concentration on the elution volume (ordinate) of polystyrene standards (Troth 1968). The average molecular weights of the standard polystyrene samples are: 1. 19,800; 2. 98,200; 3. 411,000; 4. 2,115,000 dalton

Fig. 103b. Relationship between the concentration effect and the intrinsic viscosity (Troth 1968). The abscissa shows the slope of the straight lines in *Fig.103a*. On the ordinate is plotted the intrinsic viscosity of the polystyrene solutions

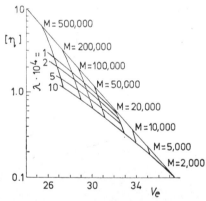

Fig. 104. Nomograph for determination of molecular weight and branching of polymer samples (Drott and Mendelson 1968)

On the abscissa is plotted the elution volume in arbitrary units (the actual value is a function of the column set used). The figures indicate the molecular weight (2,000 to 500,000 dalton) and the branching indices (0 to 10). The value of the latter is given by $\lambda = n_w/M$, where M designates the molecular weight, and n_w the weight average number of branch points

curve and the polymer concentration, usually by measurement of the refractive index. This allows a computation of the molecular weight distribution of the chromatographed polymer and illustrates the number of molecules in each molecular weight component (see *Fig. 94*). This process, however, will yield only an approximate molecular weight distribution because the elution curve for the molecules of the same molecular weight is bell shaped. This means that they are distributed over several fractions and mix with molecules of nearly the same size. This proves that the resolution of gel chromatography is far from infinite and, in the computation of the absolute molecular weight distribution, corrections must be made (Tung et al. 1966; Duerksen and Hamielec 1968; Altgelt 1968).

Kubin (1975) has published a model for the description of the gel permeation chromatographic separation of macromolecules on a packing with non-homogeneous pores.

Several authors have studied the effect of the flow rate on the elution volume of synthetic polymers, and on the resolution of the column (Smith and Killmansberger 1965; Heitz and Coupek 1968; Bilmeyer et al. 1968; Duerksen and Hamielec 1968; Altgelt 1968; Yau et al. 1968; Yau 1969). At a flow rate of less than 2 ml/min the elution volumes pertaining to the chromatographic peaks are practically constant, but the HETP may show quite considerable variations. Since the theoretical plate number of the column varies with different substances, the optimal chromatographic conditions should be determined by a series of experiments for each individual case.

Unger (1974) used a special silica packing material for the gel permeation chromatography of synthetic polymers. Porous silica beads, with a high porosity, in different pore-size ranges were prepared by hydrolytic polycondensation of polyethoxysilane (Unger 1973), and a dense layer of alkyl groups were chemically bonded on the surface by silanization; this allows polystyrenes, polyethylenes, and polypropylenes to be fractionated without any adsorption effect.

The gel permeation chromatography of polystyrene samples on macroporous silicas was studied recently by Eltekov and Nazansky (1976).

In a comprehensive publication, Muzsay (1970) has presented a number of data on the conditions for the gel chromatographic analysis of synthetic polymers.

Gel chromatography is a suitable method for the determination of total polymer in industrial materials (Hillman and Paul 1975). If a gel of small pore size is used, the polymers are eluted in a single narrow peak, separated from the low molecular weight additives, and their total quantity is easily calculated from the size of the peak.

5. GEL CHROMATOGRAPHY OF SMALL-MOLECULE ORGANIC COMPOUNDS

5.1. GEL CHROMATOGRAPHY OF ALIPHATIC AND AROMATIC HYDROCARBONS

Since n-alkanes are frequently used for the calibration of gel columns in the analysis of the molecular weight distribution of the polymers dealt with in the previous section, there is ample information available in the literature on their gel chromatographic behaviour.

Hendrickson and Moore (1966), performing chromatography from tetrahydrofuran on polystyrene gel, established that there is a linear correlation between the logarithm of the number of effective carbon atoms or the effective chain length of alkanes and their elution volume. The effective chain length was computed on the basis of the bond angles and intervals. They extended their study also to organic compounds of different atomic content. For the findings, see *Tables 37 to 40*.

When determining the effective chain length of alkanes, in addition to the carbon–carbon bonds the authors considered also the C—H bonds at the chain ends. Since the elution volume of the branching hydrocarbons is equal to that of normal hydrocarbons having the same molecular weight it was concluded that it is the molecular volume which has a decisive role and the chain length is but a useful coefficient for practical application.

Smith and Kollmansberger (1965) chromatographed normal and branching alkanes similarly from tetrahydrofurane on polystyrene gel. They established that the logarithm of the molecular volume and the elution volume are in linear relationship. The agreement between the molecular volumes determined by calculation and by gel chromatographic migration is shown in *Table 37*. The experiments of Edgards and Ng (1968) and of Cazes and Gaskill (1968) corroborate this finding. *Figure 105* illustrates the results arrived at by Cazes and Gaskill (1968) from o-dichloro benzene.

Their findings are of particular significance for compounds containing different substituents and, as proved by several publications, in the gel chromatography of aliphatic hydrocarbons the logarithm of the molecular weight is in linear relationship with the elution volume (Frank *et al.* 1968, Bombaugh *et al.* 1968b).

Accordingly, low molecular weight aliphatic hydrocarbons can be readily separated on polystyrene gel *(Fig. 106)*. Chromatographing on columns with high theoretical plate number, fractionation is feasible up to the number with approximately 30 to 40 carbon atoms but the separation of the adjacent members of the homologous series with higher molecular weight is more difficult.

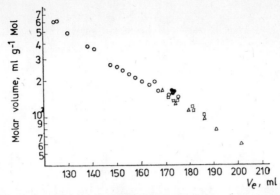

Fig. 105. The correlation of the elution volume and the logarithm of the molar volume (Cazes and Gaskill 1968). Gel chromatography performed from o-dichlorobenzene on polystyrene gel columns at 130 °C

(●) branched hydrocarbons; (□) aromatic hydrocarbons; (△) organic acids

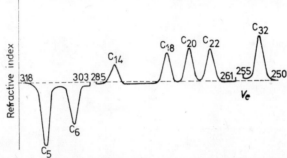

Fig. 106. Fractionation of normal hydrocarbons on polystyrene gel from tetrahydrofuran (Bombaugh et al. 1968b). On the ordinate is plotted the recorder deflection of a differential refractometer detector. On the abscissa is plotted the elution volume in arbitrary units. The numerals at the peaks denote the carbon numbers of the alkanes. The fractionation was carried out on a 160″×3/8″ column packed with Styragel of 500 Å porosity

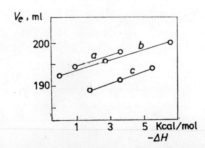

Fig. 107. Correlation of gel chromatographic elution volume with enthalpy differences for stereoisomers of perhydrophenanthrene a, perhydroanthracene b and 1,2,4,5-tetraethylcyclohycane c (Pecka et al. 1975). The chromagography was performed from tetrahydrofuran on styrene-divinylbenzene copolymer S-GEL-832 columns

Table 37

Characteristics of the behaviour of aliphatic hydrocarbons in gel chromatography

Compound	Effective C number		Chain length, Å		Molar volume, cm³ mol⁻¹		
						empirical	
	calculated*	empirical*	calculated**	empirical**	calculated**	in tetrahydrofuran	in o-dichlorobenzene+
n-Pentane	5.0	5.36 / 5.47	9.0	—	116.1	117+	148
n-Hexane	—	—	10.3	—	131.6	—	152
n-Heptane	7.0	7.14 / 7.21	11.7	—	147.5	141+	167
n-Octane	—	—	13.1	—	163.5	—	183
Octene-1	8.0	7.92	—	—	—	—	—
n-Nonane	9.0	9.08	14.4	—	179.7	174+	199
n-Decane	—	—	15.8	—	195.9	—	215
n-Dodecane	12.0	12.44	18.6	—	228.6	—	—
n-Octadecane	18.0	18.50	—	—	—	350+	357
n-Eicosane	20.0	20.14	—	—	—	382+	380
n-Octacosane	28.0	26.42	—	—	—	503+	505
2-Methylpentane	—	—	9.0	10.1	132.9	136**	—
3-Methylpentane	—	—	—	—	—	133+	—
2-Methyloctane	—	—	13.1	15.0	180.8	182**	—
3-Methyloctane	—	—	13.1	14.6	178.9	179**	—
3-Methylnonane	—	—	14.4	15.8	195.0	191**	—
3-Ethyloctane	—	—	13.1	15.5	193.3	188**	—

* Hendrickson and Moore (1966).
** Smith and Kollmansberger (1965).
+ Cazes and Gaskill (1968).

Heitz (1968) showed by his experiments that gel chromatography can be successfully applied for the fractionation of paraffin mixtures of the higher carbon atom numbers, provided that the members in the series differ by several, i.e. at least ten, carbon numbers from one another.

Hillman (1971) used gel chromatography for the analysis of various waxes.

The effects of solvents on the K_{av} values of n-alkanes using polyvinylacetate gel was studied recently by Saitoh and Suzuki (1975a, b). In all solvents except dioxane and methanol the logarithm of the molecular volume was found to be in a linear relationship with K_{av}.

The gel chromatographic behaviour on polystyrene gel in tetrahydrofuran solution of cyclo- and polycycloalkanes and the separation of the stereoiso-

Table 38

Characteristics of the behaviour of aromatic hydrocarbons and their halogen derivatives in gel chromatography

Compound	Effective C number		Molecular length, Å		Molar volume, cm^3 mol^{-1}		
						empirical	
	calculated*	empirical*	calculated**	empirical**	calculated**	in tetrahydrofuran**	in o-dichlorobenzene+
Benzene	3.55	3.52	7.1	6.4	89.2	85	105
Toluene	3.85	4.08 3.85++	—	—	107.6++	—	123
Ethyl-benzene	4.85	4.84* 4.85++	—	—	124.7++	—	142
o-Xylene	—	—	—	—	—	—	131
m-Xylene	—	—	—	—	—	—	134
p-Xylene	4.4	4.6	—	—	—	—	135
Naphthalene	5.1	4.19	—	—	—	—	—
Anthracene	7.1	5.67	—	—	—	—	—
Chlorobenzene	—	—	8.4	7.9	100.1	109	116
Bromo-benzene	—	—	8.8	7.6	103.4	100	—
Iodobenzene	—	—	9.1	7.8	109.8	103	124
p-Dichlorobenzene	—	—	9.7	9.5	113.5	122	—
p-Dibromobenzene	—	—	10.5	8.6	119.8	113	—
p-Diiodobenzene	—	—	11.5	9.5	128.7	122	—
Hexaethylbenzene	6.4	6.43	—	—	—	—	—

* Hendrickson and Moore (1966).
** Smith and Kollmansberger (1965).
+ Cazes and Gaskill (1968).
++ Larsen (1968).

mers were studied by Pecka *et al.* (1975). They found that the stereoisomer showing the higher thermodynamic stability, and a lower retention index in gas chromatography, is eluted before the stereoisomer with the higher heat content *(Fig. 107)*. The authors pointed out that this relationship between elution volumes and enthalpy is based on the relationship between heat content and the physicochemical properties directly connected with the molecular volume.

Aromatic hydrocarbons migrate on polystyrene gel generally in a way similar to paraffins, i.e. according to their molecular weights (Cazes and Gaskill 1968; *Table 38*).

Limpert and Obermiller (1968), performing column calibration with several aromatic hydrocarbons in the analysis of the asphalts of petroleum extracts, found that the elution volume is in a linear relationship with the molecular weight of the compounds.

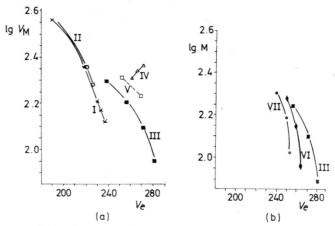

Fig. 108. The correlation between the elution volume and the logarithm of the molar volume (a), as well as the logarithm of the molecular weight (b) of aliphatic and aromatic hydrocarbons (Klimisch and Reese 1972)

The chromatography was performed from tetrahydrofuran on a Bio-Beads S-X8 column (2.45×109 cm). I. n-alkanes (hexane, heptane, octane, dodecane, eicosane); II. alkenes (2-decen, 1-dodecen, 1-eicosen); III. kata-condensed aromatic compounds (benzene, naphthalene, anthracene); IV. peri-aromatic compounds (coronene, benzo (g,h,i)pyrene, perylene); V. Kata-peri-aromatic compounds (benzo(a)pyrene, pyrene); VI. methyl-kata-aromatic compounds (toluene, 1-methylnaphthalene, 9-methylanthracene); VII. ethylkata-aromatic compounds (ethylbenzene, 1-ethylnaphthalene, 9-ethylanthracene)

Hendrickson and Moore (1966) demonstrated that the effective carbon number of the phenyl group was lower than that theoretically calculated: 2.85 instead of 3.55 for benzene *(Table 38)*. On the basis of their experiments, they had to correct the effective carbon number of the phenylene group to 2.4.

Edgards and Ng (1968) stated that, on the basis of the elution volume of aromatic hydrocarbons, molecular volumes seem to be less than those theoretically calculated. They attributed this to adsorptive aromatic interaction with the material of the polystyrene gel.

Klimisch and Reese (1972) also observed that aromatic hydrocarbons interacting with polystyrene gels pass through the gel column at a slower rate than would be expected on the basis of their molecular volume. As proved by their experiments with systematically selected compounds, the elution sequence with kata-condensed compounds corresponds to decreasing molecular weight, with peri-condensed compounds the sequence corresponds to increasing molecular weights *(Fig. 108)*. In the latter case, therefore, the adsorptive interaction with the gel is of the greatest significance. The presence of alkyl substituents causes the elution volumes to shift towards the values which characterize alkanes.

Oelert (1970) used a polyvinyl acetate type gel for the chromatography of hydrocarbons in various organic solvents and found that the investigated compounds could be divided in four types: paraffins, polyphenylenes, kata-

and peri-condensated aromatics. For each type a linear relationship of log molecular volume versus V_e was obtained, but the straight lines for different types did not coincide. This indicates that effects in addition to molecular sieving influence the migration of the compounds containing an aromatic ring.

The behaviour of alkylbenzenes on polystyrene-divinylbenzene gel in tetrahydrofuran was studied by Popl et al. (1975). These authors found that the logarithm of the relative elution volume (V_R) is a linear function of the number of non-aromatic carbon atoms (nC_a), according to the equation

$$\log V_R = 5.3 \cdot 10^{-4} x^3 - 2.51 \cdot 10^{-2} nC_a$$

where x denotes the number of substituents *(Fig. 109)*. The monoaromatic condensed ring containing the benzene derivatives forms another type of compound, in which participation of the carbon atoms is about 2.2-times that for the aliphatic ones. The compounds containing both a condensed ring and an alkyl group form a third type, the elution behaviour of which falls between those of the other two types.

The anomalies, i.e. the side-effects which influence the migration of hydrocarbons on polystyrene gels, have been discussed recently by Asche and Oelert (1975).

The interaction of aromatic hydrocarbons is particularly conspicuous on Sephadex LH-20 gel which contains a hydrophobic group. According to Wilk et al. (1966), in chromatography from isopropyl alcohol compounds containing different numbers of rings can be separated, but their elution sequence will conform to increasing molecular weight. If chloroform is used for eluant, this elution sequence will be reversed, but separation will not be satisfactory. Talarico et al. (1968) found that, on Sephadex LH-20, even cycloparaffins will migrate at a slower rate than branching open-chain hydrocarbons of the same molecular weight. They succeeded in separating dimethylnaphthalene from diphenylmethane.

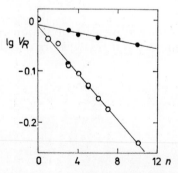

Fig. 109. Dependence of the relative elution volume ($\lg V_R$) on the number of non-aromatic carbon atoms (n) in the molecule (Popl et al. 1975)

(○) alkylbenzenes; (●) benzenes with condensed rings without side chains

Streuli (1971a) examined whether the interaction of aromatic compounds and Sephadex LH-20 is related to the resonance energy of π electrons which are characteristic of the density of the electron cloud. He ascertained that the distribution coefficient (which the author calls "adsorption value") of planar aromatic compounds, if methanol is used as solvent, can be expressed by the following equation:

$$K = 0.0199 \times (\text{resonance energy}) + 0.54$$

Consequently, the interaction between aromatic compounds and the gel is interpreted to be a Lewis-type acid–base complex formation. In the case of non-planar aromatic compounds, the values differ from those calculated from the formula. This, in the opinion of Streuli, can be attributed to the fact in such cases the rings cannot enter into simultaneous interaction with the gel and therefore these compounds are eluted faster than the planar aromatic compounds. Also in aromatic compounds containing heteroatoms or atoms of oxygen, or nitrogen as substituents, the distribution coefficient will differ from the value calculated from the equation, owing to other interactions. In another publication Streuli (1971b) proved that, for chromatography from different solvents on Sephadex LH-20 gel, various interactions became dominant. Thus, with methanol and acetonitrile, aromatic (π electron) interactions predominate, whilst with tetrahydrofuran, hydrogen bonds predominate.

Determann and Lampert (1972) only accept with reservations an interaction with π electrons. According to their experiments the elution sequence of benzene, anisole, nitrobenzene and dinitrobenzene *(Fig. 110)* does not confirm Streuli's (1971a) theory because the π electron density of anisole is higher than that of benzene; a lesser value is obtained for substances containing the nitro group. They assume a polar interaction which decreases with increasing temperature.

Determann and Lampert (1972) studied also the entropy-dependent hydrophobic interactions between apolar carbon chains and Sephadex LH-20 gel. Since these tend to take place mainly in aqueous solution, they occur in water-soluble compounds. This subject will be further considered later.

The adsorption of aromatic nitro compounds containing phenolic or basic substituents, and the influence of electrolytes on adsorption in methanol solution have been studied recently by Mori and Takeuchi (1974).

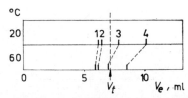

Fig. 110. Effect of temperature on the elution volume of benzene derivatives in chromatography on Sephadex LH-20 gel from n-butanol (Determann and Lampert 1972)
1. benzene; 2. anisole; 3. nitrobenzene; 4. dinitrobenzene. The arrow indicates the total volume of the column

Nakae and Muto (1976) used a methanol–water solvent mixture for the chromatography of alkylbenzenes and alkylbenzoates on polystyrene-divinylbenzene gel; the chromatographic behaviour of the compounds in this system can be interpreted in terms of the theory of partition chromatography, the elution volume increasing with increasing chain length.

5.2. GEL CHROMATOGRAPHY OF ORGANIC ACIDS, THEIR DERIVATIVES AND LIPIDS

Numerous methods have been developed for the chromatographic separation of organic acids and esters. Brewer, as long ago as 1961, in the earliest years after the introduction of gel chromatography, successfully fractionated esters from a toluene solution on slightly vulcanized natural latex. Tipton et al. (1964) fractionated lipid mixtures on styrene-divinylbenzene copolymer gel from benzene, on the assumption that micelle-forming phospholipids can be separated from triglycerides by gel chromatography. They did in fact succeed in separating the fast migrating lecithin fraction from retarded and only partly separated triglycerides, and styrene esters from extracts with different lipid content and model mixtures.

Nyström and Sjövall (1965a, b) fractionated fatty acids, cholic acids and lipids on methylated Sephadex dextran gels from organic solvent mixtures. The method they used was basically distribution chromatography, but with an inverse phase process in which the gel phase (the stationary phase) corresponded to the apolar solvent.

It is rather difficult to explain separation merely by this mechanism, because fractionation by molecular weight and adsorption on the gel also have a greater or lesser role in it. From an ethylene chloride–methanol (9 : 1) mixture, the least polar cholic acids eluted first, while from a chloroform–methanol (1 : 1) mixture, they emerged in the sequence of decreasing polarity. From a chloroform–methanol mixture, triglycerides emerged from the column according to their decreasing molecular weights. The process may therefore be regarded with equal justification as true gel chromatographic fractionation or a separation according to increasing polarity. The same system, accordingly, seems to be a true phase distribution chromatography in the separation of triglycerides but appears as an inverse phase distribution process if cholic acids are fractionated. The chromatographic zones were generally symmetric and the elution volumes did not exceed the volume of the column. For the chromatography of phospholipids (cephalines, cerebrosides, sphingomyelines) the authors found a chloroform–methanol solvent mixture to be best suited for the purpose.

The method elaborated by the authors quoted is suitable for the separation of steroids of equal molecular weight, but dissimilar polarity. From a chloroform–methanol (4 : 1) mixture, cholesterin and tetrahydroxycholane elute almost simultaneously, but from a solvent mixture of higher methanol content (1 : 2, 1 : 1, 2 : 1), tetrahydroxycholane elutes before cholesterin. From a mixture containing chloroform in the ratio of 8 : 1, tetrahydroxycholane is eluted only after cholesterin.

Bejer and Nyström (1972) fractionated free fatty acids, their methyl esters and plant steroles from organic solvent mixtures on the hydroxyalkoxypropyl ether of Sephadex gels by distribution chromatography. For the chromatography of free fatty acids they found a 3 : 7 : 1 mixture of water–methanol–ethylene chloride, for the chromatography of their methyl esters a 2 : 8 : 1 mixture of solvents, to be best suited. The elution sequence corresponded to the increasing molecular weight, which is in good agreement with inverse phase distribution chromatography.

On Sephadex LH-20 gel, which is the β-hydroxypropyl ether of dextran gel, true gel chromatographic fractionation can be performed from organic solvents (e.g. chloroform, Joustra 1967). The free hydroxyl group in the compounds retards migration in the gel column. Thus, for instance, dipalmitin (molecular weight 569) is eluted later than tricarpin (555), and, further, tributyrin (302) and triacetin (218) precede monostearate (559). Also the presence of free carboxyl groups retards migration from chloroform but the use of methanol or ethanol will eliminate interactions between both hydroxyl and carboxyl groups with the gel (Joustra 1967).

Downey et al. (1970) fractionated free fatty acids from a chloroform solution on a Sephadex LH-20 column. They were not able to separate linolenic acid which contains double bonds, from stearic acid but achieved a sharp separation of stearic acid, capronic acid, butyric acid and acetic acid *(Fig. 111)*.

Inoue et al. (1970) fractionated polymeric fatty acids and their esters on a Sephadex LH-20 column from dimethylformamide solvent. They found that elution takes place according to decreasing molecular weight.

Under similar conditions Konishi et al. (1971) chromatographed the saccharose esters of long-carbon-chain fatty acids. Saccharose, mono-, di-, and triesters separated readily and eluted according to decreasing molecular weight.

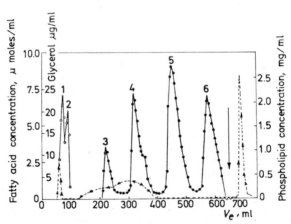

Fig. 111. Separation of glycerides and fatty acids on Sephadex LH-20 column from chloroform (Downey et al. 1970)

1. tristearin; 2. tributyrin; 3. stearic acid; 4. caproic acid; 5. butyric acid; 6. acetic acid. The dotted line shows the concentracion of the phospholipids, the arrow on the onset of the methanolic elution

Free fatty acids can be easily separated from triglycerides on Sephadex LH-20 gel. Addison and Ackman (1969) found that chloroform with 0.2% acetic acid was more suitable for this purpose than pure chloroform, because fatty acids are eluted from this mixture faster and without tailing. They applied the same method to the separation of minute quantities of fatty acids present in large amounts of triglycerides.

Aitzetmüller (1972) chromatographed monomer and dimer oleic acids and their esters from an alcoholic solution on Sephadex LH-20 gel. He experimented also with the Bio-Beads SX-1 type of polystyrene gel, in the separation of trimeric-, dimeric-, and monomeric trioleins from a benzene solution.

Determann and Lampert (1972) studied the binding of saturated and unsaturated organic acids on Sephadex LH-20 and Bio-Gel P-2 acrylamide gels. They established that, from an aqueous solution (0.002 M hydrochloric acid) there is a hydrophobic interaction between the materials of the Sephadex LH-20 gel, related to the changes in entropy. These hydrophobic bonds become increasingly distinct with rising temperature and cause an increase in the elution volume of the carboxylic acids *(Fig. 112)*.

On acrylamide gel the hydrophobic bonds are weak, the acids do not properly separate from one another and rising temperature reduces the elution volume. In organic solvents such hydrophobic interactions related to entropy changes do not occur. The authors, studying also the effect of ionic strength on benzoic acid and some other aromatic compounds, found that the elution volume increases with increasing ionic strength. The gel tends to contract and the distribution of the lipophilic substances shifts in favour of the gel phase, in a manner similar to the salting-out process.

Still more spectacular separations have been reported for chromatography on polystyrene gels from organic solvents. Under such conditions the carboxylic acid esters migrate in the same way as paraffins. Bombaugh *et al.* (1968a, b) fractionated triglycerides from tetrahydrofuran. Plotting the logarithm of their molecular weight versus the elution volume, the straight line incoincided with the points corresponding to the hydrocarbons. Mulder and Bugtenhuys (1970) similarly experienced good separation from a benzene solution on a small polystyrene gel column *(Fig. 113)*.

Klimisch and Reese (1972) studied the interaction between polystyrene gel and various compounds. They found that the esters of carboxylic acids do not adsorb and migrate from tetrahydrofuran like alkanes.

Heitz (1968) fractionated oligomethacrylic acid-butyl esters by molecular weight from tetrahydrofuran on a polystyrene gel column.

The behaviour of free organic acids is radically different from that of esters. Edgards and Ng (1968) observed that during chromatography from tetrahydrofuran on polystyrene gel, carboxylic acids emerge with a smaller elution volume than hydrocarbons of the same molecular weight. This may be attributed to the fact that, by forming hydrogen bonds, the organic acid molecules become capable of associating with the solvent molecules. Therefore, the molar volume of the compounds must be corrected in the same manner as for alcohols.

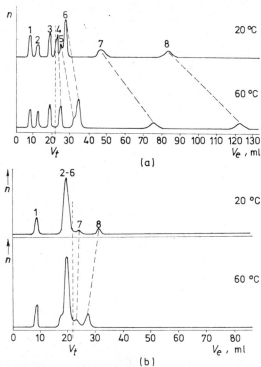

Fig. 112. Effect of temperature on the elution volume of organic acids (Determann and Lampert 1972). Chromatography from an aqueous solution on Sephadex LH-20 column (a) and on Bio-Gel P-2 column (b)

1. polyethylene glycol (void volume marker); 2. glucose; 3. formic acid; 4. propionic acid; 5. butyric acid; 6., 7. unsaturated carboxylic acids ($CH_3-CH=CH-COOH$ and $CH_3-CH=CH-CH=CH-COOH$, respectively); 8. benzoic acid. The ordinate represents the refraction index of the effluent

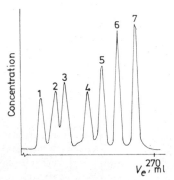

Fig. 113. Fractionation of triglycerides from benzene on polystyrene gel (Mulder and Bugtenhuys 1970)

1. tristearin; 2. trimyristin; 3. trialurin; 4. tricaprilin; 5. tricaproin; 6. hexadecane; 7. undecane

As regards the molecular volume and effective carbon number of carboxylic acids, which determine their migration rate in gel chromatography, see *Table 39*.

Chang (1968) plotted a calibration curve for the analysis of carboxylic acid samples of unknown composition by chromatography from tetrahydrofuran on polystyrene gel. Owing to association with tetrahydrofuran he had to correct the effective carbon number of the acids by three units. With the calibration curve corrected in this manner, he established the effective carbon numbers of several other organic acids (cf. *Table 39*).

Using dichlorobenzene instead of tetrahydrofuran as solvent, carboxylic acids are eluted from a polystyrene gel column at a slower rate than paraffin hydrocarbons (Frank *et al.* 1968; Hillman 1971). This may be interpreted by the adsorptive interaction between the carboxyl group and the gel. The elution volume of the dicarboxylic acids, accordingly, will be still greater (Frank *et al.* 1968).

Carboxylic acids are capable of reversibly adsorbing also on hydrophilic dextran gels; this sometimes permits fractionation. Haavaldsen and Norseth (1966) separated the keto- and enol-tautomers of p-oxyphenyl-pyruvic

Table 39

Characteristics of the gel chromatographic behaviour of carboxylic acids and esters

Compound	Effective C number		Molecular volume, $cm^3\ mol^{-1}$
	Calculated	Empirical in tetrahydrofuran	in o-dichlorobenzene+
Acetic acid	2.7*	6.22++	62
Propionic acid	—	—	80
Butyric acid	—	—	98
Valeric acid	5.7*	—	116
Hexoic acid	—	—	133
Heptoic acid	—	—	151
Octoic acid	—	—	169
Nonoic acid	9.7*	—	—
Myristic acid	14.7*	—	—
Stearic acid	18.7*	—	—
Oleic acid	—	18.7*	—
Linolenic acid	—	18.7*	—
Benzoic acid + tetrahydrofuran	8.0**	7.71**	—
Salicylic acid + tetrahydrofuran	8.67**	8.61**	—
p-Oxybenzoic acid + 2-tetrahydrofuran	12.27**	12.6++	—
Phenyl salicylate	8.67**	8.5**	—
Ethyl acetate	4.67**	4.17**	—
Ethyl formate	4.33++	4.69++	—
Methyl butyrate	5.67**	5.35**	—
Methyl benzoate	5.52**	5.18**	—

* Chang (1968).
** Hendrickson (1968).
+ Cazes and Gaskill (1968).
++ Hendrickson and Moore (1966).

and indole-pyruvic acid from a 0.01 M acetic acid solution on a Sephadex G-25 column. Bridges et al. (1969) on a Sephadex G-10 column, isolated benzoic acid from aliphatic acids from a 0.1 M phosphate buffer solution of pH 7.4. They observed that aliphatic acids precede benzoic acid under such conditions. Aliphatic acids are separated sharply from benzoic acid but their separation from each another is only partial.

Brook (1969a, b) assumed that, in their dissociated form, organic acids are excluded from the dextran gel but, in the undissociated form, they become adsorbed. Their elution volume accordingly depends on the pH. This assumption is supported by experiments. He proved also that organic acids with different dissociation constants can be separated, provided that they are dissociated to different degrees. Furthermore, if the pK values are known, the optimum pH range for separation can be calculated in advance. In a later publication Brook and Housley (1969b) stated that the pH-dependence of the K_d value of aromatic acids in gel chromatography permits the determination of their dissociation constants. *Figure 114* shows the pH-dependence of the distribution coefficient of some aromatic acids.

Similar pH-dependences for the K_d values of various carboxybenzenesulphonates were obtained recently by Vejrosta and Malek (1975).

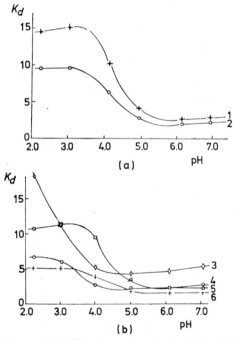

Fig. 114. The pH dependence of the K_d values of some aromatic acids chromatographed on Sephadex G-10 column from an aqueous solution (Brook and Housley 1969b)

1. m-toluic acid; 2. benzoic acid; 3. benzylic acid; 4. phenylisopropyl-hydroxy-acetic acid; 5. phenylisopropyl-acetic acid; 6. phenylacetic acid

The experiments of Laird and Synge (1971) demonstrated that the application of a 1:1:1 phenol-acetic acid-water solvent mixture does not prevent the sorption of organic acids on dextran gels. According to their measurements the K_d values of oxalic acid and citric acid are even higher in such cases than would be expected on the basis of molecular weight calculations.

5.3. FRACTIONATION OF ALCOHOLS AND ETHERS ON GEL COLUMNS

Hendrickson and Moore (1966) found that alcohols chromatographed from a tetrahydrofuran solution on polystyrene gel are eluted faster than expected on the basis of their effective carbon numbers *(Table 40)*. The elution volume of ethers, under the same conditions, is equal to that of alkanes having equal effective carbon atom numbers. This anomalous behaviour of the alcohols is attributed to linkage with a solvent molecule via hydrogen bonds, through the hydroxyl group. On the basis of chromatographic migration, the difference in the effective C number amounts to 2.5 units on average, in fair agreement with the effective carbon number of tetrahydrofuran (1.54 units). The authors could not find any tendency for association between the alcohol molecules.

According to some authors, the experimentally found effective carbon number in glycols is higher by more than 5 units from that theoretically calculated. This may be due to the binding of two tetrahydrofuran molecules, i.e. to the formation of one hydrogen bond for each hydroxyl group. Mercaptanes, which are closely related to alcohols, show practically no association with tetrahydrofuran, the number of their hydrogen bonds being less than 10% of the theoretically possible value.

Table 40

Characteristics of the gel chromatographic behaviour of alcohols and ethers

Compound	Effective C number*		Molar volume**, $cm^3 mol^{-1}$		
	calculated	empirical in tetrahydrofuran	calculated in tetrahydrofuran	empirical in tetrahydrofuran	in o-dichlorobenzene
Methanol	1.67	5.58	39	114	46
Ethanol	2.67	5.78	57	131	67
Propanol	3.67	6.57	77	152	85
Isopropanol	3.67	6.89	74	148	90
n-Heptanol	7.67	10.29	139	213	156
Ethylene glycol	3.33	8.20	62	136 210	—
Propylene glycol	4.33	7.97	—	—	—
n-Hexyl ether	12.67	—	230	—	—
Allyl ether	6.67	—	115.3	—	—
Anisole	4.52	—	105.2	—	—

* Hendrickson and Moore (1966).
** Cazes and Gaskill (1968).

The experiments of Edgards and Ng (1968) confirm these findings in relation to the association of alcohols and tetrahydrofuran. To make the points of the elution volume fall on the calibration line plotted with alkanes, the molar volume of the alcohols must be corrected by an average value of 52 ml. No correction is required for ethers on the other hand, because they do not associate with tetrahydrofuran and run in the polystyrene gel column at approximately the same rate as alkanes.

Edgards and Ng (1968) proved by experiment that benzyl alcohol shows extraordinary behaviour. It does not associate with tetrahydrofuran, but presumably by interaction with the π electrons of the aromatic ring, the hydroxyl group participates in intermolecular hydrogen bond formation.

According to Hillman (1971), in such cases alcohols adsorb on the polystyrene gel in the same way as acids do, and therefore migrate more slowly than normal hydrocarbons of corresponding molecular weight.

During chromatography from o-dichlorobenzene, alcohols do not interact with the solvent (see *Table 40*).

The adsorption on Sephadex LH-20 gel of cholesterin, and other cholestane derivatives containing hydroxyl groups from a benzene solution, was exemplified by the experiments of Brooks and Keates (1969). By chromatography from benzene–ethanol or benzene–isopropanol solvent mixtures they found that the elution volumes decreased with increasing alcohol concentration of the mixture.

As will be evident from the case of methanol and benzyl alcohol, aliphatic and aromatic alcohols are capable of binding on Sephadex LH-20 gel from both tetrahydrofuran and acetonitrile (Streuli 1971a).

Inoue *et al.* (1970) state that in the chromatography of oligomer fatty alcohols on Sephadex LH-20 gel from dimethylformamide, adsorption does not take part in the process and the K_d values are below 1. As in true gel chromatography, the elution sequence conforms to decreasing molecular weight *(Fig. 115)*. The authors studied the effect of the flow rate, temperature and concentration of the sample on the separation, and expressed the result by the following formula

$$R = \frac{2t}{b_1 + b_2}$$

where t denotes the distance between adjacent elution peaks, and b_1 and b_2 are the widths of the elution zones. The results are shown in *Figure 116*.

Klimisch and Stadler (1972), carrying out distribution chromatography on aqueous methanol–hexane and methanol–benzene solvent mixtures on Sephadex LH-20 gel, established that while ethers, when eluted with hexane (or benzene), leave the column after paraffins and terpenes, polyalcohols behave like other hydrophilic compounds and are eluted only after subsequent washing with a methanol–water mixture.

The distribution of polymer ethers by molecular weight, for instance of polyethylene glycols, tenzides or epoxy resins, on polystyrene gel columns can be analysed by chromatography from tetrahydrofuran. For this pur-

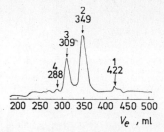

Fig. 115. Gel chromatographic fractionation of thermal-polymerized fatty alcohols on Sephadex LH-20 column from dimethylformamide (Inoue et al. 1970)
1. monomer; 2. dimer; 3. trimer; 4. tetramer fatty alcohol

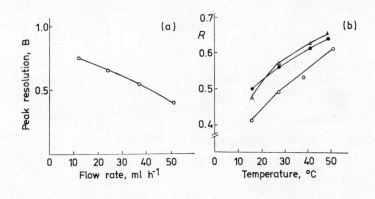

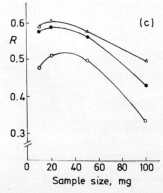

Fig. 116. The effect of (a) the flow rate, (b) the temperature and (c) the sample size on the gel chromatographic separation of polymer fatty alcohols, fatty acids and fatty acid methylesters on Sephadex LH-20 gel from dimethylformamide (Inoue et al. 1970)
(●) polymer fatty alcohols; (○) polymer fatty acids; (△) polymer fatty acid methylesters

pose the equipment can be calibrated with aliphatic and aromatic ethers (Larsen 1968) or oligomers of known molecular weight (Moore 1964; Heitz 1968). Also the oligomer members of the homologous series can be fractionated from one another. Slagt and Sipman (1972) found that fractionation from methanol on methylated Sephadex G-25 dextran gel is suitable for the thin layer gel chromatography of polyethylene glycol.

5.4. GEL CHROMATOGRAPHY OF PHENOLS

From the point of view of gel chromatography phenols occupy an exceptional place. On certain gels they may bind by aromatic adsorption, their hydroxyl group may bind solvent molecules or groups in the gel matrix by hydrogen bonding. In the dissociation phenolate ion form the molecule is surrounded by a strong ionic double-layer which prevents sorptive interaction between the π electrons and the gel matrix. During chromatography from aqueous solutions, owing to changes in acidic dissociation, the elution volume is pH-dependent. This may account for the fact that, for example, tyrosine in an alkaline solution preceedes phenylalanine and there is an inverse elution sequence if a neutral solution is applied (Gelotte 1960; Eaker and Porath 1967; Janson 1967). It was accepted even in the initial period of gel chromatography that phenols are capable of binding to dextran gels (Gelotte 1960; Porath 1960). Woof and Pierce (1967) found that, on the basis of their dissimilar adsorptivities, phenols may be fractionated on Sephadex G-25 gel even from a distilled water solution. In such cases the

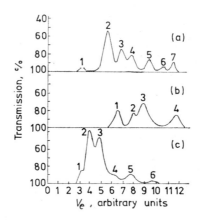

Fig. 117. Fractionation of phenols on Sephadex G-25 column (Woof and Pierce 1967). (a) elution with distilled water on a gel column of 35×2.5 cm; 1. hydroquinone; 2. phenol; 3. catechol plus resorcinol; 4. guaiacol; 5. pyrogallol; 6. cresol; 7. orcinol (b) Fractionation from 0.1 M acetic acid on a gel column of 35×2.5 cm; 1. phenol; 2. hydroquinone; 3. resorcinol plus catechol; 4. pyrogallol plus phloroglucinol (c) Gel chromatography from 0.1 M acetic acid containing 0.05 M NaCl on a 0.9×25 cm column; 1. saligenin; 2. nitrophenol plus guaiacol; 3. salicylaldehyde; 4. catechol; 5. cresol plus chlorophenol; 6. salicylic acid

phenols are eluted according to their increasing hydroxyl group content, with the exception of the fastest migrating hydroquinone *(Fig. 117)*. The authors established that, in chromatography from a 0.05 M NaCl solution, phenols fall into two groups. One of these consists of ortho-substituted catechinol and pyrogallol. The K_d value of these compounds is less by 0.7 unit than that of phenol, resorcinol and phloroglucinol, which belong to the second group. The K_d value of hydroquinone is between the values of catechinol and resorcinol. In an alkaline medium, i.e. a 0.1 M ammonium hydroxide solution, sorption — and therefore the K_d value — is lower, while in acid media, i.e. in acetic acid solution, hydrogen bonding is stronger.

The same authors examined the chromatographic migration of phenols containing hydroxyl groups in the ortho-position also in a solution which contained molybdate ion capable of forming complexes. With catechinol they found an increase in the K_d value, with pyrogallol they experienced a drop in the value of K_d compared to chromatography in an NaCl solution. Woof and Pierce (1967) used this method also for fractionation of flavonoids, with good results. For the elution of isoflavones, Nilsson (1962) used an ammonium hydroxide solution.

Brook and Housley (1969a) suggest that in the binding of phenols on dextran gel the hydrogen bond formation plays the decisive role. With undissociated phenol the hydrogen of the phenolic hydroxyl group may become bound to the oxygen of the oxypropylene bridge linking the dextran chains. With phenolate ion, a hydrogen bond may evolve between the phenolic oxygen and the hydrogen of the hydroxy group of the gel matrix. The two bonds are not equivalent, therefore the elution volume is a function of the dissociation of phenol *(Fig. 118; Table 41)*. Using phenols containing various substituents, these authors examined whether Hammett's linear free energy relationship is valid also for the K_d values which characterize the assumed interactions with gels, regarded as equilibrium constants. They determined the distribution coefficients of phenols with different substituents, and, on the basis of the distribution coefficient of phenol (K_{d0}), they plotted the values K_d/K_{d0} as a function of the Hammett constant (σ).

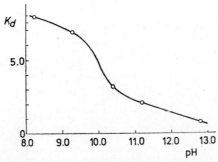

Fig. 118. The pH dependence of the K_d value of phenol in gel chromatography on Sephadex G-10 (Brook and Housley 1969a)

Table 41

The volumetric distribution coefficient (K_d) of selected phenols chromatographed on Sephadex G-10 gel*

Compound	K_d	
	in acetate buffer at pH = 4	in 0.1 M NaOH solution
Phenol	0.58	7.2
m-Cresol	0.72	10.2
p-Cresol	0.76	10.1
p-Ethylphenol	0.93	10.6
Resorcinol	—	12.7
p-Nitrophenol	5.41	16.9
p-Chlorophenol	2.85	22.8
m-Chlorophenol	2.38	25.6
m-Bromophenol	3.20	36.1
p-Bromophenol	3.90	32.5
p-Iodiophenol	6.00	57.8

* From Brook and Housley (1969a).

To a fair approximation the points fall on a straight line *(Fig. 119)*, with the exception of the halogen-substituted phenols (Brook and Housley 1969a; Brook and Munday 1970a). In these authors' opinion, this is due to interactions other than hydrogen bond formation.

Examining the binding of aromatic compounds on Sephadex LH-20 gel, Streuli (1971a) established that the distribution coefficient of phenols is

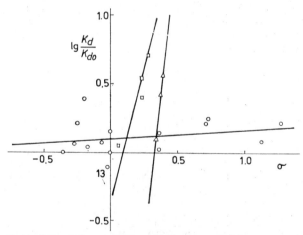

Fig. 119. Hammett-plot of the sorption of substituted phenols on Sephadex LH-20 gel (Brook and Munday 1970a)

(□) m-halogen-substituted phenols; (△) p-halogen-substituted phenols; (○) phenols containing substituents of another type

higher than the values calculated from the resonance energy of the π electrons. He attributes this effect to the hydrogen bond formed with the gel.

According to Determann and Lampert (1972), the elution volume of phenols, chromatographed from butanol on a lipophilic dextran gel, decreases with rising temperature *(Fig. 120)*. This can be explained by sorption through the assumed hydrogen bonds.

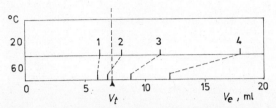

Fig. 120. The effect of temperature on the migration of benzene, benzoic acid and phenols on Sephadex LH-20 gel from butanol (Determann and Lampert 1972)
1. benzene; 2. phenol plus benzoic acid; 3. resorcinol; 4. phloroglucinol. The arrow indicates the total volume of the column

Somers (1966), in the chromatography of the colouring matter of wine from a mixture (4 : 6) of ethanol containing 0.1 % hydrochloric acid and water on Sephadex G-25 dextran gel, succeeded in partially fractionating tannins and anthocyanins. Laird and Synge (1971) chromatographed various plant substances from a phenol–acetic acid–water (1 : 1 : 1) mixture on Sephadex G-25 gel on the assumption that phenol present in the solvent will inhibit the bonding of other substances and force these out from the bonding points of the gel. However, the results failed to confirm their assumption. The K_d values of polyphenols and of several other compounds were not in agreement with the values theoretically calculated from their molecular weights.

For the chromatography of phenols, polystyrene gels are also frequently used. Cortis-Jones (1961) examined the rate of migration of phenols and other small-molecule compounds from different solvents on styrene-divinylbenzene copolymer resin. From his findings he concluded that the rate of elution depends not only on the molecular weight but also on aromatic adsorption. According to his data, in benzene the elution sequence of quinalizarin, alizarin, α-naphthol and resorcinol corresponds to decreasing molecular weight but, compared to the small molecules like acetaldehyde or ethylamine, the elution volumes are nevertheless relatively high; those of acetaldehyde and alizarin, for instance, coincide. From the faster elution of aliphatic alcohols (ethanol, n-pentanol) the author drew the conclusion that the hydroxyl groups cause exclusion. The fact that from ethylacetate or acetone, resorcinol precedes 1-naphthol and naphthylamine was explained by the higher exclusion effect caused by the larger number of hydroxyl groups present in the resorcinol molecule.

Hendrickson and Moore (1966) chromatographed phenol from tetrahydrofuran on polystyrene gel and found that its elution volume is much smaller than the value calculated theoretically from the effective carbon number

Table 42

The effective C number of selected phenols chromatographed from tetrahydrofuran on polystyrene-divinylbenzene copolymer gel*

Compound	Effective C number	
	calculated	empirical
Phenol	3.52	7.53
p-Chlorophenol	4.51	9.15
o-Cresol	4.52	8.07
m-Cresol	4.52	8.39
p-Methoxyphenol	5.19	10.44
Hydroquinone	3.73	8.61
β-Naphthol	4.68	8.88

* Hendrickson (1968).

(Table 42). From their measurements the authors concluded that phenol associates with the solvent molecules in the same way as alcohols, but unlike alcohols it is capable of binding two tetrahydrofuran molecules. This means that in addition to the phenolic hydroxyl group, hydrogen in the p-position is also capable of participating in the hydrogen bonding.

Edgards and Ng (1968) carried out chromatography from tetrahydrofuran solution, also on polystyrene gel. They verified the validity of the assumption with regard to the hydrogen bond between the phenolic hydroxyl group and the tetrahydrofuran molecule. They established that

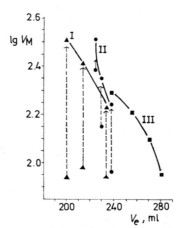

Fig. 121. Relationship between the elution volume and the logarithm of the molar volume of phenols and aromatic amines in gel chromatography from tetrahydrofuran on polystyrene gel (Klimisch and Reese 1972). The dotted lines indicate the corrections made for association with the solvent molecule

I. phenols (phenol, 2-naphthol, resorcinol, pyrogallol); II. aromatic amines (aniline, 1-aminonaphthalene, 1-aminoanthracene); III. unsubstituted kata-condensed aromatic compounds (benzene, naphthalene, anthracene)

with phenols, as in the case of alcohols, to make the elution volumes (plotted as a function of the logarithm of the molecular volume) fall on the straight line which characterizes hydrocarbons, the molar volume must be corrected by 52 ml on average for each hydroxyl group present in the phenol. That such a correction is required was confirmed by the experiments performed on various aromatic compounds by Klimisch and Reese (1972) *(Fig. 121)*.

More recently Coupek *et al.* (1974) have studied the correction values for phenols and the influence of various substituents on the ΔV_e value.

According to Klimisch and Stadler (1972), who performed distribution chromatography with methanol–water/hexane solvent mixture on Sephadex LH-20 gel, phenols emerge from the column together with the hydrophilic compounds during methanol–water elution following the hexane rinsing. The same is valid if benzene is used instead of hexane.

The separation of phenolic compounds on polyacrylamide-type gel by recycling chromatography was studied recently by Kalász *et al.* (1975) (cf. Section 3.1 of Part II).

On a Merckogel PGM 2000 column, Seki (1975) separated the three isomers of dihydroxybenzene from borate-containing phosphate buffer; the V_e value of pyrocatechol decreases in the presence of borate, probably owing to the negative charge of its borate complex.

5.5. GEL CHROMATOGRAPHY OF ORGANIC BASES

The gel chromatographic behaviour of organic bases is governed by the same factors as those for organic acids and phenols. Thus in an aqueous solution, positively charged molecules are surrounded by an ionic shell which causes exclusion from the smaller-pored gels. This effect is dependent on the pH and proportional to the degree of dissociation of the base. Performing gel chromatography from organic solvents, depending on the experimental conditions, the formation on hydrogen bonds between the solvent and the bases or between the gel matrix and the base must be reckoned with.

Examples of the chromatography of organic bases in aqueous solutions have been presented in previous sections, dealing with amino acids and nucleic acid bases.

The behaviour of aromatic bases during chromatography from aqueous solutions on Sephadex G-10 dextran gel was extensively studied by Brook and Housley (1969a). They verified by experiment that, in their protonated positively charge state, the bases examined were excluded from the gel, but, in the dissociated (deprotonated) state they are strongly sorbed on the gel. The pH-dependence of the K_d values is shown in *Figure 122*. According to the authors, the ionization constant (base exponent) of a base can be established from the pH-dependence of its behaviour in gel chromatography. If the compound contains both basic and acidic groups, the pH-dependence of the K_d value conforms to a maximum curve, as, for example, in the case of p-aminobenzoic acid.

If quaternary ammonium compounds are gel chromatographed, the hydrogen ion concentration has no effect on the elution volume. This was confirmed by the experiments of Crone (1971) who stated that the migration rate of the quaternary ammonium derivatives tested on Sephadex G-10 gel depends both on the molecular size and on the adsorption based on aromatic or hydrogen bonds. If the compound contains also a group capable of acid

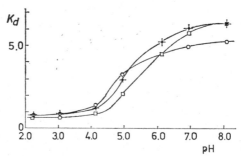

Fig. 122. The pH-dependence of the K_d values of aniline derivatives chromatographed on Sephadex G-10 gel (Brook and Housley 1969b)
(○) aniline; (+) N-methylaniline; (□) N,N-dimethylaniline

dissociation, e.g. an oxime group, as well, the K_d value will depend on the pH and decrease with increasing pH, i.e. the formation of the dipolar ion.

In chromatography from tetrahydrofuran on a polystyrene gel column the association of the amines with the solvent molecules yields smaller elution volumes. *Table 43* shows the calculated and empirical values of the effective carbon numbers of some bases according to Hendrickson (1968).

According to Edgars and Ng (1968), the effective molar volumes of primary amines calculated from their elution volumes are higher by 37 ml on average than those calculated from the composition of the molecule. The delayed elution of the primary aliphatic amines is attributed by these authors to the fact that, like water, they can be regardeds̆ poor solvents for the gel and cause less swelling than tetrahydrofuran. When the sample

Table 43

The effective C number of selected organic bases chromatographed from tetrahydrofuran on styrene-divinylbenzene copolymer gel

Compound	Effective C number	
	calculated	empirical
Methylamine	2.0	2.0
Aniline	3.76	6.68
Diphenylamine	6.51	8.59
Phenyl-β-naphthylamine	7.66	9.30
Dibenzylamine	8.71	8.79
Dibenzylphenylamine	10.36	9.95

is applied to the column the gel contracts, "closes in" the already diffused amino molecules, then, as the zone passing farther on, it swells again, releasing the entrapped molecules. This may cause tailing, similar to adsorption. Aromatic and tertiary amines do not show elongated elution zones. The effective molar volume of tertiary amines as established by gel chromatography is also not larger than, and sometimes even slightly smaller than, the theoretically calculated value. Regarding secondary aliphatic amines, the difference between experimental and calculated values varies widely. The average difference between the molar volumes for secondary aromatic amines was 18 ml.

Klimisch and Reese (1972) examined the chromatography of aromatic amines and N-heterocyclic compounds, also on polystyrene gel, from tetrahydrofuran. Their findings confirm the assumption that the tetrahydrofuran molecule is bound to aromatic amines by hydrogen bond. The authors observed a similar phenomenon in the case of carbazole-type heterocyclic compounds but they could not detect such phenomena in any other N-heterocyclic compound. The elution volume of the latter falls somewhere between the values characteristic of unsubstituted aromatic and of aliphatic hydrocarbons. This, according to the authors, indicates that the effect of the N-heteroatom — and the reduced aromatic character — causes a decrease in the interaction between the π electrons and the gel.

According to Streuli (1971a) aromatic amines when chromatographed from methanol on lipophilic Sephadex LH-20 gel show a stronger binding to gels than would be expected if it were calculated from the resonance energy of the π electrons. According to the authors the oxygen atoms of the gel matrix bind the hydrogen of the amine and contribute thus to the interaction with the π electrons. From the differences between the theoretical and experimentally detected distribution coefficients, Streuli calculated the energy of the assumed hydrogen bonds *(Table 44)*.

Table 44

The energy value of hydrogen bonds between aromatic and heterocyclic compounds and Sephadex LH-20 gel

Compound	Energy of bond, kcal
Anisole	10
Phenol	20
Aniline	17
N-methylaniline	16
N,N-dimethylaniline	13
α-Naphthylamine	18
β-Naphthylamine	16
Pyridine	11
Quinoline	8
Pyrrole	32
Indole	30
Carbazole	23

From Streuli (1971a).

Dimethylformamide as a solvent suppresses the interaction of π electrons with Sephadex LH-20 gel but does not disrupt the binding through the hydrogen bonds (Streuli 1971b). On the basis of their elution volume, a weak hydrogen bond forms under such conditions between the N-heterocyclic compounds and the gel matrix. The relatively strongly basic sec-butylamine and morpholine form a stronger bond to the gel than do aniline derivatives. The aromatic interaction with the material of the Sephadex LH-20 gel developing from tetrahydrofuran does not take place, whereby the hydrogen bond becomes predominant. According to Streuli (1971b), under such circumstances the aliphatic amines form a stronger bond than the aromatic types. Furthermore the interaction between the primary amines and the gel is stronger than in the case of secondary and tertiary amines. Aniline forms a stronger bond than naphthylamine, probably owing to the latter's larger molecular size. The N-heterocyclic compounds form a weak bond with the gel material but triazole forms a strong one.

In the course of distribution chromatography on Sephadex LH-20 gel from methanol–water/hexane solvent mixtures the amines were located in the stationary aqueous phase and their elution was performed by subsequent washing with a water–methanol solvent mixture, as in the case of amino acids and other hydrophilic compounds (Klimish and Stadler 1972).

6. GEL CHROMATOGRAPHY OF INORGANIC IONS

Previous sections have proved that inorganic salts can be separated from large-molecule substances, provided a gel of such pore size is used into which the macromolecules cannot diffuse. Gel chromatography, in addition to this purpose, can be used with advantage also for the separation of inorganic salts or of ions. The migration of inorganic ions in gel columns depends on several factors which will be surveyed below.

6.1. GEL CHROMATOGRAPHY OF INORGANIC SALTS FROM A DISTILLED WATER SOLUTION

For the gel chromatography of inorganic salts, small-pored gels are usually used in which a considerable portion of the water volume constitutes the hydrate shell of the gel matrix, forming a volume inaccessible even to small ions. Compared to the entire water space inside the gel, the quantity of water bound to the gel, in large-pored gels, is negligible, but in Bio-Gel P-2, Sephadex G-10 and G-25, it may account for as much as 20% of the water adsorbed by the gel. Gelotte (1960) demonstrated that, on a Sephadex G-25 column, the K_d value of neutral salts is around 0.8. According to Pecsok and Saunders (1968), the K_d value of acetone and ethanol, chromatographed on a Bio-Gel P-2 acrylamide gel column, is 0.81 and 0.77, respectively. Accordingly, K_d values are higher than 0.8 and indicate a sorption on this small-pored gels.

According to recent reports by Ortner and Pacher (1972), small-pored gels differ from the types with fewer crosslinks also in their lower carboxyl group content per unit weight of gel. These authors also established that there is a linear correlation between the swelling volume and the carboxyl-group content of the gel *(Table 45)*. Their measurements proved also that the treatment of dextran gels by pyridine causes o anly partial reduction in the carboxyl content, which shows that, contrary to the assumption of other authors, these gels contain not only adsorbed oleate ions which can be removed by pyridine but also covalently bound carboxylic groups.

Negatively charged carboxylate ions show an electrostatic attraction to cations and repel anions. Spitzy et al. (1961) demonstrated that small amounts of zinc and cobalt ion applied to a Sephadex G-25 column from an aqueous solution form a strong bond at the top of the column and are not

Table 45

Carboxyl content of small-pored gels

Gel	COOH content		
	xerogel*, μequ g^{-1}	xerogel washed in pyridine*, μequ g^{-1}	swollen gel**, μmol ml^{-1}
Sephadex G-10	4.4	2.1	6
Sephadex G-15	5.4	2.1	—
Sephadex G-25	7.9	3.0	—
Bio-Gel P-2	4.9	—	1.5
Sephadex LH-20	3.9	—	—

* Orther and Pacher (1972).
** Pecsok and Saunders (1968) (data refer to gels not treated with pyridine).

eluted. Under the same conditions, iodide ions are eluted at the point corresponding to the void volume of the column, i.e. they are excluded from the gel, in a similar manner to macromolecules. This exclusion, however, is not a result of reduced diffusion arising from molecular size, but the result of electrostatic repulsion, i.e. ion exclusion.

Experimental work has shown, however, that the situation is not this simple, because certain anions also form bonds with gels, and because cations too may become bound to sorbed anions, while the anions excluded from the gel reduce the diffusion of cations into the gel in conformity with the Donnan equilibrium. These effects can be partially compensated if foreign electrolytes are also present. Chromatography from electrolyte solutions and the effect of the so-called background electrolytes will be dealt with in the next section.

As proved experimentally by Saunders and Pecsok (1968), in gel chromatography from a distilled water solution salts can be characterized by K_d values which consist of two factors, Φ_k characteristic of the cation and Φ_a characteristic of the anion according to the following expression

$$K_d = \frac{s\Phi_k + t\Phi_a}{s + t}$$

where s denotes the charge of the anion and t that of the cation.

The same authors studied the chromatographic behaviour of salts containing the same anion and various cations (alkali metal, alkaline earth metal, copper, nickel, and aluminium ions), and those containing an identical cation but various anions (halide, nitrate, perchlorate, and sulphate anions) and determined the Φ_k and Φ_a values *(Table 46)*. *Table 47* shows the K_d values for some salts.

The value of K_d depends on the concentration of the sample. If salt solutions of less than 0.1 M are applied to the column, in agreement with the Donnan equilibrium the effect of the carboxylate ions of the gel causes

Table 46

Φ_k and Φ_a values of cations and anions*

Cation	Φ_k		Anion	Φ_a	
	A	B		A	B
Li^+	1.25	1.23	F^-	0.96	0.75
Na^+	1.00	1.00	Cl^-	1.63	1.44
K^+	0.88	0.90	Br^-	2.16	1.98
Cs^+	0.93	0.94	I^-	2.82	2.65
Mg^{2+}	1.76	1.63	NO_3^-	2.25	2.14
Ca^{2+}	2.48	2.28	ClO_4^-	—	2.71
Sr^{2+}	2.37	2.09	SO_4^{2-}	1.43	1.05
Ba^{2+}	2.61	2.31			
Cu^{2+}	—	2.31			
Ni^{2+}	—	2.06			
Al^{3+}	5.82	3.36			

* Saunders and Pecsok (1968).
Note: The absolute values of Φ_k and Φ_a are unknown, the values in the table are related to that of the Na^+ ions whose Φ_k value was regarded by Saunders and Pecsok as unity. Column A shows the results of chromatography of 0.005 M, column B those of the chromatography of solutions of 0.01 M concentration.

Table 47

The volumetric distribution coefficient (K_d) value of some inorganic salts chromatographed from deionized water

Salt	K_d					
	Bio-Gel P-2*				Sephadex	
	23 °C	37 °C	49 °C	65 °C	G-10**	G-20**
LiCl	1.38	—	—	—	0.56—0.62	0.92—0.95
NaF	0.91	—	—	—	—	—
NaCl	1.27	1.49	1.14	1.10	—	—
NaBr	1.44	—	—	—	—	—
NaI	1.76	1.49	1.33	1.26	—	—
$NaNO_3$	1.64	—	—	—	—	—
Na_2SO_4	1.07	1.00	0.99	0.82	0.40—0.48	0.86
KF	0.88	—	—	—	—	—
KCl	1.26	1.18	1.12	1.05	0.86—0.94	1.03—1.09
KBr	1.44	—	—	—	1.18—1.19	1.16—1.19
KI	1.63	—	—	—	1.9 —2.04	1.28—1.34
$MgCl_2$	1.53	1.35	1.39	1.18	—	—
$BaCl_2$	1.64	1.56	1.41	1.32	—	—

* Saunders and Pecsok (1968).
** Zeitler and Stadler (1972).

a higher degree of ion exclusion (Pecsok and Saunders 1968; Neddermeyer and Rogers 1968). In such cases the elution curve is asymmetric: first flat but sharp at the end. If the solutions applied are of higher concentration, the curve will be symmetric or slightly tailing backwards *(Fig. 123)*.

If two salts with different cation and anion contents are applied to the column and if the concentration of the sample is high enough to counter

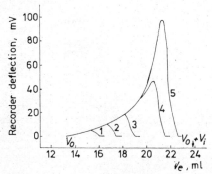

Fig. 123. The effect of the concentration of the sample on the location and shape of the chromatographic peak in chromatography from distilled water on a Sephadex G-10 column (0.61×126 cm) (Neddermeyer and Rogers 1968)

The concentrations of the sodium chloride sample solution: 1. 0.01 M; 2. 0.025 M; 3. 0.050 M; 4. 0.1 M; 5. 0.2 M
Sample volume: 0.05 ml

the effect of fixed ions, then the rate of migration in the column depends both on the relative values of Φ_k and Φ_a and the relative quantities of the salts (Saunders and Pecsok 1968). This is illustrated by the series of experiments shown in *Figure 124*; $BaCl_2$ and $KClO_4$ when separately gel chromatographed, emerge at practically the same elution volume and the difference of their experimentally established K_d values is negligible (1.64 and 1.77, respectively). In the chromatography of a stoichiometric mixture of the two salts, a KCl peak will appear earlier, while with larger elution volume will be a $Ba(ClO_4)_2$ peak because the K_d value of KCl is 1.26, which is lower than that of the salts $KClO_4$ and $BaCl_2$. At the same time the K_d value of the $Ba(ClO_4)_2$ salt is higher, namely 2.63, as proved by the experiment.

This phenomenon favours the application of gel chromatography for preparative purposes in inorganic chemistry: from salts applied in stoichiometric quantities, salts of other, different, composition will emerge from the column. If the mixing ratio of the salts is not stoichiometric, a third chromatographic peak will also be observed. This contains the excess salt, following the exchange reaction ($BaCl_2$) in the case illustrated in *Figure 124*.

The fact that the individual cations as well as anions can be characterized by different migration rates is a consequence of the various interactions between the gel material and the ions. Using the chromatographic term, the selectivity of the gels varies with the different ions. The elution volume of the various ions is not a unique function of the size of the hydrated ion. This is clearly shown by Egan's experiments (1968). The elution sequence *(Fig. 125)* of the cations examined by Egan on a Bio-Gel P-2 column, when applied in the form of chloride salts was as follows

$$K^+, Na^+, Li^+, Mg^{2+}, Ca^{2+}$$

On the basis of decreasing ion diameter this should have been

$$Mg^{2+}, Ca^{2+}, Li^+, Na^+, K^+$$

The elution sequence from the chromatography of nitrate salts on an acrylamide gel column was

$$Cs^+, Na^+, Tl^+, Mn^{2+}, Ag^+$$

The sequence on the basis of the hydrated size of the ions should have been

$$Mn^{2+}, Na^+, Ag^+, Tl^+, Cs^+$$

Egan (1968), performing chromatography on the larger pored Bio-Gel P-100 acrylamide gel, found that the elution volume of the salts is not completely equal and that the elution sequence is similar to that in small-pored gels.

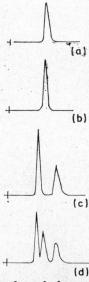

Fig. 124. Exchange reaction during the gel chromatography of a mixture of $BaCl_2$ and $KClO_4$ from distilled water (Saunders and Pecsok 1968)

(a) The chromatogram of $BaCl_2$ alone; (b) the chromatogram of $KClO_4$ alone; (c) the chromatogram obtained when a stoichiometric mixture of the two salts was applied to the column; the first peak contained KCl, and the second $Ba(ClO_4)_2$; (d) the chromatogram obtained when an equimolar mixture of the two salts was applied to the column; the first peak contained KCl, the second (in the middle) $BaCl_2$ and the third $Ba(ClO_4)_2$

These facts indicate that there are interactions between the gel matrix and the cations which can be attributed to ionic or complex bonds. If small-pored dextran gels and acrylamide gels are used, the effects are different (Egan 1968; Ortner and Pacher 1972). Although dextran gels contain generally a larger number of carboxyl groups than do acrylamide gels, from solutions of lower concentration the latter are capable of binding cations more strongly. This is supported by the experiments of Ortner and Pacher (1972). These authors attribute this phenomenon to the combined effect of the carboxyl and carboxamide groups of the gel and the formation of complex bonds if zinc and cobalt ions are concerned, which are susceptible to the formation of complex compounds. It might be of interest that Sephadex G-25 pretreated with pyridine behaves differently, a stronger bond is formed with metal ions on this gel compared with both acrylamide and untreated dextran gels.

Pecsok and Saunders (1968) assume a weak physical bond rather than a chemical one; i.e. that, during sorption of the gel, there is no change taking place in the first coordination sphere of the metal ion. As shown by the determination of the Φ_k values, this weak physical sorption depends first and foremost on the charge of the metal ion. The Φ_k values of ions with the same number of charges show little difference, those for ions with two or three times greater charge are two or three times higher (see *Table 46*).

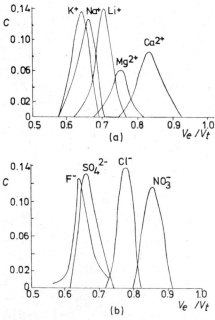

Fig. 125. The position of the elution of (a) some cations and (b) some anions chromatographed from distilled water on a 2.5×223 cm Bio-Gel P-2 column (Egan 1968). Cations were applied to the column in the form of chloride salts, anions in the form of sodium salts

In their spectrophotometric and NMR (nuclear magnetic resonance) experiments, the authors observed no changes which would have indicated the formation of a chemical bond between soluble polyacrylamide and metal ions. Of course, this does not preclude with complete certainty the possibility of the binding by complex bonds of some metal ions to gel surfaces containing carboxyl and carboxamide groups in close proximity.

Owing to the cation-binding capacity of gels, the products commercially available might be contaminated by metal ions. Morgan *et al.* (1972) found a few micrograms of zinc in one gram of Sephadex product. They assume that this is caused by bonding of the Zn^{2+} ion on glycopyranosides.

Henry and Rogers (1968) observed an interesting case of interaction with the gel. In the chromatography of iron(III) nitrate, hydrolyzed to different degrees, on dextran gels, they found one form with a high molecular weight (polymerized) excluded from the gel, and another, of lower molecular weight which eluted at the point corresponding to the total volume of the column. By increasing the pore size of the gel, they obtained invariably the same two peaks, while separation became increasingly sharper, and the height of the first peak became lower. They presume that what fractionation reflects is not separation according to the actual molecular size but a separation process caused by adsorption and depolymerization processes which take place in the gel pores.

Gels are selective also to anions. Here two opposite effects may exist: the repulsion of the negative ions on the one hand and the forming of non-ionic bonds with the gel material on the other. The former appears mostly if dilute salt solutions are applied. On the basis of the Donnan equilibrium, Pecsok and Saunders (1968) calculated the expected variations of the K_d values, applying samples in different concentration to Sephadex G-10 and Bio-Gel P-2 gels, and found that the experimental points agree with the theoretical curve calculated for Sephadex G-10 but deviate considerably in the case of the acrylamide gel. A possible explanation being that the carboxyl groups of the gel are not randomly distributed in the volume accessible to the inorganic ions in the gel.

Observations of the sorption of some anions on the gel began quite some time ago. Gelotte reported as early as 1960 the slow migration of hydroxide and tetraborate ions on Sephadex columns. According to Egan (1968), sodium salts containing sulphate, fluoride, chloride and nitrate anions show different elution sequences when chromatographed on Sephadex G-10 and Bio-Gel P-2 columns, which was certainly not what would be expected on the basis of the sizes of the hydrated anions — namely sulphate, fluoride, nitrate and chloride. *Figure 125b* illustrates chromatography on a Bio-Gel P-2 column.

Zeitler and Stadler (1972) observed that, in dextran gel columns, alkali-halide salts migrate at different rates. They indicated also that thiocyanate ion is absorbed on dextran gels and is eluted with considerable delay.

The selectivity of gels towards different anions varies according to the different matrix materials.

From NMR measurements, Pecsok and Saunders (1968) established that the proton of amide in the "trans" position is capable of producing a hydrogen bond with perchlorate anion in polyacrylamide. This may explain the sorption in acrylamide gels of anions able to participate in the formation of hydrogen bonds. Hydrogen bonds may form between the hydroxyl groups of the gel and individual anions even on dextran gels.

The chromatographic migration in aqueous solutions of V(VI), Mo(VI) and W(VI) anions is affected by the pH-dependence of polyacids and that of formation of chelate complexes with the dextran gel matrix (Ortner et al. 1973; Ortner and Dalmonego 1974, 1975; Ortner 1975a). Rhenium(VI) forms a chelate complex only at pH 3, but the K_d values of the perrhenate anion in neutral and alkaline media are also surprisingly high, similar to that of the perchlorate anion, indicating a similarity in the specific sorption of these anions (Oertner 1975b).

Considering what has been discussed above, it should be stressed that, in the course of gel chromatography from a distilled water medium, the effects on cations and anions act in combination. We can speak, therefore, on the migration rate of *salts*, and not on that of individual ions. This will be evident from *Table 47*.

The interactions between the gel material and the ions depend on temperature. This is clearly indicated by the decreasing K_d values of salts at higher temperatures (see *Table 47*). From a thermodynamic analysis of the experimental results, Pecsok and Saunders (1968) established that the change in enthalpy caused by the sorption of BaI_2 is -3 ± 1 kcal/mol. According to these authors, since hydrogen bond formation between the anions and the acrylamide gel is characterized by a value of -5 kcal/mol, and that of the physical sorption of cations by an enthalpy change of less than 6 kcal/mol, the gross sorption energy for the various salts should represent a value of -3 to -4 kcal/mol.

The changes of K_d values with concentration are explained by various authors on different theoretical bases. It is a fact that the gel chromatographic migration of inorganic ions in distilled water is not a simple molecular sieving, but a more complex function of such additional effects as ion exchange, ion exclusion, Donnan distribution, complex formation and weak adsorption. The dominant role of any one of these effects depends on the conditions.

6.2. THE INFLUENCE OF BACKGROUND ELECTROLYTES ON THE BEHAVIOUR OF INORGANIC IONS DURING GEL CHROMATOGRAPHY

The interference of the carboxylate ions of the gels may be eliminated by chromatography in electrolyte solutions instead of distilled water. In such cases the gel column is washed with an eluant containing 0.01—0.1 M NaCl, or another electrolyte, and the sample is dissolved in the same solution.

With a correctly selected background electrolyte, the cations and anions will pass through the column without binding at a rate corresponding to the diameter of the hydrated ion. This means that they are fractionated according to the size of the hydrated ions. A proper choice of the background electrolyte means the selection of one whose ions diffuse deeper into the gel pores than the ions to be chromatographed. Yoze and Ohashi (1969) called attention to the fact that, if background electrolytes are used, the inner space of the gel can be divided into two phases — one containing the ions of the electrolyte, the other inaccessible to them. This latter can therefore be regarded as a distilled water phase. Accordingly, the gel has a "polyfunctional" character. In the chromatography of ions which are capable of diffusing into the "distilled water" phase of the gel, the behaviour will be anomalous insofar as there will be a change in the position of the elution peak, the elution curve becoming asymmetric and tailing taking place.

In the chromatography of alkaline earth metal ions on a Sephadex G-25 column, Yoza and Ohashi (1969) found that, using a background electrolyte of 0.1 M NaCl, the only elution curve which is symmetric is that for the Mg^{2+} ion. Under these conditions, the K_d values of the cations will increase in the following sequence

$$Mg^{2+} < Na^+ < Sr^{2+} \approx Ca^{2+} < Ba^{2+}$$

which means that the Mg^{2+} ion is the only one which is unable to penetrate into the space inaccessible to the Na^+ background cation. When these authors used 0.01 M hydrochloric acid as eluant, the elution volumes decreased, the related elution position of the Sr^{2+} and Ca^{2+} ions altered, and the elution curves became generally more symmetric. The sequence of the K_d values became

$$Mg^{2+} < Ca^{2+} \approx Sr^{2+} < Na^+ < Ba^{2+}$$

which shows that, except for the Ba^{2+} ion, alkaline earth metal ions move only in the phase which contains the Na^+ ion. The experimentally determined K_d values can be correlated with the radius of the hydrated ions *(Fig. 126)*.

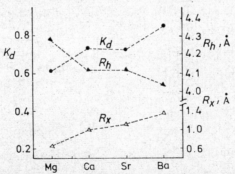

Fig. 126. The relationship between the K_d values and the ionic radius of alkali earth metal ions (Yoza and Ohashi 1969). The K_d values were obtained by chromatography from 0.1 M NaCl solution containing 0.01 M HCl on Sephadex G-15 gel. On the right-hand side ordinate the radius (R_x) of the ions in crystals, and that of the hydrated ion (R_h) are illustrated in Å

According to Ueno *et al.* (1970a, b, c), the migration rate of cations depends also on the properties of the anions of the background electrolyte. As eluant the authors used alkali chlorides, sulphates or nitrates, or perchlorate salt solutions in a concentration of 0.1 M; the eluants contained also the appropriate mineral acids in a concentration of 0.01 N. The K_d values of the chromatographed metal ions, determined in the presence of different background electrolytic solutions, is shown in *Figure 127*.

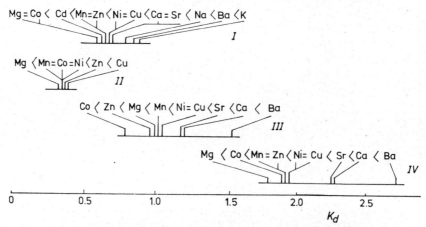

Fig. 127. The K_d values of various metal ions chromatographed on Sephadex G-15 gel from electrolyte solutions containing *I*. chloride, *II*. sulphate, *III*. nitrate and *IV*. perchlorate anion (Ueno *et al.* 1970)

As will be evident, in the system containing sulphate anion, the metal ions migrate at a faster rate than if chromatographed with chloride-containing eluant. The sulphate ion can diffuse into a smaller portion of the inner gel volume than would be expected on the basis of its ion radius, and from a background electrolyte solution containing chloride ion it migrates faster than the metal ions. When metal ions associate with sulphate ion, as counterion, they cannot diffuse into the "distilled water" part of the inside space of the gel and their K_d value decreases, compared to that of the system containing the chloride anion.

On the other hand, according to Ueno *et al.* (1970a, b, c) nitrate and perchlorate ions migrate more slowly from background electrolytes containing chloride than chloride ions do. This may be explained by a specific sorption, since the ion radius of these ions is larger than that of the chloride ion.

The relatively high K_d values of metal ions in the presence of eluants containing nitrate or perchlorate anions can be attributed to a reversible binding of cations to anions bound to the gel. This phenomenon is analogous to ion exchange chromatography. The fact that the K_d value of Mg^{2+}, from a background electrolyte solution containing perchlorate anions, on Sephadex G-15 and G-25 gels with different pore sizes, is nearly the same, is in good agreement with this assumption.

The effect of counteranion on the migration of Mg^{2+} ions on Sephadex G-15 has been further studied by Ogata et al. (1971).

In chromatography from a background electrolyte containing chloride anion, the migration rate of K^+, NH_4^+, Na^+, Ni^{2+} and Mg^{2+} ions corresponds to the radius of the hydrated ions *(Fig. 128)*. If metal ions can form chloro-complexes with chloride ion, then their ion radius must not be used to interpret the migration rate. Although calcium, strontium and barium

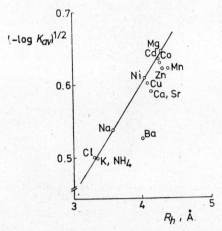

Fig. 128. The correlation between the K_{av} values and the hydrated ionic radius (R_h) in chromatography from KCl—HCl electrolyte solution on Sephadex G-15 gel (Ueno et al. 1970a)

ions do not form stable complexes with the chloride ion, their migration rate is anomalous: Yoza et al. (1970) studied in detail the gel chromatographic behaviour of alkaline earth metal ions and found that the concentration of the sample influences the elution rate, even from a background electrolyte of 0.1 M NaCl *(Fig. 129)*. In the low concentration range the elution curve will be symmetric only for the Mg^{2+} ion, and its K_d value will be independent of the concentration. The elution curve of the Sr^{2+} and Ba^{2+} ions, in such cases, are at the front elongated and, with increasing sample concentration, the K_d value decreases. In the high sample concentration range, the K_d value of all the three ions increases with increasing concentration and tailing takes place. The authors explain this phenomenon by the hypothesis that, at high ion concentrations, the size of the hydrate shell is reduced. Since the variation of the K_d value is greatest with Mg^{2+} and smallest with Ba^{2+}, they presume that, by increasing the concentration of the ions with larger hydrate shells, a greater decrease in hydration will be achieved.

Although increasing ion concentration may reduce the thickness of the hydrate shell of the gel, according to Ueno et al. (1970a, b, c) this does not provide an explanation for the variations in the elution volume. Increasing the inner volume of the gel by dV_i would merely cause an increase of

dV_iK_d in the elution volumes, i.e. the changes in the elution volume of metal ions of smaller diameter should be greater, which is exactly opposite to the experimental findings.

Neddermeyer and Rogers (1968) also examined the effect of background electrolytes. They found that both the shape of the elution curve of an NaCl sample applied to a Sephadex G-10 column and the elution maximum depend on the concentration of the NaCl solution used for the eluant, and that, to obtain a symmetric elution curve *(Fig. 130)*, a salt concentration of approximately 0.01 M would be required. In a later work the same authors studied the gel chromatography of inorganic polyphosphates and found the appearance of two elution peaks regarding electrolyte concentration,

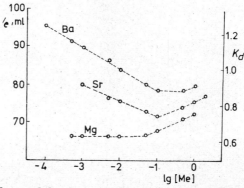

Fig. 129. The dependence of the elution volume and the K_d value of alkali earth metal ions on the sample concentration in chromatography on Sephadex G-15 gel from 0.1 M sodium chloride solution (Yoza et al. 1970)

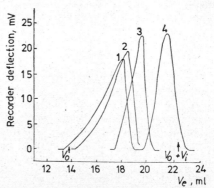

Fig. 130. Changes in the shape of the elution curve of NaCl by the changes of the concentration of the background electrolyte solution in chromatography on Sephadex G-10 (Neddermeyer and Rogers 1968). Column size 0.61×126 cm

1. chromatography from distilled water; 2. chromatography from 10^{-4} M NaCl; 3. that from 10^{-3} M NaCl; 4. that from 10^{-2} M NaCl-background electrolyte solution; 0.05 ml 0.05 M sodium chloride samples were applied to the gel column

when the background electrolyte was not identical to the sample chromatographed. The second peak represented the eluant–electrolyte, the height of which depended on both the sample and eluant concentrations.

Yoza et al. (1971b) have shown the appearance of pseudo-peaks when various sodium salts were chromatographed from sodium chloride background electrolyte solutions. In the gel chromatography of trisodium phosphate in 0.1 M NaCl, a "negative" sodium chloride concentration peak appeared first and later a positive one. This phenomenon is explained by exclusion of sodium chloride from the sample zone, thus giving a pair of negative and positive peaks. The phosphate ions move together with the negative NaCl peak. Another, third, lower migrating peak was found which contained OH^- ion originating from retention of the PO_4^{3-} ion with water, yielding HPO_4^{2-} ion. The reaction of the pseudo-peak by exclusion is independent of the sample concentration, but its height is greatly dependent on this factor.

Neddermeyer and Rogers (1969) found that the elution sequence of various oligophosphates on acrylamide and dextran gels corresponds with decreasing molecular weight. For chromatography on Sephadex G-10 from 0.01 M NaCl solution, the oligophosphates were eluted close to the void volume. On Sephadex G-25, using a 0.025 M Na_2HPO_4 eluant solution, the phosphate salts emerged somewhere between the void volume and the total volume of the column.

The gel chromatography of oligophosphates from 0.1 M KCl solution on Sephadex G-25 was studied earlier by Ohashi et al. (1966) and proved the validity of fractionation according to the size of the molecules.

In later publications Ueno et al. (1970a, b, c) studied the effect of background electrolytes, and found that, from distilled water, the K_d values are extremely small, and further, that these are independent of the pH of the solution, if 0.1 M KCl is used as eluant. That increasing KCl concentration causes a rise in the K_d values is interpreted to be the result of changes in the degree of ion hydration. The linear correlation, found with a 0.1 M KCl background electrolyte solution, between the logarithm of the degree of polymerization and the K_{av} values permits a determination of the average degree of polymerization of samples of unknown composition by gel chromatography.

For the gel chromatography of longer-chained polymers containing 10 to 40 phosphate units, Felter et al. (1968) found that Sephadex G-100 large-pored gels are best suited.

Experiments by Zeitler and Stadler (1972) proved that the elution curves of orthophosphate ions with different charges differ in chromatography on a Sephadex G-10 column: the PO_4^{3-} ion is excluded to a large extent from 0.1 M NaOH solution, the elution volume of HPO_4^{2-} being greater and the $H_2PO_4^-$ ion attaining the highest value if chromatography is performed in an acetate buffer of pH = 4.3.

Deguchi (1975) studied the migration of various halide anions in 0.1 M NaCl and 0.1 M $NaNO_3$ eluants, and, in the latter, a good separation could achieve with the order of elution $F^- < Cl^- < Br^- < I^-$ *(Fig. 131)*. The radii of the hydrated ions are in the order $F^- < Cl^- < I^- < Br^-$, therefore the migration of the I⁻ ion is affected by specific adsorption. This is further supported by the fact that the K_d value for iodide is greater than unity.

According to Wilson and Greenhouse (1975), the elution volume of radio-iodide in tris-HCl buffer on Sephadex G-10 or G-15 is higher than that of sodium or potassium iodide. The increasing concentration of carrier NaI results in a change of V_e for ^{131}I⁻ or ^{125}I⁻ to that of sodium iodide. This

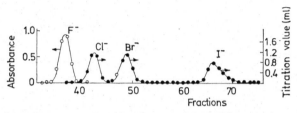

Fig. 131. Fractionation of halide anions in 0.1 M NaNO₃ solution (Deguchi 1975)

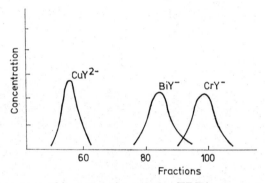

Fig. 132. Gel chromatographic separation of the EDTA complexes of Cu²⁺, Bi³⁺ and Cr³⁺ metal ions from sodium chloride containing background electrolyte solution (Deguchi 1976)

probably reflects the concentration-dependence of the adsorption of iodide ion on the gel matrix.

Yoza et al. (1971a) have shown that the retention of anions in sodium chloride eluant increases in the order $SO_4^{2-} < NO_3^- < ClO_2^- < OH^-$.

The adsorption of various anions on Sephadex G-25 and LH-20 was recently studied by Sinibaldi and Lederer (1975).

The chromatographic behaviour of the EDTA complexes of Co(II), Ni(II), Cu(II), Cr(III), Fe(III), Co(III), and Bi(III) on Sephadex G-10,-15, and -25 in NaCl-containing solvent has been studied by Deguchi (1976). The $M^{II}Y^{2-}$ complexes were easily separated from the $M^{III}Y^-$ complexes; however, apart from these two types, fractionation is more difficult. In some cases, e.g. in the chromatography of BiY⁻ and CrY⁻, the use of a long column results in a good separation *(Fig. 132)*. Increase of the NaCl concentration increases the V_e values, indicating a reduction of the effective size of the complexes to that of other charged solutes (cf. Eaker and Porath 1967).

Further data on various polymeric anions, complexes and on the side-effects of the migration of simple inorganic ions have been published in a summary by Yoza (1973).

6.3. GEL CHROMATOGRAPHY OF INORGANIC COMPOUNDS USING ORGANIC SOLVENT

Ortner and Spitzy (1968a, b) studied the gel chromatography of alkali metal ions from methanol–water mixtures. In such cases a gel phase, rich in water, is formed and the water content of the gel, depending on the methanol content of the mixture, varies according to a maximum-containing curve. The authors found that the use of an eluant with a 75% methanol content offers optimal results. They pointed out that the use of organic solvent

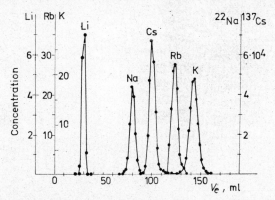

Fig. 133. Separation of alkali chloride salts on Sephadex G-25 column from tris buffer of 0.05 ionic strength containing 75% methanol (Ortner and Spitzy 1968b). Column size: 42×1.07 cm; flow rate 2.64 ml h^{-1}

mixtures enhances the ion exchange properties of the gels, the susceptibility for forming ion pairs increases, and, in addition, the attainment of the equilibrium state may be delayed. Owing to the high solubility of the alkali salts in methanol, this solvent should be used because, from mixtures containing ethanol or propanol, only very small amounts of samples can be chromatographed.

Using 75% methanol as eluant in chromatography on Bio-Gel P-2, Ortner and Spitzy (1968a, b) found that alkali chlorides are eluted in a broad zone as a result, probably, of the lower hydrophilic character of the acrylamide gels, compared to dextran gels. The elution sequence on Sephadex G-25 was as follows

LiCl, NaCl, CsCl, RbCl, KCl

which corresponds to the ratio of their solubility in water to that in methanol, respectively. Using a Sephadex G-10 column, NaCl and CsCl change places. This phenomenon may be explained by fractionation according to the ionic radius. On the larger-pored G-25 column, separation is achieved mainly by distribution on the basis of solubility.

Ortner and Spitzy also studied in detail the effect of the factors which influence separation, i.e. temperature, flow rate, sample concentration, and established the most favourable conditions for the separation of different pairs of alkali metal salts. *Figure 133* shows the fractionation of a mixture containing all the alkali metal ions.

Saitoh and coworkers (1974; Saitoh and Suzuki 1975a,b) have studied the gel chromatography of the acetylacetone complexes of Cr(III), Fe(III), Co(III), Al(III), Cu(II), Ni(II) and Be(II) metal ions in organic solvents on the polyvinyl acetate type, Merckogel OR-2000. In tetrahydrofuran the K_{av} values showed a linear decreasing relationship with the molar volume of the complexes. This line, however, does not coincide with that obtained for n-alkanes under the same conditions, which may be explained by some interaction between the gel matrix and the solute.

The use of organic solvents extends greatly the applicability of gel columns in the fractionation of inorganic salts.

BIBLIOGRAPHY

Ackers, G. K., Biochemistry *3*, p. 723 (1964)
Ackers, G. K., J. Biol. Chem. *242*, p. 3237 (1967a)
Ackers, G. K., J. Biol. Chem. *242*, p. 3026 (1967b)
Ackers, G. K., Advances in Protein Chemistry. Elsevier, New York, London, Vol. 24, pp. 343—446 (1970)
Ackers, G. K. and R. L. Steere, Biochim. Biophys. Acta *59*, p. 137 (1962)
Ackers, G. K. and I. E. Thompson, Proc. Natl. Acad. Sci. U. S. *53*, p. 342 (1965)
Acrilex Gel Filters, Reanal Fine Chemicals Factory, Budapest (1971)
Adams, H. E., K. Farhat and B. L. Johnson, Ind. Eng. Chem. Prod. Res. Devel. *5*, p. 126 (1966)
Addison, R. F. and R. G. Ackman, Anal. Biochem. *28*, p. 515 (1969)
Agostini, A., C. Vergani and B. Lomanto, J. Lab. Clin. Med. *69*, pp. 522—529 (1967)
Aitzetmüller, K., J. Chromatogr. *71*, p. 355 (1972)
Akabosi, M., K. Kavai and A. Waki, Anal. Biochem. *47*, p. 687 (1972)
Akai, H., K. Yokorayashi, A. Misaki and T. Harada, Biochim. Biophys. Acta *237*, p. 422 (1971a)
Akai, H., K. Yokorayashi, A. Misaki and T. Harada, Biochim. Biophys. Acta *252*, p. 427 (1971b)
Alberts, B. M., Methods Enzymol. *12A*, pp. 566—581 (1967)
Albertsson, P. A., Partition of Cell Particles and Macromolecules. Almquist and Wiksell, Uppsala (1960)
Albertsson, P. A., Biochim. Biophys. Acta *103*, pp. 1—12 (1965)
Albertsson, P. A., Methods Virol. *2*, pp. 303—321 (1967)
Albertsson, P. A. Advances in Protein Chemistry. Elsevier, New York, London, Vol. 24, pp. 309—341 (1970a)
Albertsson, P. A., Science Tools *17*/3, pp. 53—67 (1970b)
Albertsson, P. A. and G. D. Baird, Exp. Cell Res. *28*, pp. 296—322 (1962)
Albertsson, P. A. and E. J. Nyns, Nature *184*, pp. 1465—1468 (1959)
Albertsson, P. A. and E. J. Nyns, Ark. Kemi *17*, pp. 197—206 (1961)
Albertsson, P. A., S. Sasakawa and H. Walter, Cross partition and isolectric points of proteins, Nature *228*, pp. 1329—30 (1970)
Altgelt, K. H., Makromol. Chem. *88*, p. 75 (1965a)
Altgelt, K. H., J. Appl. Polymer Sci. *9*, p. 3389 (1965b)
Altgelt, K. H., Adv. in Chromatogr., *7* (1968)
Altgelt, K. H., Preprints Div. Petroleum Chem. Am. Chem. Soc. *15* (2), p. A 115 (1970)
Alvord, C. E., M. W. Kies, F. N. Le Baron and R. E. Martenson, Science *151*, p. 821 (1966)
Alvsaker, J. O., Scand. J. Clin. Lab. Invest. *17*, p. 467 (1965)
Ames, B. N., R. G. Martin and B. J. Garry, J. Biol. Chem. *236*, p. 2019 (1961)
Amsterdam, A., Zvi Er-el and S. Shaltiel, Arch. Biochem. Biophys. *171*, p. 673 (1975)
Anderson, D. M. W., A. Hendrie and A. C. Munro, J. Chromatogr. *44*, p. 178 (1969)
Anderson, D. M. W., I. C. M. Dea, S. Rahman and J. F. Stoddart, Chem. Commun. 145 (1965)
Anderson, D. M. W. and J. F. Stoddart, Anal. Chim. Acta *34*, p. 401 (1966a)
Anderson, D. M. W. and J. F. Stoddart, Carbohyd. Res. *2*, p. 104 (1966b)

Anderson, J. H. and C. E. Carter, Biochemistry *4*, p. 1102 (1965)
Andrasko, J., Biophys. J. *15*, p. 1235 (1975)
Andrews, P., Nature *196*, p. 36 (1962)
Andrews, P., Biochem. J. *91*, p. 223 (1964)
Andrews, P., Biochem. J. *96*, pp. 595—606 (1965)
Andrews, P., Lab. Pract. *16*, p. 851 (1967a)
Andrews, P., Prot. Biol. Fluids *14*, p. 573 (1967b)
Andrews, P., Methods of Biochemical Analysis (Ed. D. Glick). Interscience Publ., New York, London, Vol. 18, pp. 1—53 (1970)
Andrews, P., R. C. Bray and P. Edwards, Biochem. J. *93*, p. 627 (1964)
Anfinsen, C. H. B. and E. Haber, J. Biol. Chem. *236*, p. 1361 (1961)
Araki, C., J. Chem. Soc. Japan *58*, pp. 1338—1350 (1937)
Araki, C., Bull. Chem. Soc. Japan *29*, p. 543 (1956)
Armstrong, A., H. Hagopiam, M. Ingram and E. K. Wagner, Biochem. J. *5*, p. 3027 (1965)
Arnott, S., A. Fulner, W. E. Scott, I. C. M. Dea, R. Moorhouse and D. A. Rees, J. Mol. Biol. *90*, pp. 269—284 (1974)
Asche, W. and H. H. Oelert, J. Chromatogr. *106*, p. 490 (1975)
Aspberg, K. and J. Porath, Acta Chem. Scand. *24*, pp. 1839—1841 (1970)
Aspinall, P. T. and J. N. Miller, Anal. Biochem. *53*, pp. 509—513 (1973)
Auerswald, W., W. Doleschel and E. Zweymüller, Z. Kinderheilk. *109*, pp. 262—270 (1971)
Auricchio, F. and C. B. Bruni, Biochem. Z. *340*, p. 321 (1964)
Auricchio, F. and C. B. Bruni, Biochem. J. *98*, pp. 290—292 (1966)
Ayad, S. R., R. W. Bonsall and S. Hunt, Science Tools *14*/3, p. 40 (1967)
Baba, N., T. Hara and Y. Meda, Shimadzu Hyoron *26*/3, pp. 279—284 (1969)
Baba, N. and S. Sato, Bunseki Kagaku *20*/2, pp. 208—214 (1971)
Bachmann, R. C. and B. F. Burnham, J. Chromatogr. *41*, pp. 394—399 (1969)
Baghurst, P. A., L. W. Nichol, A. G. Ogston and D. J. Winzor, Biochem. J. *147*, pp. 575—583 (1975)
Barral, E. M. and J. H. Cain, J. Polymer. Sci. C *21*, p. 253 (1968)
Barrer, R. M. and D. Brook, Trans. Faraday Soc. *49*, p. 9 (1953)
Bartoli, F. and C. Rossi, J. Chromatogr. *28*, p. 30 (1967)
Basedow, A. M., K. H. Ebert, H. Ederer and H. Hunger, Makromol. Chem. *177*, p. 1501 (1976)
Bathgate, G. N., J. Chromatogr. *47*, pp. 92—96 (1970)
Battle, A. M. del C., J. Chromatogr. *28*, p. 82 (1967)
Behal, F. J., G. H. Little and R. A. Klein, Arylamidase of human liver, BBA *178*, p. 118 (1969)
Beiss, U. and R. Marx, Naturwiss. *49*, p. 142 (1962)
Bejer, K. and E. Nyström, Anal. Biochem. *48*, p. 1 (1972)
Belenkii, B. G., E. S. Gankina, P. P. Nefedov, M. A. Lazareva, T. S. Savitskaya and M. D. Volchikhina, J. Chromatogr. *108*, p. 61 (1975a)
Belenkii, B. G., L. Z. Vilenchik, V. V. Nesterov, V. J. Kolegov and S. Y. Frenkel, J. Chromatogr. *109*, p. 233 (1975b)
Beling, C. G., Acta Endocrinol. *43*, Suppl. 79, p. 98 (1963)
Beling, C. G., Nature *192*, pp. 326—327 (1961)
Bengtsson, S. and L. Philipson, Biochim. Biophys. Acta *79*, p. 399 (1964)
Bennich, H., Biochim. Biophys. Acta *51*, p. 265 (1961)
Benoit, H., 5th International Seminar Gel Permeation Chromatography, London (1968)
Benoit, H., Z. Grubisic, P. Rempp, D. Decker and J. G. Zilliox, J. Chim. Phys. *53*, p. 1507 (1966)
Berger, K. C., Makromol. Chem. *175*, pp. 2121—2132 (1974)
Berger, K. C., Makromol. Chem. *176*, pp. 399—410 (1975)
Bergquist, P. L., B. C. Baguley and J. M. Robertson, Biochem. Biophys. Acta *108*, pp. 531—539 (1965)
Bergquist, P. L. and J. M. Robertson, Biochim. Biophys. Acta *95*, p. 357 (1965)
Bergström, K., Acta Med. Scand. *445*, pp. 127—131 (1966)
Bernal, J. D. and J. Mason, Nature *188*, pp. 910—911 (1960)

Bernofsky, C., Anal. Biochem. *68*, p. 311 (1975)
Bertonière, N. R., L. F. Martin and S. P. Rowland, Carbohyd. Res. *19*, pp. 189—196 (1971)
Billmeyer, F. W. Jr., G. W. Johnson and R. N. Kelley, J. Chromatogr. *34*, p. 316 (1968)
Bio-Rad Laboratories, Catalogue U/V, Bio Rad Materials for Ion Exchange, Gel Filtration, Adsorption 1970, Richmond, California (1973)
Birch, N. J. and I. Goulding, Anal. Biochem. *66*, p. 293 (1975)
Biserte, G., M. Bonte and P. Sautière, J. Chromatogr. *35*, pp. 168—178 (1968)
Bjorklund, B., Anal. Biochem. *41*/1, pp. 287—290 (1971)
Bloemendal, H., W. S. Bout, J. F. Jonkind and J. H. Wisse, Nature *193*, p. 437 (1962)
Blue Dextran 2000, Pharmacia Fine Chemicals AB, Uppsala (1964)
Blume, K. G., R. W. Hoffbauer and D. Busch, Biochim. Biophys. Acta *227*, p. 364 (1971)
Blundell, D. J., A. Keller and I. M. Ward, J. Polymer Sci. *B 4*, p. 781 (1966)
Bly, D. D., Science, *168* (No. 3931), pp. 527—533 (1970)
Bly, D. D., K. A. Boni, M. J. R. Cantow, J. Cazes, D. J. Harmon, J. N. Little and E. D. Weir, J. Polymer Sci. *B 9*/6, pp. 401—411 (1971)
Boman, H. G. and S. Hjertén, Arch. Biochim. Biophys. Suppl. 1, p. 276 (1962)
Bock, H. G., Science *191*, p. 380 (1976)
Bogardt, R. A. Jr., F. E. Dvulet, L. D. Lehman, B. N. Jones and F. R. N. Gurd, Biochemistry *15*, p. 2597 (1976)
Bombaugh, K. J., The practice of gel permeation chromatography in Modern Practice of Liquid Chromatography (Ed. J. J. Kirkland). J. Wiley and Sons Inc., New York—London—Sydney—Toronto, Ch. 7, pp. 237—285 (1971)
Bombaugh K. J., W. A. Dark and R. N. King, J. Polymer Sci. Part C *21*, p. 131 (1968a)
Bombaugh K. J., W. A. Dark and R. F. Levangie, 5th International Seminar Gel Permeation Chromatography, London, 1968 (1968b)
Bombaugh, K. J., W. A. Dark and J. N. Little, Anal. Chem. *41*/10, p. 1337 (1969a)
Bombaugh, K. J., W. A. Dark and R. F. Levangie, J. Chromatogr. Sci. *7*, p. 42 (1969b)
Bombaugh, K. J. and R. F. Levangie, Anal. Chem. *41*, p. 1357 (1969c)
Bombaugh, K. J. and R. F. Levangie, J. Chromatogr. Sci. *8*, pp. 560—566 (1970a)
Bombaugh, K. J. and R. F. Levangie, Separ. Sci. *5*, p. 892 (1970b)
Boni, K. A., D. E. Nemzer and F. A. Sliemers, 6th International Seminar Gel Permeation Chromatography, Miami Meach, Florida, p. 445 (1968)
Bonilla, C. A., Anal. Biochem. *32*, p. 522 (1969)
Bonilla, C. A., J. Chromatogr. *47*, p. 499 (1970)
Boutoux, J., J. P. Bali and A. Dauplan, Ann. Phys. Biol. Med. *4*, pp. 151—158 (1970)
Boross, L., Ioncserés kromatográfia a szerves és biokémiában (Ion Exchange Chromatography in Organic Chemistry and Biochemistry), Műszaki Könyvkiadó, Budapest (1968)
Boross, L., Biochim. Biophys. Acta *96*, p. 52 (1965)
Bosch, L., G. van der Wende, M. Senyser, H. Bloemendal (Sluyser), Biochim. Biophys. Acta *53*, p. 44 (1961)
Boschetti, E., R. Tixier and J. Uriel, Biochimie *54*/4, pp. 439—444 (1972)
Boschetti, E., R. Tixier and R. Garelle, Science Tools *21*, pp. 35—38 (1974)
Bowden, J. A. and M. Fried, Comp. Biochem. Physiol. *32*, pp. 391—400 (1970)
Braun, T. and A. B. Farag, Anal. Clin. Acta *62*, p. 476 (1972)
Brewer, P. I., Nature *188*, p. 934 (1960); *190*, p. 625 (1961)
Brewer, P. I., Polymer *6*, p. 603 (1965)
Bridges, R. L., L. R. Finn and S. L. Tinkler, J. Chromatogr. *39*, p. 519 (1969)
Brizová, E., M. Popl and J. Coupek, J. Chromatogr. *139*, pp. 15—20 (1977)
Broman, L. and K. Kjellin, Biochim. Biophys. Acta *82*, pp. 101—109 (1964)
Brook, A. J. W., J. Chromatogr. *39*, p. 328 (1969)
Brook, A. J. W. and S. Housley, J. Chromatogr. *41*, p. 200 (1969a)
Brook, A. J. W. and S. Housley, J. Chromatogr. *42*, p. 112 (1969b)
Brook, A. J. W. and K. G. Munday, J. Chromatogr. *47*, pp. 1—8 (1970a)
Brook, A. J. W. and K. C. Munday, J. Chromatogr. *51*, pp. 307—310 (1970b)
Brooks, C. J. W. and R. A. B. Keates, J. Chromatogr. *44*, p. 500 (1969)
Brown, W., J. Chromatogr. *52*, p. 273 (1970)
Brown, W., J. Chromatogr. *59*, pp. 335—342 (1971)

Brown, W., Chemica Scripta *2*, pp. 25—29 (1972)
Brown, W. and O. Andersson, J. Chromatogr. *57*, pp. 255—263 (1971)
Brown, W. and O. Andersson, J. Chromatogr. *67*, p. 163 (1972)
Brown, W. and K. Chitumbo, J. Chromatogr. *66*, p. 370 (1972)
Brown, W. and K. Chitumbo, J. Chem. Soc. Faraday Transactions I. *71*, pp. 1—11 and 12—21 (1975)
Brönsted, J. N., I. Z. Phys. Chem., Bodenstein-Festband, pp. 257—266 (1931)
Brumbaugh, E. E. and G. K. Ackers, J. Biol. Chem. *243*, p. 6315 (1968)
Bryan, S. E. and E. Frieden, Biochemistry *6*, p. 2728 (1967)
Bryce, C. F. A. and R. R. Crichton, J. Chromatogr. *63*, pp. 267—280 (1971)
Burstein, M., H. R. Scholnick and R. Morfin, J. Lip. Res. *11*, p. 583 (1970)
Calderon, M. and W. J. Baumann, Biochim. Biophys. Acta *210*, pp. 7—14 (1970)
Cameron, B. F., Separ. Sci. *6/2*, pp. 229—237 (1971)
Candido, E. P. M. and G. H. Dixon, J. Biol. Chem. *247*, p. 3868 (1972)
Cann, J. R. and W. B. Goad, J. Biol. Chem. *240*, p. 148 (1965)
Cantow, M. J. R. and J. F. Johnson, J. Appl. Polymer Sci. *11*, p. 1851 (1967)
Cantow, M. J. R. and J. F. Johnson, J. Polymer Sci. Part A-1, *5*, p. 2835 (1967)
Cantow, M. J. R., R. S. Porter and J. F. Johnson, J. Polymer Sci., Part A-1, *5*, p. 987 (1967a)
Cantow, M. J. R., R. S. Porter and J. F. Johnson, J. Polymer Sci., Part A-1, *5*, p. 1301 (1967b)
Cantow, M. J. R., R. S. Porter and J. F. Johnson, J. Polymer Sci., Part B, *4*, p. 707 (1967c)
Carman, P. C., Flow of Gases through Porous Media, Academic Press, New York (1956)
Carmichael, J. B., J. Polymer Sci., Part A-2, *6*, p. 517 (1968)
Carnegie, P. R., Nature *206*, p. 1128 (1965a)
Carnegie, P. R., Biochem. J. *95*, p. 9P (1965b)
Carnegie, P. R. and C. E. Lumsden, Immunology *12*, pp. 133—145 (1967)
Carnegie, P. R. and G. Pacheco, Proc. Soc. Exp. Biol. Med. *117*, p. 137 (1964)
Carter, D. B., P. H. Efird and C. B. Chee, Biochemistry *15*, p. 2603 (1976)
Carter, J. H. and E. Y. C. Lee, Anal. Biochem. *39*, pp. 521—542 (1971)
Carter, P. M. and J. R. Hobbs, Brit. Med. J. *2*, pp. 260—261 (1971)
Casassa, E. F., J. Polymer Sci. *B5*, p. 773 (1967)
Casassa, E. F., J. Phys. Chem. *75*, pp. 3929—3939 (1971)
Catsimpoolas, N. and J. Kenney, J. Chromatogr. *64*, pp. 77—83 (1972)
Cazes, J., J. Chem. Educ. *43*, p. A 567 (1966)
Cazes, J. and D. R. Gaskill, Separ. Sci. *2*, p. 421 (1967)
Cazes, J. and D. R. Gaskill, 6th International Seminar Gel Permeation Chromatography, Miami Beach, Florida (1968)
Cazes, J. and D. R. Gaskill, Separ. Sci. *4*, p. 15 (1969)
Čech, M., M. Jelinková and J. Čoupek, J. Chromatogr. *135*, pp. 435—440 (1977)
Chang, T. L., Anal. Chem. *40*, p. 989 (1968)
Chaterjee, A. K., G. J. Durcint, H. Hendrickson, I. C. Lee and R. Montgomery, Biochim. Biophys. Res. Comm. *4*, p. 425 (1961)
Cheetham, N. W. H. and D. J. Winzor, J. Chromatogr. *48*, pp. 400—405 (1970)
Cherkin, A., F. E. Martinez and M. S. Dunn, J. Am. Chem. Soc. *75*, p. 1244 (1953)
Chersi, A., Science Tools *11*, p. 1 (1964)
Chiancone, E., L. M. Gilbert, G. A. Gilbert and G. E. Kellett, J. Biol. Chem. *243*, p. 1217 (1968)
Chitumbo, K. and W. Brown, J. Chromatogr. *80*, p. 187 (1973a)
Chitumbo, K. and W. Brown, J. Chromatogr. *87*, p. 17 (1973b)
Chow, C. D., J. Chromatogr. *114*, p. 486 (1975)
CH-Sepharose 48, Ah-Sepharose 48 for Affinity Chromatography, Pharmacia Fine Chemicals AB, Uppsala (1973)
Chun, P. W., S. J. Kim, C. A. Stanley and G. K. Ackers, Biochemistry *8*, p. 1625 (1969)
Claesson, J. and S. Claesson, Arkiv. Kemi *19A*, No. 5, p. 1 (1944)
Claesson, J. and S. Claesson, Phys. Rev. *73*, p. 1221 (1948)
Claesson, S., Arkiv. Kemi *26A*, No. 24, p. 1 (1948)
Clark, R. T., Anal. Chem. *30*, p. 1676 (1958)

Cleland, W. W., Biochemistry *3*, p. 480 (1964)
Clenn, W. L., J. Biol. Chem. *246*, p. 9 (1971)
Clifford, J. and T. F. Child, Proc. 1st Eur. Biophys. Cong. Baden, Vol. III, p. 461 (1971)
Clifford, J., L. Oakes and H. J. T. Tiddy, Special Discussions Faraday Soc. *1*, p. 175 (1970)
Clinton, B. A., N. C. Palczuk and L. A. Stauber, J. Immunol. *108*, pp. 1570—1577 (1972)
Coleman, W. H., J. Kaur and M. E. Iwert, J. Bacteriol. *96*, pp. 1137—1143 (1968)
Colman, R. F., Anal. Biochem. *46*, pp. 358—363 (1972)
Collins, R. C. and W. Haller, Anal. Biochem. *54*, p. 47 (1973)
Constantopoulos, G., A. S. Dehuban and W. R. Caroll, Anal. Biochem. *31*, p. 59 (1969)
Cooper, A. R., A. R. Bruzzone, J. H. Cain and E. M. Barrall, J. Appl. Polym. Sci. *15*/e, pp. 571—588 (1971)
Cooper, A. R. and J. F. Johnson, J. Appl. Polym. Sci. *13*, p. 1487 (1969)
Cooper, A. R. and S. Wood, J. Pharm. Pharmacol. SO Suppl. p. 1505 (1968)
Cortis-Jones, B., Nature *191*, p. 272 (1961)
Coupek, J., S. Prokorny and J. Propisil, J. Chromatogr. *95*, p. 103 (1974)
Coxeter, H. S. M., Illlinois J. Math. *2*, p. 746 (1958)
Craft, H. J., Biochim. Biophys. Acta *54*, p. 611 (1961)
Craven, G. R. Jr., E. Steers and C. B. Anfinsen, J. Biol. Chem. *240*, p. 2468 (1965)
Crone, H. D., J. Chromatogr. *60*, pp. 185—194 (1971)
Crone, H. D., R. M. Dawson and E. M. Smith, J. Chromatogr. *103*, p. 71 (1975)
Cruft, H. J., Biochim. Biophys. Acta *54*, p. 609 (1961)
Cuatrecasas, P., Proc. Nat. Acad. Sci. U. S. A. *63*, p. 450 (1969)
Cuatrecasas, P., J. Biol. Chem. *245*, pp. 3059—3065 (1970a)
Cuatrecasas, P., Nature *228*, pp. 1327—1328 (1970b)
Cuatrecasas, P. and C. B. Anfinsen, Ann. Rev. Biochem. *40*, pp. 259—278 (1971)
Culling, C. F. A., Handbook of Histopathological Techniques. Butterworth and Co., London (1963)
Cunningham, W. L. and J. L. Simkin, Biochem. J. *99*, p. 434 (1966)
Curtain, C. C. and W. G. Nayler, Biochem. J. *89*, p. 69 (1963)
Daisley, K. W., Nature *191*, p. 868 (1961)
Dawkins, J. V. and M. Hemming, Makromol. Chem. *176*, p. 1777 (1975)
 1. Universal calibration for cyclohexane at 35 °C;
 2. Separation mechanism, p. 1795;
 3. Temperature dependence of universal calibration for trans-decalin, p. 1815
Davis, B. J. and L. Ornstein, Deliv. at the Soc. for the Study of Blood at the N. Y. Acad. Med. (1959)
Davison, R., Science *161*, p. 906 (1968)
Dawson, R. M. C., D. C. Elliott, W. H. Elliott and K. M. Jones (Eds), Data for Biochemical Research. Clarendon Press, Oxford, pp. 475—508 (1969)
Deelder, R. S., J. Chromatogr. *47*, pp. 307—312 (1970)
Deguchi, T., J. Chromatogr. *108*, p. 409 (1975)
Deguchi, T., J. Chromatogr. *120*, p. 159 (1976)
Deibler, G. E., R. E. Marbuson and M. W. Kies, Biochim. Biophys. Acta *200*, p. 342 (1970)
De la Llosa, P., C. Tertrin and M. Justiz, Biochim. Biophys Acta *115*, p. 464 (1966)
Demassieux, S. and J. P. Lachance, J. Chromatogr. *89*, pp. 251—258 (1974)
De Mets, M. and A. Lagasse, J. Chromatogr. *47*, pp. 486—489 (1970)
Denamur, R. E. and P. J. B. Gaye, Eur. J. Biochem. *19*, p. 23 (1971)
Determann, H., Experientia *18*, pp. 430—432 (1962)
Determann, H., Angew. Chem. *3*, p. 608 (1964a)
Determann, H., Angew. Chem. *76*, p. 635 (1964b)
Determann, H., Gel Chromatography. Springer-Verlag, Berlin—Heidelberg—New York (1969)
Determann, H. and K. Lampert, J. Chromatogr. *69*, pp. 123—128 (1972)
Determann, H., G. Lüben and H. Wieland, Makromol. Chem. *73*, p. 168 (1964)
Determann, H., and K. Mätter, FEBS Letters *2*, p. 163 (1969)
Determann, H. and W. Michel, Z. Anal. Chem. *212*, pp. 211—218 (1965)

Determann, H. and W. Michel, J. Chromatogr. *25*, pp. 303—313 (1966)
Determann, H. and I. Walter, Nature *219*, pp. 604—605 (1968)
Deuel, H., J. Solms and L. Auyas-Weisz, Helv. Chim. Acta *33*, p. 2171 (1950)
Deuel, H. and H. Neukom, Natural Plant Hydrocolloids, in Advances in Chemistry Series *11*, p. 51 (1954)
Deutsch, B., R. D. Levere and J. Levine, J. Clin. Path. *16*, p. 183 (1963)
De Vries, A. J., M. Le Page, R. Bean and C. L. Guillemin, Anal. Chem. *39*, pp. 935—939 (1967)
Di Marzio, E. Z. and C. M. Guttman, J. Polymer Sci. *7*, p. 267 (1969)
Dintzis, F. R. and R. Tobin, J. Chromatogr. *88*, p. 77 (1974)
Dizik, N. S. and F. W. Knapp, J. Food Sci. *35*, pp. 282—285 (1970)
Dobos, D., Elektrokémiai táblázatok (Electrochemical Tables). Műszaki Könyvkiadó, Budapest (1965)
Doley, S. G., M. J. Harvey and P. D. G. Dean, FEBS Letters *65*, p. 87 (1976)
Done, J. N. and J. H. Knox, J. Chromatogr. Sci. *10*, p. 606 (1972)
Dorrington, K. K., M. H. Zarlengo and C. Tanford, Proc. Nat. Acad. Sci. U. S. A. *51*, p. 996 (1967)
Downey, W. K., and P. Andrews, Biochem. J. *94*, p. 642 (1965)
Downey, W. K. and P. Andrews, Biochem. J. *101*, p. 651 (1966)
Downey, W. K., R. F. Murphy and M. K. Keogh, J. Chromatogr. *46*, p. 120 (1970)
Drott, E. E. and R. A. Mendelson, 5th International Seminar Gel Permeation Chromatography, London (1968)
Duerksen, J. H. and A. E. Hamielec, 5th International Seminar Gel Permeation Chromatography, London (1968)
Dvulet, F. E., R. A. Bogardt, B. N. Jones and L. D. Lehman, Biochemistry *14*, p. 5336 (1975)
Eaker, D. and J. Porath, Separ. Sci. *2*, pp. 507—550 (1967).
Edgards, G. D. and Q. Y. Ng, J. Polymer Sci., Part C, *21*, p. 105 (1968)
Edmond, E., S. Farquhar, J. R. Dunstone and A. G. Ogston, Biochem. J. *108*, pp. 755—763 (1968)
Edmond, E. and A. G. Ogston, Biochem. J. *109*, pp. 569—576 (1968)
Edmond, E. and A. G. Ogston, Biochem. J. *117*, pp. 85—89 (1970)
Edwards, V. H. and J. M. Helft, J. Chromatogr. *47/3*, pp. 490—493 (1970)
Egan, B. Z., J. Chromatogr. *34*, p. 382 (1968)
Ehrlich, S. D., J. P. Thiery and G. Bernardi, Biochim. Biophys. Acta *246*, pp. 161—168 (1971)
Eichenberger, W., Chimia *23*, p. 85 (1969)
Eipper, B. A. and R. E. Mains, Biochemistry *14*, p. 3836 (1975)
Ekman, R., B. G. Johansson and U. Ravnskov, Anal. Biochem. *70*, 628 (1976)
Ellingboe, J., E. Nyström and J. Sjövall, J. Lip. Res. *11/3*, pp. 266—273 (1970)
Eltekov, Y. A. and A. S. Nazansky, J. Chromatogr. *116*, p. 99 (1976)
Emnéus, A., J. Chromatogr. *32*, p. 243 (1968)
Engel, J., J. Kurtz, E. Katchalski and A. Berger, J. Mol. Biol. *17*, p. 255 (1966)
Epstein, W. V. and M. Tan, J. Chromatogr. *6*, pp. 258—263 (1961)
Epton, R., S. R. Holding and J. V. McLaren, J. Chromatogr. *110*, p. 327 (1975)
Erikson, R. L. and J. A. Gordon, Biochem. biophys. Res. Commun. *23*, pp. 422—428 (1966)
Fairclough, G. F. Jr. and J. S. Fruton, Biochemistry *5*, pp. 673—683 (1966)
Fasold, H., H. G. Gundlach and F. Turba, in Chromatography (Ed. E. Heffman), p. 406, Reinhold, New York (1961)
Fawcett, J. S. and C. I. O. R. Morris, Separ. Sci. *1*, p. 9 (1966)
Fehrenbach, F. J., Eur. J. Biochem. *18*, pp. 94—102 (1971)
Fejes Tóth, L., Regular Figures. Pergamon Press, Oxford—London—New York—Paris, pp. 288—300 (1964)
Fejes Tóth, L., Lagerungen in der Ebene auf der Kugel und im Raum. Springer Verlag, Berlin—Heidelberg—New York, pp. 171—185 (1972)
Felter, S., O. Dirheimer and J. B. Ebel, J. Chromatogr. *35*, p. 207 (1968)
Ferrier, B. M., D. Jarvis and V. du Vigneaud, J. Biol. Chem. *240*, pp. 4264—4266 (1965)
Fish, W. W., K. G. Mann and C. Tanford, J. Biol. Chem. *244*, p. 4989 (1969)

Fischer, L., An introduction to gel chromatography, in Laboratory Techniques in Biochemistry and Molecular Biology (Eds T. S. Work and E. Work). North Holland Publ. Co., Amsterdam—London, Vol. I, part II (1969)
Fishman, M. L. and R. A. Barford, J. Chromatogr. *52*/3, pp. 494—496 (1970)
Flodin, P., J. Chromatogr. *5*, pp. 103—115 (1961)
Flodin, P., Dextran gels and their applications in gel filtration. Dissertation, Pharmacia AB, Uppsala, Sweden pp. 1—85 (1962)
Flodin, P. and K. Aspberg, Biological Structure and Function. Academic Press, New York, Vol. 1, p. 345 (1961)
Flodin, P. and K. Granath, Symposium über Makromoleküle, Wiesbaden 1959, Verlag Chemie, Weinheim (1959/60)
Flodin, P., B. Gelotte and J. Porath, Nature *188*, p. 493 (1960)
Flodin, P., J. D. Gregory and L. Rodén, Anal. Biochem. *8*, p. 424 (1964)
Flodin, P. and J. Killander, Biochim. Biophys. Acta *63*, pp. 403—410 (1962)
Flodin, P. and J. Porath, Molecular sieve processes, in Chromatography (Ed. E. Heftmann), Reinhold, New York, p. 328 (1961)
Fox, J. B. Jr., R. G. Calhour and W. J. Eglinton, J. Chromatogr. *43*, pp. 48—65 (1969)
Franek, F. and R. S. Neslin, Biokhimiya *28*, p. 193 (1963)
Franek, M. and K. J. Hruska, J. Chromatogr. *119*, p. 167 (1976)
Frank, F. C., I. M. Ward and T. Williams, 5th International Seminar Gel Permeation Chromatography, London (1968)
Frankland, B. T. B., M. D. Hollenberg and D. B. Hope, Brit. J. Pharmacol. *26*, p. 502 (1966)
Frenkel, M. J. and R. J. Blagrove, J. Chromatogr. *111*, p. 397 (1975)
Fritsche, P. and V. Gröbe, IUPAC Symposium on Macromolecular Chemistry, Preprint 422. Prague (1965)
Fudano, S. and K. Konishi, J. Chromatogr. *71*, pp. 93—100 (1972)
Gamble, L. W., L. Westerman and E. A. Krupp, Rubber Chem. Technol. *38*, p. 823 (1965)
Gelotte, B., J. Chromatogr. *3*, p. 330 (1960)
Gelotte, B., Naturwiss. *48*, p. 554 (1961)
Gelotte, B., in New Biochemical Separations (Eds A. T. James and L. J. Morris), Van Nostrand, London, Ch. 6 (1964a)
Gelotte, B., Acta Chem. Scand. *18*, p. 1283 (1964b)
Gelotte, B. and A. Emnéus, Chem. Ing.-Techn. *38*, pp. 445—451 (1966)
Gelotte, B., P. Flodin and J. Killander, Arch. Biochem. Biophys. Suppl. *1*, p. 319 (1962)
Gelotte, B. and A. B. Krantz, Acta Chem. Scand, *13*, p. 2127 (1959)
Gelotte, B. and J. Porath, in Chromatography (Ed. E. Heftmann). Reinhold, New York (1966)
Gerlich, W., H. Determann and T. Wieland, Z. Naturforsch. *25.B*, pp. 1235—1239 (1970)
Giddings, J. C., Dynamics of Chromatography. Marcel Dekker, New York, Vol. 1. Ch. 1 (1965)
Giddings, J. C., Anal. Chem. *39*, p. 1027 (1967)
Giddings, J. C. and K. Dahlgren, Separ. Sci. *5*, pp. 717—722 (1970)
Giddings J. C., E. Kucera, C. P. Russel and M. N. Myers, J. Phys. Chem. *72*, p. 4397 (1968)
Giddings, J. C. and K. L. Mallik, Anal. Chem. *38*, pp. 997—1000 (1966)
Gilbert, G. A., Proc. Roy. Soc. *A 276*, p. 354 (1963)
Gilbert, G. A., Nature *210*, pp. 299—300 (1966a)
Gilbert, G. A., Nature *212*, pp. 296—297 (1966b)
Gilbert, G. A., Anal. Chim. Acta *38*, p. 275 (1967)
Ginzburg, B. L. and D. Cohen, Trans. Faraday Soc. *60*, p. 185 (1964)
Giorgio, N. A. Jr., A. T. Yip, J. Fleming and G. W. E. Plant, J. Biol. Chem. *245*, p. 5469 (1971)
Glazer, A. N. and D. Wellner, Nature *194*, pp. 862—863 (1962)
Glueckauf, E., Ion Exchange and its Applications, Soc. Chem. Ind., London, p. 34 (1955a)
Glueckauf, E., Trans. Faraday Soc. *51*, p. 34 (1955b)

Glueckauf, E., Trans. Faraday Soc. *51*, p. 1540 (1955c)
Goldman, L. P., de L. Ballivian and E. Melgar, Clin. Chem. *16*, pp. 740—742 (1970)
Goodson, J. M., V. Distefano and J. C. Smith, J. Chromatogr. *54*, pp. 43—53 (1971)
Gordon, W., R. T. Havran and I. L. Schwartz, Proc. Nat. Acad. Sci. U. S. A. *60*, pp. 1353—1355 (1968)
Granath, K. A., J. Colloid. Sci. *13*, p. 308 (1958)
Granath, K. A., in New Biochemical Separations (Eds A. T. James and L. J. Morris). Van Nostrand, London, p. 112 (1964)
Granath, K. A. and P. Flodin, Makromol. Chem. *48*, p. 160 (1961)
Granath, K. A. and B. E. Kvist, J. Chromatogr. *28*, pp. 69—81 (1967)
Griebel, R. and Z. Smith, Biochemistry *7*, pp. 3676—3681 (1968)
Griffith, I. P., J. Chromatogr. *109*, p. 399 (1975)
Grubisic-Gallot, Z. and H. Benoit, Proc. 7th Intern. GPC Seminar, p. 65. Monte Carlo, 1969. Publ. by Waters Ass. Inc. Franningham, Mass. (1969)
Gupta, D., Nature *206*, p. 575 (1963)
Haavaldsen, R. and T. Norseth, Anal. Biochem. *15*, p. 535 (1966)
Habeeb, A. F. S. A., Biochim. Biophys. Acta *121*, p. 21 (1966)
Haeckel, R., Hess, B. and Lauterborn, W., Hoppe-Seyler's Z. Physiol. Chem. *349*, p. 699 (1968)
Haller, W., Nature *206*, p. 693 (1965)
Haller, W., J. Chromatogr. *32*, p. 676 (1968)
Haller, W., J. Chromatogr. *85*, p. 129 (1973)
Hallick, R. B., C. Lipper, O. C. Richards and W. J. Rutter, Biochemistry *15*, p. 3039 (1976)
Hamilton, P. B., Anal. Chem. *30*, p. 914 (1958)
Hanai, T., P. D. S., Wood, G. D. Michaelis and L. W. Kinsell, Proc. Soc. Exp. Biol. Med. *129*, p. 226 (1968)
Hanson, L. A., B. G. Johansson and L. Rymo, Clin. Chim. Acta *14*, pp. 391—398 (1966)
Hanson, L. A., B. G. Johannsson and L. Rymo, Protides Biol. Fluids *14*, p. 579 (1967)
Hanus, J. and J. Kucera, J. Chromatogr. *97*, pp. 270—272 (1974)
Harada, T., A. Misaki, H. Akai and K. Yokorayashi, Biochim. Biophys. Acta *268*, p. 497 (1972)
Hardingham, T. E., S. Assan-Jackson and H. Muio, Biochem. J. *129*, p. 101 (1972)
Harrison, J. F. and B. E. Northam, Clin. Chim. Acta *14*, pp. 679—688 (1966)
Havran, R. T., I. L. Schwartz and R. Walter, J. Am. Chem. Soc. *91*, pp. 1836—1840 (1969)
Havran, R. T. and V. du Vigneaud, J. Am. Chem. Soc. *91*, pp. 3626—3628 (1969)
Hayes, F. N., E. Hanstury and V. E. Mitchell, J. Chromatogr. *16*, p. 410 (1964)
Hawker, C. D., H. Rasmussen and J. Glass, Am. J. Med. *43*, pp. 656—661 (1967)
Heftmann, E., Chromatography, Reinhold Publ. Co., New York, p. 406 (1961)
Heftmann, E., Chromatography, Reinhold Publ. Co., New York, Ch. 14, pp. 343—369 (1967)
Heinegard, J., Biochim. Biophys. Acta *285*, pp. 181, 193 (1972)
Heinz, F. and W. Prosch, Anal. Biochem. *40*, pp. 327—330 (1971)
Heitz, W., 6th International Seminar Gel Permeation Chromatography, Miami Beach, Florida, p. 130 (1968)
Heitz, W., Angew. Makromol. Chem. *10*/117, pp. 115—125 (1970a)
Heitz, W., J. Chromatogr. *53*/1, pp. 37—49 (1970b)
Heitz, W., Z. Anal. Chem. *277*, pp. 323—333 (1975)
Heitz, W., B. Bomer and H. Ullner, Makromol. Chem. *121*, pp. 102—116 (1969)
Heitz, W., and J. Coupek, 5th International Seminar Gel Permeation Chromatography, London, 1968 (1968a)
Heitz, W. and J. Coupek, 6th International Seminar Gel Permeation Chromatography, Miami Beach, Florida, p. 203, 1968 (1968b)
Heitz, W. and J. Coupek, Makromol. Chem. *112*, p. 286 (1968c)
Heitz, W. and W. Kern, Angew. Makromol. Chem. *1*, p. 150 (1967)
Heitz, W., H. Ullner and H. Höcker, Makromol. Chem. *98*, p. 42 (1966)
Helfferich, F., Ionenaustauscher, Band 1. Verlag Chemie GmbH, Neinheim/Bergstr. (1959)

Hellsing, K., Acta Chem. Scand. *19*, pp. 1791–1792 (1965)
Hellsing, K., J. Chromatogr. *36*, pp. 170–180 (1968)
Hendrickson, J. G., Anal. Chem. *40*, p. 59 (1968)
Hendrickson, J. G. and J. C. Moore, J. Polymer Sci. A. *174*, 167 (1966)
Henn, S. W. and G. K. Ackers, Biochemistry *8*, p. 3829 (1969)
Henry, R. A. and L. B. Rogers, Separ. Sci. *3*, p. 11 (1968)
Hermier, C. and M. Jutisz, Biochim. Biophys. Acta *175*, pp. 402–408 (1969)
Hersch, C. K., Molecular Sieves, Reinhold, New York (1961)
Heufer, G. and D. Braun, Polymer Letters *3*, p. 495 (1965)
Hiatt, C. W., A. Shelokov, E. J. Rosenthal and J. M. Galimore, J. Chromatogr. *56*, p. 362 (1971)
Hibberd, G. E., A. G. Ogston and D. J. Winzor, J. Chromatogr. *48*, pp. 393–399 (1970)
Hillman, D. E., Anal. Chem. *43*, p. 1007 (1971)
Hillman, D. E. and J. I. Paul, J. Chromatogr. *108*, p. 397 (1975)
Hjertén, S., Biochim. Biophys. Acta, *53*, p. 514 (1961)
Hjertén, S., Arch. Biochem. Biophys. Suppl. *1*, p. 147 (1962a)
Hjertén, S., Biochim. Biophys. Acta *62*, p. 445 (1962b)
Hjertén, S., Arch. Biochem. Biophys. *99*, pp. 466–475 (1962c)
Hjertén, S., Biochim. Biophys. Acta *79*, pp. 393–398 (1964)
Hjertén, S., Chromatography on agarose spheres. In Methods in Immunology and Immunochemistry (Eds M. W. Chase, M. W. and C. A. Williams), Academic Press Inc., New York, pp. 149–154 (1968)
Hjertén, S., J. Chromatogr. *50*, pp. 189–208 (1970)
Hjertén, S. and R. Mosbach, Anal. Biochem. *3*, p. 109 (1962)
Hodgson, R. and P. Sewell, J. Med. Lab. Techn. *22*, p. 130 (1965)
Hoffmann, L. G., J. Chromatogr. *40*, pp. 39–52 (1969)
Hoffmann, L. G. and P. W. McGivern, J. Chromatogr. *40*, pp. 53–61 (1969)
Hofsten, B. V., Exp. Cell Res. *41*, pp. 117–123 (1966)
Hohn, Th. and W. Pollmann, Z. Naturforsch. *18*b, p. 919 (1963)
Hohn, Th. and H. Schaller, Biochim. Biophys. Acta *138*, p. 466 (1967)
Holking, J. D., L. D. Jolly and R. D. Batt, Biochem. J. *128*, p. 69 (1972)
Hollman, M., J. Papenberg and W. Piper, Klin. Wschr. *48*, pp. 493–497 (1970)
Höglund, W., Virology *32*, pp. 662–677 (1967)
Hruby, V. J., C. W. Smith and D. K. Linn, J. Am. Chem. Soc. *94*, pp. 5478–5480 (1972)
Hruby, V. J. and V. du Vigneaud, J. Am. Chem. Soc. *91*, pp. 3624–3626 (1969)
Hummel, J. P. and W. J. Dreyer, Biochim. Biophys. Acta *63*, p. 530 (1962)
Inman, F. P. and A. Nisonoff, J. Biol. Chem. *241*, p. 322 (1966)
Inoue, H., K. Konishi and N. Taniguchi, J. Chromatogr. *47*, p. 348 (1970)
Ishizaka, T., K. Ishizaka and H. Bennich, J. Immunol. *104*, pp. 854–862 (1970)
Iverius, P. H. and T. C. Laurent, Biochim. Biophys. Acta *133*, pp. 371–372 (1967)
Iwama, M., N. Tagata and T. Homma, Kogyo Kagaku Zasshi *74*/2, pp. 277–281 (1971)
Jacobsson, L., Clin. Chim. Acta. *7*, p. 180 (1962)
Jacobsson, L. and G. Widström, Scand. J. Clin. Lab. Invest. *14*, p. 285 (1962)
Jamaluddin, M., S. Kim and W. K. Paik, Biochemistry *15*, p. 3077 (1976)
James, A. T. and A. J. P. Martin, Analyst *77*, p. 915 (1952)
James, A. N., E. Pickard and P. G. Shotton, J. Chromatogr. *32*, pp. 64–74 (1968)
Janado, M., J. Biochem. *76*, p. 1183 (1974)
Janson, J. C., J. Chromatogr. *28*, pp. 12–20 (1967)
Janson, J. C., J. Agr. Food Chem. *19*, pp. 581–588 (1971)
Jaworek, D., Chromatographia *2*, p. 289 (1970a)
Jaworek, D., Chromatographia *3*, pp. 414–417 (1970b)
Jefferis, R., Science Tools *22*, p. 33 (1975)
Johannsson, B. G. and L. Rymo, Acta Chem. Scand. *16*, pp. 2067–2068 (1962)
Johannsson, B. G. and L. Rymo, Acta Chem. Scand. *18*, pp. 217–223 (1964)
Johansson, G., BBA *221*, pp. 390–393 (1970a)
Johansson, G., BBA *222*, pp. 381–389 (1970b)
John, M., Technicon Symposium 70, Bad Homburg V. D. H., pp. 29–30 (1970)
Jones, O. W., G. E. Townsend, H. A. Sober and C. A. Heppel, Biochem. J. *3*, p. 238 (1964)

Joustra, M. K., Protides Biol. Fluids *14*, p. 533 (1967)
Joustra, M. K. Gel filtration on agarose. In Modern Separation Methods of Macromolecules and Particles (Ed. T. Gerritsen), Vol. 2, John Wiley and Sons, New York (1970)
Joustra, M., B. Söderquist and L. Fischer, J. Chromatogr. *28*, p. 21 (1967)
Junowicz, E., S. E. Charm and H. E. Blair, Anal. Biochem. *47*, p. 193 (1972)
Kado, C. I. and Y. Yin, Anal. Biochem. *39*, p. 339 (1971)
Kaiser, R., Chromatographia *1*, pp. 38—40 (1970)
Kakiuchi, K., S. Kato and A. Imanishi, J. Biochem. (Tokyo) *55*, p. 102 (1964)
Kalász, H., J. Chromatogr. *78*, p. 233 (1973)
Kalász, H., J. Nagy and J. Knoll, J. Chromatogr. *107*, pp. 35—42 (1975)
Kalász, H., K. Magyar and W. T. Barnes, J. Chromatogr. *128*, p. 300 (1976)
Kalász, H., T. Lengyel, J. Timár, G. Jóna and J. Knoll, Abstr. Commun. 7th Meet. Eur. Biochem. Soc., p. 339 (1971)
Karger, B. L., The Relationship of Theory to Practice in High-Speed Liquid Chromatography. Ch. 1. In Modern Practice of Liquid Chromatography (Ed. J. K. Kirkland). J. Wiley and Sons, Inc., New York, London, pp. 3—53 (1971)
Karlstam, B. and P. A. Albertsson, FEBS Letters *5*, pp. 360—363 (1969)
Katz, S., Anal. Biochem. *5*, p. 7 (1963)
Kecskés, E., I. Sures and D. Gallwitz, Biochemistry *15*, p. 2541 (1976)
Keil, B. and Z. Sormová, Biokémiai Laboratóriumi Technika (Biochemical Laboratory Techniques), Műszaki Könyvkiadó, Budapest (1968)
Kelley, R. N. and F. W. Billmeyer, Separ. Sci. *5/3*, pp. 291—296 (1970)
Kenedy, G. J. and J. H. Knox, J. Chromatogr. Sci. *10*, p. 549 (1972)
Kennedy, J. F., J. Chromatogr. *69*, p. 325 (1972)
Keresztes-Nagy, S., R. F. Mais, Y. T. Oester and J. F. Zarashinski, Anal. Biochem. *48*, p. 80 (1972)
Khym, J. X. and M. Uziel, J. Chromatogr. *47*, pp. 9—14 (1970)
Kibukamusoke, J. W. and N. E. Wilks, Lancet *1*, p. 301 (1965)
Killander, J., Biochim. Biophys. Acta *93*, pp. 1—14 (1964)
Killander, J., S. Bengtsson and L. Philipson, Proc. Soc. Exp. Biol. Med. *115*, p. 861 (1964)
Killander, J., J. Pontén and F. Rodén, Nature *192*, p. 182 (1961)
King, T. P., Biochemistry *11*, p. 367 (1972)
Kingsbury, D. W., J. Mol. Biol. *18*, pp. 195—203 (1966)
Kirkland, J. J. (ed.), Modern Practice of Liquid Chromatography. J. Wiley and Sons, Inc., New York, London, Sydney, Toronto (1971a)
Kirkland, J. J., Anal. Chem. *43*, pp. 36A—48A (1971b)
Kirkland, J. J., J. Chromatogr. *9*, pp. 206—214 (1971c)
Kirkland, J. J., J. Chromatogr. Sci. *10*, p. 129 (1972)
Kirret, O., I. Arro and H. Heinlo, Trans. Acad. Sci. Estonian SSR *15*, p. 414 (1966)
Klaus, G. G. B., D. E. Nitecki and J. W. Goodman, Anal. Biochem. *45*, pp. 286—297 (1972)
Klimisch, H. J. and D. Reese, J. Chromatogr. *67*, p. 299 (1972)
Klimisch, H. J. and L. Stadler, J. Chromatogr. *67*, pp. 291—297 (1972)
Koh, T. Y. and B. T. Khouw, Can. J. Biochem. *48*, pp. 225—227 (1970)
Kohn, J., Immunology *15*, pp. 863—865 (1968)
Koike, T., T. Uchida and F. Egami, J. Biochem. (Tokyo) *69*, p. 111 (1971)
Konig, K., W. Osterle and H. Rutly, Hoppe-Seyler's Z. Physiol. Chem. *352*, p. 977 (1971)
Konishi, K., H. Inoue and N. Taniguchi, J. Chromatogr. *54*, p. 367 (1971)
Kremmer, T., Gélszűrés. A biokémia modern módszerei, 3 (Gel filtration. Modern Biochemical Methods, 3), Magyar Kémikusok Egyesülete Biokémiai Szakosztálya, Kossuth Könyvkiadó, Budapest (1968)
Kremmer, T., D. Gaál and L. Holczinger, Protides of the Biol. Fluids (Ed. H. Peeters). Pergamon Press, Oxford, New York, Toronto, Sydney, Braunschweig, Vol. 19, pp. 71—74 (1972)
Kresse, H., H. Heidel and E. Buddecke, Eur. J. Biochem. *22*, p. 557 (1971)
Krieger, M., R. E. Koeppe and R. M. Stroud, Biochemistry *15*, p. 3458 (1976)
Kubin, M., J. Chromatogr. *108*, p. 1 (1975)
Kun, K. A. and R. Kunin, J. Polymer. Sci. *132*, 587 (1964)

Kunin, R. and R. J. Meyers, Discuss. Faraday Soc. 7, p. 114 (1949)
Kusch, P. and H. Zahn, Angew. Chem. 77, p. 720 (1965)
Laas, T., J. Chromatogr. 111, p. 373 (1975)
Laird, W. M. and R. L. M. Synge, J. Chromatogr. 54, p. 433 (1971)
Lakshmi, T. S. and P. K. Nandi, J. Chromatogr. 116, p. 177 (1976)
Lampert, K. and H. Determann, J. Chromatogr. 56, pp. 140—142 (1971a)
Lampert, K. and H. Determann, J. Chromatogr. 63, pp. 420—422 (1971b)
Langhammer, G. and K. Quitzsch, Makromol. Chem. 43, p. 160 (1961)
Langley, T. J. and E. L. Smith, J. Biol. Chem. 246, p. 3789 (1971)
Larsen, F. D., 6th International Seminar Gel Permeation Chromatography, Miami Beach, Florida, p. 111 (1968)
Lathe, G. H. and C. R. Y. Ruthven, Biochem. J. 60, p. XXXIV (1955)
Lathe, G. H. and C. R. Y. Ruthven, Biochem. J. 62, p. 665 (1956)
Laurent, T. C., Acta Chem. Scand. 17, p. 2664 (1963a)
Laurent, T. C., Biochem. J. 89, p. 253 (1963b)
Laurent, T. C., Biochim. Biophys. Acta 136, pp. 199—205 (1967)
Laurent, T. C. and K. A. Granath, Biochim. Biophys. Acta 136, pp. 191—198 (1967)
Laurent, T. C., K. Hellsing and B. Gelotte, Acta Chem. Scand. 18, p. 274 (1964)
Laurent T. C. and J. Killander, J. Chromatogr. 14, p. 317 (1964)
Laurent, T. C. and A. Pietruszkiewich, Biochim. Biophys. Acta 49, p. 258 (1961)
Lay, C. Y., Arch. Biochim. Biophys. 128, p. 202 (1968)
Lea, D. J. and A. H. Sehon, Canad. J. Chem. 40, p. 159 (1962)
Leach, A. A. and P. C. O'Shea, J. Chromatogr. 17, pp. 245—251 (1965)
Lederer, E. and M. Lederer, Chromatography. A review of principles and applications. Elsevier Publ. Co., Amsterdam, London, New York, Princeton (1957)
Ledvina, M., Chem. Listy 65/4, pp. 439—440 (1971)
Lee, M. and J. R. Debro, J. Chromatogr. 10, p. 68 (1963)
Lee, E. Y. C., C. Mercier and W. Shelan, Arch. Biochim. Biophys. 125, p. 1028 (1968)
Lehmann, F. G., Clin. Chim. Acta 28, p. 335 (1970)
Lehner, J. and A. I. Schepartz, J. Chromatogr. 39, p. 132 (1969)
Leibnitz, E. and H. G. Struppe, Handbuch der Gas-Chromatographie. Akademische Verlagsgesellschaft, Geest and Portig K. G., Leipzig, p. 109 (1970)
Lengyel, B., Műanyagipari Kutató Int. Közl. (Bulletin of the Plastic Research Inst.), 15, p. 3 (1968)
Le Page, M., R. Beau and A. J. De Vries, J. Polymer Sci. C 21, p. 119 (1968)
Levinson, J. W., A. Resostov, I. F. Liebes and J. J. McCormick, Biochim. Biophys. Acta 447, p. 260 (1976)
Lift, M. and V. M. Ingram, Biochem. J. 3, p. 560 (1964)
Limpert, R. J. and E. L. Obermiller, 6th International Seminar Gel Permeation Chromatography, Miami Beach, Florida, p. 285 (1968)
Lindner, E. B., A. Elmiquist and J. Porath, Nature 184, p. 1565 (1959)
Lindquist, I., Acta Chem. Scand. 21, p. 2564 (1967)
Lindquist, B. and T. Störgards, Nature 175, p. 511 (1955)
Lionetti, F. J. et al., J. Lab. Clin Med. 64, p. 519 (1963)
Lissitzky, S. and J. Bismuth, Clin. Chim. Acta 8, p. 269 (1963)
Lissitzky, S., J. Bismuth and M. Rolland, Clin. Chim. Acta 7, p. 183 (1962)
Lissitzky, S., J. Bismuth and C. Simon, Nature 199, p. 1002 (1963)
Little, G. H. and F. J. Behal, J. Chromatogr. 60, p. 137 (1971)
Little, J. N., J. L. Waters, K. L. Bombaugh and W. J. Pauplis, J. Polymer Sci. 7, p. 1775 (1969)
LKB Ultrogel, Pre-swollen polyacrylamide/agarose gel beads for high-speed gel filtration. Stockholm, Sweden (1975)
Lloyd, K. O., Biochemistry 11, p. 3884 (1972)
Locascio, G. A., H. A. Tigier and A. M. del C. Batlle, J. Chromatogr. 40, p. 453 (1969)
Loeb, J. E., Biochim. Biophys. Acta 157, pp. 424—426 (1968)
Loeb, J. E. and J. Chauveau, Biochim. Biophys. Acta 182, p. 225 (1969)
Louie, A. J., M. T. Sung and G. H. Dixon, J. Biol. Chem. 248, p. 355 (1973)
Lurquin, P. F., G. Tshitenge, G. Delaunoit and L. Ledeux, Anal. Biochem. 65, p. 1 (1975)
Lynn, K. R., J. Chromatogr. 66, pp. 375—376 (1971)

Lyon, H. and J. A. Singer, J. Biol. Chem. *246* p. 277 (1971)
Mach, B. and E. L. Tatum, Proc. Nat. Acad. Sci. U. S. A. *52*, p. 876 (1964)
Maggi, S., V. Fasano, N. De Ceglie, M. Colonna and N. Padolecchia, Bol. Soc. Ital. Biol. Sper. *44*, p. 18 (1968)
Mahowald, T. A., Biochemistry *4*, p. 732 (1965)
Makarova, S. B. and E. V. Yegorov, J. Chromatogr. *49*/1, p. 40 (1970)
Makowetz, E., K. Müller and H. Spitzy, Mikrochem. J. *10*, p. 194 (1966)
Malchow, D. and O. Lüderitz, Eur. J. Biochem. *2*, p. 469 (1967)
Male, C. A., Laboratory Practice *16* (1967)
Maley, L. E., J. Polymer Sci. *C8*, p. 253 (1965)
Manipol, V. and H. Spitzy, Int. J. Appl. Radiat. *13*, p. 647 (1962)
Margolis, S., J. Lip. Res. *8*, p. 501 (1967)
Marrink, J. and M. Gruber, FEBS Letters *2/4*, p. 242 (1969)
Marsden, N. V. B., Ann. N. Y. Acad. Sci. *125*, p. 428 (1965)
Marsden, N. V. B., J. Chromatogr. *58*, p. 304 (1971)
Marsden, N. V. B. and B. Wieselblad, JEOL News *13*, p. 11 (1976)
Marshall, J. J., J. Chromatogr. *53*, p. 379 (1970)
Marshall, R. L., W. C. Jones, R. A. Vigna and F. R. Gund, Z. Naturforsch. *C 29*, p. 90 (1974)
Martin, A. J. P. and R. L. M. Synge, Biochim. J. *35*, p. 1358 (1941)
Martin, L. F., N. R. Bertonière and S. P. Rowland, J. Chromatogr. *64*, p. 263 (1972)
Maruyama, H. and D. Mizuno, Biochim. Biophys. Acta *108*, p. 593 (1965)
Masamichi, I. and H. Terutake, Kogyo Kagaku Zasshi *74*, p. 277 (1971)
Maxwell, M. A. B. and J. P. Williams, J. Chromatogr. *31*, p. 62 (1967)
Mayhew, S. G. and L. G. Howell, Anal. Biochem. *41*, p. 466 (1971)
Mäkinen, K. K. and I. K. Paunio, Anal. Biochem. *39*, p. 202 (1971)
McGrath, R., Anal. Biochem. *16*, p. 402 (1966)
Malgunov, I. V., J. Chromatogr. *109*, p. 204 (1975)
Mercier, C. and W. I. Whelan, Eur. J. Biochem. *16*, p. 579 (1970)
Meyerhof, G., Makromol. Chem. *89*, p. 282 (1965a)
Meyerhof, G., Ber. Bunsen Ges. phys. Chem. *69*, p. 866 (1965b)
Meyerhof, G., Makromol. Chem. *134*, p. 129 (1970)
Mikes, J. A., J. Polymer Sci. *30*, p. 615 (1958)
Mikes, O., Laboratory Handbook of Chromatographic Methods. Van Nostrand Reinhold Co., London, New York, Toronto, Melbourne (1970)
Miller, J. N., O. Erinle, J. M. Roberts and C. Thirkettle, J. Chromatogr. *105*, p. 317 (1975)
Miranda, F., J. Chromatogr. *7*, p. 142 (1962)
Miranda, F., C. Kupeyan and H. Rochat, Eur. J. Biochem. *16*, p. 514 (1970)
Miranda, F., H. Rochat and S. Lissitzky, J. Chromatogr. *7*, p. 142 (1962)
Mizutani, T. and A. Mizutani, J. Chromatogr. *111*, p. 214 (1975)
Mizutani, T. and A. Mizutani, J. Chromatogr. *120*, p. 206 (1976)
Molselect 72. Gélszűrők és ioncserélők (Gel Filters and Ion Exchangers). Reanal Finomvegyszergyár, Budapest (1972)
Moore, J. C., J. Polymer Sci. *A2*, p. 835 (1964)
Moore, J. C., 3rd Int. Seminar on Gel Permeation Chromatography, Geneva (1966)
Moore, J. C. and J. G. Hendrickson, Polymer Preprints *5*, p. 706 (1964)
Moore, J. C. and J. G. Hendrickson, J. Polymer, Sci *C8*, 233 (1965)
Morávek, L., J. Chromatogr. *59*, p. 343 (1971)
Morgan, R. S., N. H. Morgan and R. A. Guinavan, Anal. Biochem. *45*, p. 668 (1972)
Mori, S. and T. Takeuchi, J. Chromatogr. *95*, p. 159 (1974)
Morris, C. I. O. R., J. Chromatogr. *16*, pp. 167—175 (1964)
Morris, C. I. O. R. and P. Morris, Separation Methods in Biochemistry. Interscience Publ. Inc., New York (1963)
Morris, C. I. O. R. and P. Morris, Separation Methods in Biochemistry. Interscience, London, Ch. 11 (1964)
Muench, K. H. and P. Berg, Biochemistry *5*, pp. 970—981 (1966)
Mulder, J. L. and F. A. Bugtenhuys, J. Chromatogr. *51*, p. 459 (1970)
Murphy, B. E. P., J. Clin. Endocrinol. Metab. *27*, p. 973 (1967)
Murphy, B. E. P., Nature N. Biol. *232*, pp. 21—24 (1971)

Muzsay, A., Műanyagipari Kut. Int. Közl. (Bulletin of the Plastic Research Inst.), *18*, p. 1, Budapest (1970)
Müller, K., Clin. Chim. Acta *17*, p. 21 (1967)
Nagy, B. S., Dielektrometria (Dielectrometry). Műszaki Könyvkiadó, Budapest (1970)
Nakae, A. and G. Muto, J. Chromatogr. *120*, p. 47 (1976)
Nandi, P. K., J. Chromatogr. *116*, p. 93 (1976)
Nathenson, S. G., Biochemistry, *4*, p. 2412 (1965)
Neddermeyer, P. A. and L. B. Rogers, Anal. Chem. *40*, pp. 755—762 (1968)
Neddermeyer, P. A. and L. B. Rogers, Anal. Chem. *41*, pp. 94—102 (1969)
Nelsestuen, G. L. and J. W. Suttie, J. Biol. Chem. *247*, p. 6096 (1972)
Nichol, L. W., A. G. Ogston and D. I. Winzor, Arch. Biochem. Biophys. *121*, pp. 517—521 (1967a)
Nichol, L. W., A. G. Ogston and D. I. Winzor, J. Phys. Chem. *71*, p. 726 (1967b)
Nichol, L. W., W. H. Sawyer and D. I. Winzor, Biochem. J. *112*, pp. 259—260 (1969)
Nichol, L. W. and D. J. Winzor, J. Phys. Chem. *68*, p. 2455 (1964)
Nilsson, A., Acta Chem. Scand. *16*, p. 31 (1962)
Nilsson, G. and K. Nilsson, J. Chromatogr. *101*, p. 137 (1974)
Nordin, P., Arch. Biochem. Biophys. *99*, p. 101 (1962)
Novotny, J., FEBS Letters *14*, p. 7 (1971)
Nyström, E. and J. Sjövall, J. Chromatogr. *17*, p. 574 (1965a)
Nyström, E., Anal. Biochem. *12*, p. 235 (1965b)
Nyström, E. and J. Sjövall, J. Chromatogr. *24*, pp. 208—212 (1966a)
Nyström, E., J. Chromatogr. *24*, pp. 212—214 (1966b)
Nyström, E. and J. Sjövall, Arkiv Kemi *29*, pp. 107—115 (1968)
Nyström, E. and J. Sjövall, Methods in Enzymology (Ed. J. M. Löwenstein). Academic Press, New York, Vol. XXXV, p. 378 (1975)
Octyl-Sepharose CL-48; Phenyl-Sepharose Cl-48 for hydrophobic interaction chromatography; Pharmacia Fine Chemicals AB, Uppsala, Sweden (1976)
Oelert, H. H., J. Chromatogr. *53*, p. 241 (1970)
Ogata, T., N. Yoza and S. Ohashi, J. Chromatogr. *58*, pp. 267—276 (1971)
Ogston, A. G., Trans. Faraday Soc. *54*, p. 1754 (1958)
Ogston, A. G. and P. Silpananta, Biochem. J. *116*, p. 171 (1970)
Ogston, A. G. and J. D. Wells, Biochem. J. *119*, p. 119 (1970)
Ogston, A. G. and J. D. Wells, Biochem. J. *128*, p. 685 (1972)
Ohashi, S., N. Yoza and Y. Uene, J. Chromatogr. *24* (1966)
Oka, S., S. Shigeta and S. Sato, Agr. Biol. Chem. *35*, p. 1216 (1971)
Olesen, H. and P. O. Pedersen, Protides Biol. Fluids *14*, p. 607 (1966)
Ortner, H. M., Anal. Chem. *47*, p. 162 (1975a)
Ortner, H. M., J. Chromatogr. *107*, p. 335 (1975b)
Ortner, H. M. and H. Dalmonego, J. Chromatogr. *89*, p. 287 (1974)
Ortner, H. M. and H. Dalmonego, J. Chromatogr. *107*, p. 341 (1975)
Ortner, H. M., H. Krainer and H. Dalmonego, J. Chromatogr. *82*, p. 249 (1973)
Ortner, H. M. and O. Pacher, J. Chromatogr. *71*, p. 55 (1972)
Ortner, H. and H. Spitzy, Z. Anal. Chem. *238*, p. 167 (1968a)
Ortner, H. and H. Spitzy, Z. Anal. Chem. *238*, p. 251 (1968b)
Oster, H., S. Van Damme and E. Ecker, Chromatographia *4*, p. 209 (1971)
Öberg, B. and L. Philipson, Arch. Biochem. Biophys. *119*, pp. 504—509 (1967)
Öbrink, B., T. C. Lawrent and R. Rigler, J. Chromatogr. *31*, p. 48 (1967)
Örgen, S. and U. Lindahl, Biochem. J. *125*, p. 1119 (1971)
Östling, G., Acta Soc. Med. Upsaliensis *65*, p. 222 (1962)
Pacco, J. M., J. Chromatogr. *55/1*, p. 99 (1971)
Pagé, M. and C. Godin, J. Chromatogr. *50*, p. 66 (1970)
Painter, R. H. and G. A. McVicar, Can. J. Biochem. *41*, p. 2269 (1963)
Papenberg, J., W. Piper and Y. Knobloch, Z. Anal. Chem. Fresenius *252*, p. 248 (1970)
Partridge, S. M., Nature *169*, p. 496 (1952)
Patrick, R. L. and R. E. Thiers, Clin. Chem. *9*, p. 283 (1963)
Patterson, P. H. and W. J. Lennarz, Biochim. Biophys. Res. Commun. *40*, p. 408 (1970)
Pearson Murphy, B. E. and C. J. Pattee, J. Clin. Endocrinol. Metabol. *24*, p. 187 (1964)
Pecka, K., S. Hal, J. Chlebek, M. Karas and B. Kremanova, J. Chromatogr. *104*, p. 91 (1975)

Pecsok, R. L. and D. Sauders, Separ. Sci. *3*, p. 325 (1968)
Pedersen, K. O., Arch. Biochem. Biophys. Suppl. *1*, p. 157 (1962)
Pertroft, H. and A. Hnallén, J. Chromatogr. *128*, p. 125 (1976)
Peterson, E. A., Celluloic Ion Exchangers. In Laboratory Techniques in Biochemistry and Molecular Biology (Eds T. S. Work and E. Work), North-Holland Publ. Co., Amsterdam, London, pp. 297—314 (1970)
Peterson, E. A. and J. Rowland, J. Chromatogr. *5*, p. 330 (1961)
Peterson, E. A. and H. A. Sober, Anal. Chem. *31*, p. 857 (1959)
Petrovič, S., M. Novakovič and J. Petrovič, Biochim. Biophys. Acta *254*, p. 493 (1971)
Pettersson, G., E. B. Vowling and J. Porath, Biochim. Biophys. Acta *67*, p. 1 (1963)
Pettersson, U., L. Philipson and S. Höglund, Virology *35*, p. 204 (1968)
Pettijohn, D., Eur. J. Biochem. *3*, p. 25 (1967)
Pfleiderer, G., Mechanismen enzymatischer Reaktionen. Springer Verlag, Berlin, p. 300 (1964)
Philipson, L., Virology *28*, p. 36 (1966)
Pierce, J. G. and P. L. Zubhoff, Biochim. Biophys. Acta *87*, p. 343 (1964)
Piukovich, E. and L. Boross, Proc. XIIIth Meet. Hung. Biochem. Szombathely (1972)
Polson, A., Biochim. Biophys. Acta *19*, p. 53 (1956)
Polson, A., Biochim. Biophys. Acta *50*, p. 565 (1961)
Polson, A. and W. Katz, Biochem. J. *112*, p. 387 (1969)
Popl, M., C. Coupek and S. Pokorny, J. Chromatogr. *104*, p. 135 (1975)
Popovic, D. A., Anal. Biochem. *67*, p. 462 (1975)
Porath, J., Clin. Chim. Acta *4*, p. 776 (1959)
Porath, J., Biochim. Biophys. Acta *38*, p. 193 (1960)
Porath, J., Advances in Protein Chemistry. Academic Press, New York, Vol. 17, p. 209 (1962a)
Porath, J., Nature *196*, p. 47 (1962b)
Porath, J., J. Pure Appl. Chem. *6*, p. 233 (1963)
Porath, J., Metabolism *13*, p. 1004 (1964)
Porath, J., Laboratory Practice *16*, p. 838 (1967)
Porath, J., J. Chromatogr. *60*, p. 167 (1971)
Porath, J., Biotechnol. Bioeng. Symp. *3*, p. 145 (1972)
Porath, J., Biochimie *55*, p. 943 (1973)
Porath, J. and H. Bennich, Arch. Biochem. Biophys. *29*, Suppl. 1, p. 152 (1962)
Porath, J. and P. Flodin, Nature *183*, p. 1657 (1959)
Porath, J. and P. Flodin, Protides Biol. Fluids *10*, p. 290 (1963)
Porath, J., J. C. Janson and T. Laas, J. Chromatogr. *60*, p. 167 (1971)
Porath, J., T. Laas and J. C. Janson, J. Chromatogr. *103*, p. 49 (1975)
Porath, J. and E. B. Lindner, Nature *191*, p. 69 (1961)
Porath, J. and N. Ui, Biochim. Biophys. Acta *90*, p. 324 (1964)
Porter, H., Biochim. Biophys. Acta *229*, p. 143 (1971)
Porter, L. J. and R. D. Wilson, J. Chromatogr. *71*, p. 570 (1972)
Preston, J. F., E. Lapis and J. E. Gander, Arch. Biochem. Biophys. *134*, p. 324 (1969a)
Preston, J. F., E. Lapis, S. Westerhouse and J. E. Gander, Arch. Biochem. Biophys. *134*, p. 316 (1969b)
Přistoupil, T. I., J. Chromatogr. *19*, p. 64 (1965)
Přistoupil, T. I. and S. Ulrych, J. Chromatogr. *28*, p. 49 (1967)
Ptizin, H. and Y. Eizner, Sovjet Phys. Tech. Phys. *4*, p. 1020 (1966)
Pusztai A. and W. B. Watt, Eur. J. Biochem. *10*, p. 523 (1969)
Pusztai, A. and W. B. Watt, Biochim. Biophys. Acta *214*, p. 463 (1970)
Quarfordt, S. N., A. Nathans and M. Dowdee, J. Lip. Res. *13*, p. 435 (1972)
Radola, B. J., J. Chromatogr. *38*, pp. 61—77 (I) (1968a)
Radola, B. J., J. Chromatogr. *38*, pp. 78—90 (II) (1968b)
Raftery, M. A., Anal. Biochem. *30*, p. 427 (1969)
Rapp, W. and H. E. Lehmann, Clin. Chim. Acta *31*, p. 45 (1971)
Ratcliff, A. P. and J. Hardwicke, J. Clin. Path. *17*, p. 676 (1964)
Rauen, H. M. and K. Felix, Z. Physiol. Chem. *283*, p. 139 (1948)
Raymond, S. and Y. J. Wang, Anal. Biochem. *1*, p. 391 (1960)
Raymond, S. and L. Weintraub, Science *130*, p. 710 (1959)

Reanal's Acrilex Gel Filters, Reanal, Budapest (1972)
Reanal's Molselect Gel Filters, Reanal, Budapest (1972)
Reiner, J. M. and B. Reiner, Anal. Biochem. *4*, p. 1 (1962)
Renkin, E. M., J. Gen. Physiol. *38*, p. 225 (1955)
Revell, P. A. and H. Muir, Biochem. J. *130*, p. 597 (1972)
Rice, O. K., J. Chem. Phys. *12*, p. 1 (1944)
Richardson, R. W., Nature *164*, p. 916 (1949)
Richardson, R. W., J. Chem. Soc. *910* (1951)
Ringer, D., S. Chludek and J. Ofengand, Biochemistry *15*, p. 2759 (1976)
Rivera, J. V., E. Toro-Goyco and M. L. Matos, Amer. J. Med. Sci. *249*, p. 371 (1965)
Roark, D. W., T. E. Geoghegan, G. H. Keller, K. V. Matter and R. L. Engle, Biochemistry *15*, p. 3019 (1976)
Roberts, G. P., J. Chromatogr. *22*, p. 90 (1966)
Roberts, G. P. and R. A. Gibbons, Biochem. J. *98*, p. 426 (1966)
Rogers, J., D. Boldt, S. Kornfeld, S. A. Skinner and C. R. Valeri, Proc. Nat. Acad. Sci. U.S. A. *69*, p. 1685 (1972)
Rogers, K. S., L. Hellerman and T. G. Thompson, J. Biol. Chem. *240*, p. 198 (1965)
Romanowska, R., Anal. Biochem. *33*, p. 383 (1970)
Roncari, G. and H. Zuber, Int. J. Protein Res. *1*, p. 45 (1969)
Ross, J. H. and M. E. Casto, J. Polymer Sci. Part C, *21*, p. 143 (1968)
Rothstein, F., J. Chromatogr. *18*, p. 36 (1965)
Roubal, W. T. and A. L. Tappel, Anal. Biochem. *9*, p. 211 (1964)
Röschenthaler, R. and P. Fromageot, J. Mol. Biol. *11*, p. 358 (1965)
Russel, B., T. H. Mead and A. Polson, Biochim. Biophys. Acta *86*, p. 169 (1964)
Russel, J. and J. M. Osborn, Nature *220*, p. 1125 (1968)
Ruttenberg, H., Ph. D. Thesis. Cit.: Janson (1967)
Sabbagh, N. K. and J. S. Fagerson, J. Chromatogr. *120*, p. 55 (1976)
Sachs, D. H. and E. Painter, Science *175*, p. 781 (1972)
Saitoh, K., M. Satoh and N. Suzuki, J. Chromatogr. *92*, p. 291 (1974)
Saitoh, K. and N. Suzuki, J. Chromatogr. *109*, p. 333 (1975a)
Saitoh, K. and N. Suzuki, J. Chromatogr. *111*, p. 29 (1975b)
Sajgó, M., Acta Biochim. Biophys. Acad. Sci. Hung. *4*, p. 385 (1969)
Samuelson, O., Cit. W. Lautsch: Angew. Chem. *57*, p. 149 (1944)
Samuelssson, E. G., A. Emneus and B. Hallström, Svesnka Mejeritidn *59*, p. 3 (1967a)
Samuelsson, E. G., P. Tibbling and S. Holm, Food Technol. *21*, p. 121 (1967b)
Sanfelippo, P. M. and J. G. Surak, J. Chromatog. *13*, p. 148 (1964)
Sargent, R. N. and D. L. Graham, Anal. Chim. Acta *30*, p. 101 (1964)
Sata, T., D. L. Estrich, P. D. S. Wood and L. W. Kinsell, J. Lip. Res. *11*, p. 331 (1970)
Sato, S. and Y. Otaka, Chem. Econ. Eng. Rev. *3*, p. 40 (1971)
Saunders, D. and R. L. Pecsok, Anal. Chem. *40*, p. 44 (1968)
Scatchard, G., Ann. N. Y. Acad. Sci. *51*, p. 660 (1949)
Schales, O. and S. S. Schales, J. Biol. Chem. *140*, p. 879 (1941)
Schell, H. and V. Ghetie, Stud. Cercet. Biochim. *11*, p. 69 (1968)
Schell, P. L., J. Biol. Chem. *240*, p. 472 (1971)
Schleich, T. and J. Goldstein, Proc. Nat. Acad. Sci. (Wash.) *52*, p. 744 (1964)
Schmidt-Kastner, G., Naturwiss. *51*, p. 38 (1964)
Schulsatz, H. W., R. Kisters, A. Wollner and H. Gyeiling, Z. Physiol. Chem. *352*, p. 289 (1971)
Schutton, W. D., Biochim. Biophys. Acta *240*, p. 522 (1971)
Schwartz, A. N., A. W. Yee and B. A. Zetin, J. Chromatogr. *20*, p. 154 (1965)
Scott, G. D., Nature *188*, p. 908 (1960)
Scott, R. P. W., J. Chromatogr. Sci. *9*, p. 449 (1971)
Scott, T. A. and E. H. Melvin, Anal. Chem. *25*, p. 1656 (1953)
Segal, L., Polymer Letters *4*, p. 1011 (1966)
Segal, L., J. Polymer Sci. C *21*, p. 267 (1968)
Seki, T., J. Chromatogr. *115*, p. 262 (1975)
Selby, K. and C. C. Maitland, Biochem. J. *94*, p. 578 (1965)
Senior, A. E. and D. H. McLennan, J. Biol. Chem. *245*, p. 5080 (1970)

Separation News, Pharmacia Fine Chemicals AB, Uppsala. January—March—May—September 1972, 1973, 1974, 1—2—3—1975, 1—2—1976, 1—1977

Sephadex 1. — A Unique Substance for Modern Chromatography. Sephadex 2. — Theory and Experimental Technique, Pharmacia Fine Chemicals AB, Uppsala, Sweden (1963)

Sephadex Laboratory Columns and Accessories, Pharmacia Fine Chemicals AB, Uppsala (1964)

Sephadex-G-10 and G-15 for Fractionation of Low Molecular Weight Solutes, Pharmacia Fine Chemicals AB, Uppsala (1965)

Sephadex-G-150 for Increased Separation Possibilities, Pharmacia Fine Chemicals AB, Uppsala (1965)

Sephadex, Upward-Flow Adaptor, Pharmacia Fine Chemicals AB, Uppsala (1965)

Sephadex — Solvent Resistant Column, Pharmacia Fine Chemicals AB, Uppsala (1966)

Sephadex G-25 Medium and G-50 Medium in Bead Form, Pharmacia Fine Chemicals AB, Uppsala (1967)

Sephadex — DEAE Anion Exchanger; Sephadex — QUAE a Fully Quaternized Ion Exchanger, Pharmacia Fine Chemicals AB, Uppsala (1968)

Sephadex — Thin-Layer Gel Filtration, Superfine Beads, Pharmacia Fine Chemicals AB, Uppsala (1964)

Sephadex — Column K 100/100 and New K 215/100 for Preparative and Plot Plant Gel Filtration, Pharmacia Fine Chemicals AB, Uppsala (1968)

Sephadex — Industrial Gel Filtration with the Sephamatic System, Pharmacia Fine Chemicals AB, Uppsala (1968)

Sephadex — and Other Separation Products, Pharmacia Fine Chemicals AB, Uppsala (1969)

Sephadex — Ion Exchangers, an Outstanding Aid in Biochemistry, Pharmacia Fine Chemicals AB, Uppsala (1969)

Sephadex Laboratory Columns, Pharmacia Fine Chemicals AB, Uppsala (1969)

Sephadex — Gel Filtration in Theory and Practice, Pharmacia Fine Chemicals AB, Uppsala (1969)

Sephadex — Separation Bulletin, Pharmacia Fine Chemicals AB, Uppsala (1969)

Sephadex Gel and Eluant Reservoirs, Pharmacia Fine Chemicals AB, Uppsala (1969)

Sephadex Ion Exchangers, Pharmacia Fine Chemicals AB, Uppsala (1970)

Sephadex LH-20, Chromatography in Organic Solvents, Pharmacia Fine Chemicals AB, Uppsala (1970)

Sephadex LH-60, Chromatography in Organic Solvents, Pharmacia Fine Chemicals AB, Uppsala (1976)

Sephadex Gel Filtration in Theory and Practice, Pharmacia Fine Chemicals AB, Uppsala (1970)

Sephadex
Dextran Fractions, Dextran Sulphate,
DEAE-Dextran, Defined Polymers for Biological Research,
Pharmacia Fine Chemicals AB, Uppsala (1971)

Sephadex
Thin-Layer Gel Filtration with the Pharmacia TLG Apparatus, Pharmacia Fine Chemicals AB, Uppsala (1971)

Sephadex
3-Way Valve LV-3,
4-Way Valve LV-4,
Tubing Connectors, for Liquid Chromatographic Systems, Pharmacia Fine Chemicals AB, Uppsala (1971)

Sephadex-Literature References,
Sephadex Ion Exchanger, Sepharose, Dextran Fractions, Dextran Derivatives, Ficoll, Volumes: 1958—1968, 1969, 1970, 1971, 1972, 1973, 1974, Pharmacia Fine Chemicals AB, Uppsala

Sepharose 2B—4B—6B,
Pharmacia Fine Chemicals AB, Uppsala (1969)

Sepharose CL for Gel Filtration and Affinity Chromatography,
Pharmacia Fine Chemicals AB, Uppsala, Sweden (1975)

Sepharyl S-200 Superfine for High Performance Gel Filtration, Pharmacia Fine Chemicals AB, Uppsala (1976)
Shapiro, B. and J. L. Rabinowitz, J. Nucl. Med. *3*, p. 417 (1962)
Sheikh, M. I. and J. V. Möller, Biochim. Biophys. Acta *158*, p. 456 (1968)
Srivastava, P. K., H. Goch and K. Zakreewaki, Biochim. Biophys. Acta *271*, p. 310 (1972)
Siakotos, A. N. and G. Rouser, J. Am. Oil Chem. Soc. *42*, p. 913 (1965)
Siegel, L. M. and K. J. Monty, Biochem. Biophys. Res. Commun. *19*, p. 494 (1965)
Siegel, L. M. and K. J. Monty, Biochem. Biophys. Acta *112*, p. 346 (1966)
Simkin, P. A., J. Chromatogr. *47*, p. 103 (1970)
Sinibaldi, M. and M. Lederer, J. Chromatogr. *107*, p. 210 (1975)
Sjövall, J. and R. Vihko, Acta Chem. Scand. *20*, p. 1419 (1966a)
Sjövall, J. and R. Vihko, Steroids *7*, p. 447 (1966b)
Slagt, C. and W. A. Sipman, J. Chromatogr. *74*, p. 352 (1972)
Smith, J. K., R. H. Eaton and L. G. Whitby, Anal. Biochem. *23*, p. 84 (1968)
Smith, W. B. and A. Kollmansberger, J. Phys. Chem. *69*, p. 4157 (1965)
Smith, W. B., S. A. May and C. W. Kim, J. Polymer Sci. *4*, Part *A-2*, p. 365 (1966)
Smith, W. V., J. Polymer Sci. Part A 2, *8*, p. 207 (1970)
Smithies, O., Biochem. J. *61*, p. 629 (1955)
Snyder, L. R., J. Chromatogr. Sci. *7*, p. 352 (1969)
Snyder, L. R., J. Chromatogr. Sci. *10*, p. 200 (1972a)
Snyder, L. R., J. Chromatogr. Sci. *10*, p. 369 (1972b)
Somers, T. C., Nature *209*, p. 368 (1966)
Spenzer, M. E. and I. B. Walker, Eur. J. Biochem. *19*, p. 451 (1971)
Spitzy, H., H. Skrube and K. Müller, Microchim. Acta *296* (1961)
Squire, P. G., Arch. Biochem. Biophys. *107*, p. 471 (1964)
Staal, G. E. I., J. F. Koster and H. Kamp, Biochim. Biophys. Acta *227*, p. 86 (1971)
Staal, G. E. I., J. Visser and C. Veeger, Biochim. Biophys. Acta *185*, p. 39 (1969)
Steere, R. L. and G. K. Ackers, Nature *196*, p. 475 (1962)
Stevanson, G. T., J. Chromatogr. *37*, p. 116 (1968)
Stracher, A., Biochim. Biophys. Res. Commun. *35*, p. 519 (1969)
Strain, H., Anal. Chem. *32*, 3R (1971)
Streuli, C. A., J. Chromatogr. *56*, p. 219 (1971a)
Streuli, C. A., J. Chromatogr. *56*, p. 225 (1971b)
Sun, K. and A. H. Sehon, Canad. J. Chem. *43*, p. 969 (1965)
Susskind, H. and W. Becker, Nature *212*, p. 1564 (1966)
Svensson, H., J. Chromatogr. *25*, p. 266 (1966)
Swart, A. C. W. and H. C. Hemker, Biochim. Biophys. Acta *222*, p. 692 (1970)
Sweetman, L. and W. L. Nyhan, J. Chromatogr. *32*, p. 662 (1968)
Sweetman, L. and W. L. Nyhan, J. Chromatogr. *59*, p. 349 (1971)
Synge, R. L. M. and A. Tiselius, Biochem. J. *78*, p. 31P (1950)
Synge, R. L. M. and M. A. Youngson, Biochem. J. *78*, p. 31P (1961)
Szepesy, L., Gázkromatográfia (Gas Chromatography). Műszaki Könyvkiadó, Budapest, p. 218 (1970)
Takagi, T., 3rd International Seminar on Gel Permeation Chromatography. Geneva, May (1966)
Takagi, T. and S. Iwanage, Biochim. Biophys. Res. Commun. *38*, p. 129 (1970)
Talarico, P. C., E. W. Albaugh and R. E. Snyder, Anal. Chem. *40*, p. 2192 (1968)
Tanaka, H. and K. Konishi, J. Chromatogr. *64*, p. 61 (1972)
Tanaka, K., H. H. Richards and G. L. Cantoni, Biochim. Biophys. Acta *61*, p. 846 (1962)
Tanford, C., Physical Chemistry of Macromolecules. John Wiley and Sons Inc., New York, pp. 193—221 (1963)
Tangen, O., H. J. Berman and P. Marfey, Thromb. Diath. Haemorrh. *25*, p. 268 (1971)
Taniguchi, N., Clin. Chim. Acta *30*, p. 801 (1970)
Tentori, L., G. Vivaldi and S. Carta, J. Chromatogr. *28*, p. 55 (1967)
Terner, C., E. I. Szabó and N. L. Smith, J. Chromatogr. *47*, p. 15 (1970)
Texter, J. A., R. Kellerman and K. Klier, Carbohydrate Res. *41*, p. 191 (1975)
Thang, M. N., B. Beltcher, M. Grunberg and C. Manago, Eur. J. Biochem. *19*, p. 184 (1971)

Therriault, D. G. and R. H. Poe, Can. J. Biochem. *43*, p. 1427 (1965)
Thieme, T. R. and C. E. Ballen, Biochemistry *11*, p. 1115 (1972)
Thompson, A. R., Nature *169*, p. 495 (1952)
Tipton, C. L., J. W. Paulis and M. D. Pierson, J. Chromatogr. *14*, p. 486 (1964)
Tiselius, A., Nature *133*, p. 212 (1934)
Tiselius, A., Am. Rev. Biochem. *37*, 1 (1968a)
Tiselius, A., Bull. Soc. Chim. Biol. *50*, 12, p. 2201 (1968b)
Tiselius, A., J. Porath and P. Albertsson, Science *141*, p. 13 (1963)
Toro-Goyco, E., M. Martirez-Maldonado and M. Matos, Proc. Soc. Exptl. Biol. Med. *122*, p. 301 (1966)
Tortelani, G. and E. Romegudi, Anal. Biochem. *66*, p. 29 (1975)
Troth, H. G., 5th International Seminar Gel Permeation Chromatography, London (1968)
Troy, F. A., F. E. Frerman and E. C. Heath, J. Biol. Chem. *246*, p. 118 (1970)
Tsai, C. S., Anal. Biochem. *36*, p. 116 (1970)
Tsiganos, C. P., T. E. Hardingham and H. Muir, Biochim. Biophys. Acta *229*, p. 529 (1971)
Tung, F., P. Kantesaria and P. Morfey, Biochim. Biophys. Acta *335*, p. 318 (1974)
Tung, L. H., J. Appl. Polymer Sci., *10*, p. 375 (1966)
Turner, M. W. and H. Bennich, Biochem. J. *107*, p. 171 (1968)
Turner, M. W., L. Martensson and J. B. Natvig, Nature *221*, p. 1166 (1969)
Udvarhelyi, K., Kém. Közl. *32*, p. 41 (1969)
Ueno, Y., N. Yoza and S. Ohashi, J. Chromatogr. *52*, p. 321 (1970a)
Ueno, Y., Yoza and S. Ohashi, J. Chromatogr. *52*, p. 469 (1970b)
Ueno, Y., N. Yoza and S. Ohashi, J. Chromatogr. *52*, p. 481 (1970c)
Ulrych, S. and T. I. Přistoupil, J. Chromatogr. *59*, 437 (1971)
Unger, K., J. Chromatogr. *83*, p. 5 (1973)
Unger, K., J. Chromatogr. *99*, p. 435 (1974)
Ungerer, K., cit. Lengyel, B., Műanyagipari Kut. Int. Közl. (Bulletin of the Plastic Research Inst.), *15* (1968)
Uriel, J., I. Bergés, E. Boschetti and R. Tixier, C. R. Acad. Sci. (Paris) Série C *273*, p. 2358 (1971)
Uziel, M. and W. E. Cohn, Biochim. Biophys. Acta *103*, p. 539 (1965)
Vámos, E., Kromatográfia. Mőszaki Könyvkiadó, Budapest (1959)
Van Deemter, J. J., P. J. Ziderweg and A. Klinkenberg, Chem. Eng. Sci. *5*, p. 271 (1956)
Vándor, E., Dextrán-epiklórhydrin kopolimerek előállítása és tulajdonságainak vizsgálata (Production and Properties of Dextran-Epichlorhydrin Copolymers, Thesis). Budapest (1965)
Van Tilburg, A. M. J. and Ch. J. Muller, Clin. Chim. Acta *29*, p. 5 (1970)
Vaughan, M. F., Nature *188*, p. 55 (1960)
Vaughan, M. F., Nature *195*, p. 801 (1962)
Vavruch, I., Kolloid-Z. *205*, p. 32 (1965)
Veatch, W. R. and E. R. Blut, Biochemistry *15*, p. 3026 (1976)
Vejrosta, J. and J. Malek, J. Chromatogr. *109*, p. 101 (1975)
Vilko, R., Acta Endocrin. 52, Suppl. *109*, p. 15 (1966)
Vink, H., J. Chromatogr. *52*, p. 205 (1970)
Vink, H., J. Chromatogr. *69*, p. 237 (1972)
Vondruška, M., M. Šudřich and M. Mládek, J. Chromatogr. *116*, p. 457 (1976)
Wadström, T. and K. Hisatsune, Biochem. J. *120*, p. 735 (1970)
Wagner, E. and V. H. Ingram, Biochem. J., p. 3019 (1966)
Waldmann-Meyer, H., Biochim. Biophys. Acta *261*, p. 148 (1972)
Wallach, D. F. and V. B. Kamat, Proc. Nat. Acad. Sci. (Wash.) *52*, p. 721 (1964)
Walter, H., E. J. Krob and R. Garza, Biochim. Biophys. Acta *165*, p. 507 (1968a)
Walter, H., R. Garza and R. P. Coyle, Biochim. Biophys. Acta *156*, p. 409 (1968b)
Walter, H., E. J. Krob and R. Garza, Exp. Cell. Res. *55*, p. 57 (1969)
Walter H. and S. Sasakawa, Biochemistry *10*, p. 108 (1971)
Walter, H., F. W. Selby and R. Garza, Biochim. Biophys. Acta *136*, p. 148 (1967)
Walter, H., R. Winge and F. W. Selby, Biochim. Biophys. Acta *109*, p. 293 (1965)

Ward, D. N. and M. S. Arnott, Anal. Biochem. *12*, p. 296 (1965)
Warshaw, H. S. and G. K. Ackers, Anal. Biochem. *42*, p. 405 (1971)
Wasternack, C., Pharmazie *27*, p. 67 (1972a)
Wasternack, C., J. Chromatogr. *71*, p. 67 (1972b)
Wasternack, C. and H. Reinbothe, J. Chromatogr. *48*, p. 551 (1970)
Wasteson, A., Biochim. Biophys. Acta *117*, p. 152 (1969)
Wasteson, A., J. Chromatogr. *59*, p. 87 (1971)
Wasteson, A., U. Lindahl and A. Hallen, Biochem. J. *130*, p. 729 (1972)
Wasyl, Z., E. Luchter and W. Bielanski, Biochim. Biophys. Acta *243*, p. 11 (1971)
Waters, J. L., J. N. Little and D. F. Horgan, J. Chromatogr. Sci. *7*, p. 293 (1969)
Weeke, J., H. Ørskov and H. J. G. Gundersen, J. Lab. Clin. Med. *80*, p. 839 (1972)
Weiss, P., J. Herz, P. Rempp, Z. Gallot and H. Benoit, Makromol. Chem. *145*, p. 105 (1971)
Welling, W., Science Tools *15*, p. 24 (1968)
Wells, M. A. and J. C. Dittmer, Biochemistry *2*, p. 1259 (1963)
Westman, A. E. R. and H. R. Hugill, J. Amer. Ceramic Soc. *13*, p. 767 (1930)
Wheaton, R. M. and W. C. Bauman, Ann. N. Y. Acad. Sci. *57*, p. 159 (1953)
Whitaker, J. R., Anal. Chem. *35*, p. 1950 (1963)
White, H. D. and W. R. Jencks, Amer. Chem. Soc. Meeting Abstr. *43* (1970)
White, M. L. and G. H. Dorion, J. Polymer Sci. *55*, p. 731 (1961)
Widén, R. and K. E. Eriksson, J. Chromatogr. *15*, p. 429 (1964)
Wiegner, G., J. Soc. Chem. Ind. *50*, p. 65 T (1931)
Wieland, T. and H. Bende, Chem. Ber. *98*, p. 504 (1965)
Wieland, T. and H. Determann, J. Chromatogr. *28*, p. 2 (1967)
Wieland, Th., P. Duesberg and H. Determann, Biochem. Z. *337*, p. 303 (1963)
Wilk, M., J. Rochlitz and H. Bende, J. Chromatogr. *24*, p. 414 (1966)
Williams, J. P. and P. A. Merrilees, Lipids *5*, p. 367 (1970)
Williams, K. W., Lab. Practice *21*, p. 667 (1972)
Williamson, J. and A. C. Allison, Lancet *2*, p. 123 (1967)
Wilson, N. and V. Y. Greenhouse, J. Chromatogr. *107*, p. 210 (1975)
Winkler, L., T. Heim and H. Schenk, J. Chromatogr. *70*, p. 164 (1972)
Winzor, D. J., Biochem. J. *101*, p. 30/c (1966)
Winzor, D. J., J. P. Loke and W. L. Nichol, J. Phys. Chem. *71*, p. 4492 (1967)
Winzor, D. J. and L. W. Nichol, Biochim. Biophys. Acta *104*, p. 1 (1965)
Winzor, D. J. and H. A. Scheraga, Biochemistry *2*, p. 1263 (1963)
Winzor, D. J. and H. A. Scheraga, J. Phys. Chem. *68*, p. 338 (1964)
Wolkoff, A. W. and R. H. Larose, J. Chromatogr. Sci. *14*, p. 51 (1976)
Wood, G. C. and P. F. Cooper, Chromatogr. Rev. *12*, p. 88 (1970)
Woof, J. B. and J. S. Pierce, J. Chromatogr. *28*, p. 94 (1967)
Wuthier, R. E., J. Lip. Res. *7*, p. 558 (1966)
Yamashiro, D., Nature *201*, p. 76 (1964)
Yamashiro, D., H. L. Aanning and V. du Vigneaud, Proc. Nat. Acad. Sci. U.S.A. *54*, p. 166 (1965)
Yamashiro, D., D. Gillesen and V. du Vigneaud, J. Am. Chem. Soc. *88*, p. 1310 (1966)
Yau, W., J. Polymer Sci., Part A-2, *7*, p. 483 (1969)
Yau, W. W. and D. P. Malone, J. Polymer Sci. Part B-5, p. 663 (1967)
Yau, W. W., H. L. Suchan and C. P. Malone, J. Polymer Sci. Part A-2, *6*, 1349 (1968)
Yogo, T., A. Fújimok and H. Mizuno, Biochim. Biophys. Acta *240*, p. 564 (1971)
Yoshiaki, Y., H. Fujimoto and D. Mizuno, Biochim. Biophys. Acta *240*, p. 564 (1971)
Yoshinaga, K. and M. Shimomura, J. Biochem. (Tokyo) *69*, p. 425 (1971)
Yoza, N., J. Chromatogr. *86*, p. 325 (1973)
Yoza, N., T. Ogata and S. Ohashi, J. Chromatogr. *52*, p. 329 (1970)
Yoza, N., T. Ogata, and Y. Ueno, J. Chromatogr. *61*, p. 295 (1971b)
Yoza, N. and S. Ohashi, J. Chromatogr. *41*, p. 429 (1969)
Yoza, N., H. Matsumo and S. Ohashi, Anal. Chim. Acta *54*, p. 538 (1971a)
Zachau, H. G., Biochim. Biophys. Acta *108*, p. 355 (1965)
Zadrazil, S., Z. Šormova and F. Šorm, Collection Czech. Acad. Sci. *26*, p. 2643 (1961)
Zeitler, H. J. and E. Stadler, J. Chromatogr. *74*, p. 59 (1972)
Zeleznick, L. D., J. Chromatogr. *14*, p. 139 (1964)

Zhdanov, S. P., E. V. Koromaldi, R. G. Vinogradova, M. B. Ganetsku, O. M. Golvnko, N. E. Zhiltsova, B. G. Belenkii, L. Z. Vilenehik and P. P. Nefedov, J. Chromatogr. 53, p. 77 (1970)
Zimmerman, J. K., Biochemistry 13, p. 384 (1974)
Zimmerman, J. K., D. J. Cox and G. K. Ackers, J. Biol. Chem. 246, pp. 1078, 4242 (1971)
Ziska, P., J. Chromatogr. 53, p. 385 (1970)
Ziska, P., J. Chromatogr. 60, p. 139 (1971)
Zuber, H., Z. Physiol. Chem. 349, p. 1337 (1968)
Zwaan, J. and A. F. van Dam, Acta Histochem. 11, p. 306 (1961)
Zweig, G. and I. Sherma, Anal. Chem. 44, p. 42R (1972)

SUBJECT INDEX

Acetic acid
— in eluants 201, 202
—, separation of 240
Acetylacetone chelates 271
Acidic dissociation
—, effect on aromatic adsorption of acids 244
—, effect on aromatic adsorption of phenols 201, 248, 251
Acidic polysaccharides 216
Ackers's equation 81
Acrylamide-acrylate gels 33
Acrylate gels 38
Acrylex gels 38
Adenylic acid 211
Adsorption 42, 101, 167
—, preventing of 42
Adsorption of
— acids 243
— alcohols 246
— aromatic acids 200—202, 244
— cellulose nitrate 219
— hydrocarbons 236
— organic bases 253, 254, 255, 256
— proteins 183, 193, 194
— purin and pyrimidine bases 208—209, 211, 212
— ribonucleosides 212
Adsorption value 238
Aerogels 18
Affinity number 202
Agarose-polyacrylamide gels 46
Albumin 186, 193
Alcanes 232—234
Aldolase 201
Aliphatic hydrocarbons 232—234
Alizarin 251
Alkaline
— earth metal ions 265, 268
— halid salts 263, 271
— metal ions 261, 271
Alkylbenzenes 237, 239
Alkylbenzoates 239
Aluminium oxide gels 44
Amino acids 200—202

Amylase 192, 195
Amylose 216
Aniline, aniline derivates 254, 256
Anionic polymerization 40
Anthocyanins 251
Anthrone reaction 146
Aromatic
— acids 244
— amines 252, 255
— amino acids 200—202
— hydrocarbones 235—238
Aromatic adsorption, see Adsorption
Association constant of complexes 199

Background electrolytes 264—271
Basic proteins 185
Basket centrifuges 169
Batch procedures 169
Benzene 238
Benzoic acid 244
Benzylalcohol 246
Bio-Beads S 41
Bio-Gel
— A types 25
— P types 35
Bio-Glas 43
Blue Dextran 2000 142, 146, 167
— complex formation with enzymes 192
Borate, complex formation with pyrocatechol 253
Branched polymer molecules 228, 230, 232, 233
Bromocyanide splitting products of proteins 200—201
Buffer change 171
— of protein solutions 183—184
Buffers 117
Butylamide 256
Butyric acid 240

Calibration line, for determination of mol. weight of protein subunits 198
Calibration curves
— of synthetic polymers 226, 267
—, universal 227

293

Capacity factor 58, 71
Carbohydrates 213—223
—, permethylated 220
— sorption on gels 220—222
Carboxylate ion content of gels 257, 258, 262
Cellodextrines 219, 221—222
Cellulose, molecular weight distribution 218
Cellulose oligosaccharides 215
Charge effects, see pH effects
Chelate complex formation 264
Chloretone 115
Chloride
—, anion 263, 266, 269—272
—, titration 145
Cholic acids 239
Chondroitin sulfate 216—217, 219—220
Chromosomes 207
Chymotrypsin 197
Citric acid 245
Coagulation factor 167
Cobalt ion 257, 262, 266, 270
Column chromatography 119
— instruments and auxiliaries 125
Columns
—, checking 131
—, packing 131
—, selection of 120
— sizes and proportion 122
—, thermostating 130
— types 120
Complex formation 164
Complexes of
— aromatic amino acids with small molecules 201
— nucleotides with ions 212
— proteins 195—200
Concentration
— by xerogels 169
—, effect of 103, 190—192, 228, 230, 260
— of proteins and protein solutions 171, 187—188
Controlled Porosity Glass 44
Convective currents 133
Coordination numbers 51
Countercurrent distribution (see also Distribution Chromatography) 207
Cresol 248, 250, 252
Cycle number 156
Cycloalkanes 234, 237
Cytidylic acid 199, 211
Cytochrome C 88, 193
— retention factor 93

Darcy law 135
Density
—, solvent of 59
—, swollen gels 59, 62

Desalting 171
— of nucleic acids 203
— of nucleotides 211
— of polysaccharides 213
— of proteins 183
Detection
—, enzymatic 179
— of substances 149
Determann equations 89
Dextran 26
— dissolving from the gels 213
— oligosaccharides 215
Dextran gels 26
— chemical properties 29
—, structure of 28
Dextran preparations
— fractionations 214—216
— mol. weight distribution 214—215
Dextrins 216
Dinitrobenzene 238
Disaccharides 220
Dissociation constant, determination by GELFAC 199
Distribution by molecular weight 214—215, 218, 225, 231
Distribution chromatography 201, 207—208, 239, 246, 253, 256
Distribution coefficient 58, 61
DNS
— fragments 207
—, native 203, 204
Documentation of results 152
Dodecyl sulphate
— in solvent 201
— separation from proteins 184
Donnan equilibrium 184, 258, 263, 264
Dopamine 155
Drying of gels 115
Dyed polysaccharides 215
Dyeing of proteins 189
Dyes, removal from proteins 184

EDTA complexes 270
Effect of
— adsorption 101
— concentration 103
— temperature 101
Effective C numbers 212, 226, 227, 232, 234, 235, 236, 243, 245, 251, 254
Electrolite concentration effect in chromatography of amino acids 201
Electrostatic interactions 257
Elution volume 56
—, relative 57
Empirical equations 95
Enthalpy 102
Enthropy 102
— dependent hydrophobic interactions 238
Enzyme inhibitors 192

Epoxy resines 246
Equations of
 — Ackers 81
 — Brönsted 98
 — Determann 89
 — Eichenberger 158
 — empirical 95
 — enthalpy-enthropy 102
 — Giddings 76
 — Kozeny-Carmen 139
 — Laurent and Killander 66
 — Ogston 66
 — Porath and Squire 63
 — Renkin 80
 — Stokes radius 67
Equilibration of gel column 135
Equivalent sharp boundary 196
Esters 239—241, 246
Ethers 245—248
Euglobulins 162, 186
Exclusion limit 59, 254
Extensions 133

Fatty
 — acids 239—243
 — alcohols 246
Ferritin 88
Ferry's factor 80
Fetoprotein 187
Fibrinolysin 188
Fibrous proteins 195
Flavonoids 249
Flow
 —, hysteresis curves 137
 —, linear 135
 —, rate 246
 —, regulation of 135
 —, resistance 142
Fluiditation of gels 110
Fluoride anion, chromatography 263, 269—270
Formic acid, in eluants 201, 202
Fraction
 — collecting 148
 — collectors 149
 — detection 147
 — volume 148
Fractionation 107
Friction coefficient 86
Frontal analysis 195, 196 (see also GELFAC)

Galactomannane 217
Gel bed 49
 — formation 17
 — forming substances 20
  — natural 21
  — semi-synthetic 24
 — forming synthetic 32
 —, inner volume of 50

Gel bed
 —, outer volume of 50, 56
 —, structure of 18
 —, terminology of 17
 —, total volume of 56
 —, volumetric distribution of 50
Gel chromatography
 —, general processes in 147
 —, hydrodinamic parameters of 69
 —, moving phase of 116
 —, resolution in 69
 —, special techniques in 155
 —, terminology of 56
Gel chromatography of
 — alcohols 245
 — aliphatic and aromatic hydrocarbons 232
 — carbohydrates 232
 — ethers 245
 — inorganic ions 257
 — lipids 239
 — mononucleotides 206
 — nucleic acids 203
  — bases 208
 — nucleosides 208
 — oligo nucleotides 203
 — oligo saccharides 219
 — organic acids 239
 — organic bases 253
 — peptides and amino acids 207
 — phenols 248
Gel chromatography of
 — polysaccharides 213
 — protein complexes 195
 — salts 257
 — small molecule organic compounds 232
 — synthetic polymers 224
 — transfer ribonucleic acids 200
Gel permeation chromatography 40, 219, 225, 229, 231
Gels
 —, Acrilex 38
 —, acrylate-acrylamide 33
 —, acrylate-hydrophylic 38
     -organophylic 38
 —, Bio-Beads S 41
 —, Bio-Gel P 34
 —, drying 115
 —, hydrophylic 19
 —, hydrophobic 19
 —, macroreticular 19
 —, microreticular 19
 —, mol. select 30
 —, organophilic dextran 30
 —, osmotic properties of 100
 — particle 49
  — size, shape 49
  —, spheric orientation of 54
 —, polystyrene acetate 45

Gels
— preparation 108
— preservation 114
— purification 115
— selection 29
—, Sephadex 26
 — LH-20 31
 — LH-60 31
— sterilization 114
—, Styragels 41
— swelling 112
GELFAC method 199—200
Geometric models 50
— of molecular exclusion 63
Gibbs-Donnan equilibrium 100
Giddings equation 76
Glass beads, porous 42
Globulins 186
Glutamate dehydrogenase 201
Glutathion reductase 167
D-glyceraldehyde-3-phosphate dehydrogenase 200
Glycerin 188
Glycols 245
Glycopeptides 223
Glycoproteins 195, 223
Gradient elution 158
Gramicidin C 202
Graphic method 139
Group separation 107
Guanidine hydrochloride 198, 200—201
Gyration radius 216

Haemoglobin 188
Halide anions 263, 269—270
Halogen substituted phenols 250
Hammett's linear free energy relationship 249
Haptoglobin 166
Heparin 218
HETP 72, 231
—, influence of flow rate on 75
— in chromatography 77
Hexahedron model 53
Hexoses 220
Hibitane 115
Histones 185
Historical survey 11
Hummel and Dreyer method 165
Hyaluronic acid oligosaccharides 219
Hydrated ions (see also ionic strength effects, pH effects) 261, 263, 264, 265, 267
Hydrocarbons 241
Hydrodynamic
— parameters 48
— radius 66
— volume 227—228
Hydrogen bonds 219, 220, 241, 246, 249, 252, 253, 255, 256, 264

Hydrophobic interactions 193, 202, **241**
Hydroxide ion, migration of 263
Hrydroxyalkylmethacrylate gels 39
Hydoxy-steroid dehydrogenase 186

Immiscible-phase polymer solutions 98
Immunglobulin 194
Indubiose AcA 46
Inner volume 50
Inorganic ions
—, chromatography of 257—272
— polyphosphates 268
— salts, removal (see desalting)
Insulin, binding 166
Interaction with gel matrix 192, 193, 222, 240, 260, 262, 272
Iodide ion 258, 259, 269—270
Ion exclusion 183, 201, 208, 258, 263
Ionic strength effects 201, 207, 222
Intrinsic viscosity 227, 228, 230
Iron(III) nitrate 263
Isoflavones 249
Isomaltoses 219

Johnston — Ogston effect 192

Kinetic theories 69

Lactate dehydrogenase 192, 199
Laurent and Killander's model 66
Lecithin 239
Linolenic acid 240
Lipase 192
Lipids 168, 239
Lipopolysaccharides 239
Lipoproteins 164, 187, 193
—, separation of 164
Longitudinal diffusion 75
Lyzozyme 192, 193

Macroglobulins 179
Macroporous silicas 231
Mannodextrines 221—222
Mariotte flasks 127
Matrix 18
Mercaptanes 245
Merck-o-gel
— -OR 45
— -Si 45
Merthiolate 115
Metal ions, binding on gels (see also inorganic ions, alkali metal ions) 263
Microreticular gels 19
Migration rate 93
Mixing volume 124
Models
—, electric analogue 68
— geometric 50
— geometric of molecular exclusion 63
— spheric of Laurent — Killander 66
Molecular-kinetic theories 48

Molecular weight
— determination 82
— sieves 11
— sieving theory 68
— size and shape 86
— thin layer gel chromatography 92
Molecular weight determination of
— proteins 189
— protein subunits 198
Molecular weight distribution of
— cellulose 218
— dextran preparations 214, 215
— polymer samples 225, 231
Molselect gels 30
Monosaccharides 221
Mucopolysaccharides 216, 217

Naphtol 251
N-heterocyclic compounds 255
Nicotinamide adenine dinucleotide 199, 200
Nitrate anion 263, 266
Nucleic acids 203—208
Nucleosides 209—212
Nucleotides 203, 209—212

Octahedron model 52
Octyl-Sepharose CL-4B 32
Oestrogens 162
Ogston equation 66
Oleic acid 241
Oligoadenylic acids 211
Oligomer fatty alcohols 246
Oligomethacrylic acid butyl esters 241
Oligonucleotides 208—209, 210
Oligopeptides 200—202
Oligophosphates 269
Oligosaccharides 215, 219—223
Oligothymidylic acids 210, 211
Organic acids 239
Organophilic dextran gels 30
— acrylate gel 38
Osmotic properties of gels 100
Outer volume 50
—, determination of 146
—, measuring of 63
Ovalbumin 190, 191
Oxalic acid 244
Oxytocin 202

Packing of gels 132
—, checking of 142
— by pressure 134
— by sedimentation 132
Packings for gel chromatography 42
Partition chromatography (see also Distribution chromatography) 167
Pentoses 220
Pepsin 171
Peptides 200—202

Perchlorate anion, sorption on gels 266
Permeability, specific 136
pH effects 201, 244, 249, 253, 264
Phenol-resitol gels 45
Phenols 201, 203, 248—253
Phenyl mercury salts 115
Phenyl-Sepharose CL-4B 32
Plate number 73
Phosphofructokinase 192
Phospholipides 239
Phosphomannane 217
Plasma proteins 199
Polyanions 163
Polyacrylamide-agarose gels 46
Polydisperse systems 93
Polyfunctional character of gels 265
Polyisobuten 224, 228, 229
Polymer fatty acids 240
Polynucleotides 205, 206, 209
Polyphenols 251
Polyphenylenes 236
Polypropylen glycols 224
Polysaccharides 213—219
Polystyrene 224, 226, 228, 229, 230
Polystyrene gels 39
Polyvinyl acetate gels 45
Porapak P, Q, R, S 45
Porasil 44
Porath's equation 63
Pore size of gels 18, 59
— and chromatographic properties 89
— and molecular dimensions 82
Porous glass beads 42
Preparative techniques 169
Preservation of gels 114
Protein complexes 185, 195—198
Protein gel interactions 185
Proteins 183—200
—, characterization of isolated 188
—, molecular weight determination of 85, 189
— subunits 198
Proteoglycanes 217
Prothrombin 217
Pseudo globulins 162, 186
Pseudo peaks 269
Pullulane 215
Pumps 128
Purification of gels 115
Purin
— bases 208, 210, 211
— nucleosides 209
Pyridine treatment of gels 257
Pyrimidine bases 208, 211
Pyrocatechol 253
Pyruvate kinase 167, 192

Quaternary ammonium compounds 254
Quinalizarin 251

297

Recycling gel chromatography 155
Relative migration 177
Replica technique 178
Resolution 231
Resolution in gel chromatography 69
Resorcinol 248, 250, 251
Restricted diffusion theory 79
Retention volume 56
—, constant 57
Ribonuclease 193, 199, 205
Ribonucleic acids 203—208
— adsorption of r-RNA on agar gel 207
— products of enzymic splitting 205
— r-RNAs 205, 206
— separation of soluble RNAs from nucleotides 206
— t-RNAs 204—206
Ribosome subunits 205, 206
Rubber latex 224

Saccharose esters 240
Safety loop 125
Salting-in chromatography 186
Salting-in, -out chromatography 161
Salts
—, chromatography of 257—272
—, effects of (see ionic strength effects)
—, removal of (see desalting)
Sample
— application 143
—, concentration of 144
— preparation 143
—, viscosity of 144
— volume 57
Sedimentation of gels 110
Selectivity curves
—, Bio-Gel P 37
— of gel columns 71
—, Sephadex G 29
—, Ultrogel AcA 47
Separation volume 57
— of gel particles 110
Sephacryl S-200 32
Sephadex
— G gels 26
— LH-20, LH-60 31, 102
Sepharose types 25
—, Octyl CL-4B 32
—, Phenyl CL-4B 32
Serum albumin 186, 193—195, 199
Serum proteins 185, 186, 187
Sieving of gels 109
Silochrom 45
Snake venom neurotoxin 193
Sodium azide 114
Solubilization chromatography 186
Solubility chromatography 161
Solubilizers 118
Solvat shell, effect on adsorption of amino acids (see also hydrated ions) 201

Solvent reservoirs 128
— regain 59
Special techniques 155
Spheron gels 38
Spherosil 45
Stearic acid 240
Stereoisomers 235
Steric-volumetric
— model 66
— theory 55
Sterilization of gels 114
Steroids 239
Stoke radius 67
Styragels 41
Swelling
— of gels 112
— time 113
— volume 59
Synthetic polymers 224—231

Tannins 251
Temperature effects 222, 229, 238, 241, 242, 246, 251
Tenzides 246
Tetraborate ion 263
Tetrahydrofuran, association with
— alcohols 245
— amines 255
— organic acids 241, 242
— phenols 252
Thermodynamic aspects 97
Thermostating 130
Thin-layer gel chromatography 172
—, evaluation of 178
Thrombin 197
Thymidine oligonucleotides 210, 211
Thyocyanate ion 263
Thyramine 155
Thyrocidines 201
Thyroid hormones 166
Total volume 56
Transfer ribonucleic acids (see ribonucleic acids)
Tributirinase 192
Triglicerides 239—242
Triolein 241
Trypsin 166, 192
Tryptophane peptides 199, 201
Tubings and joints 125
Tyrosine 201
— peptides 201

Ultrogel AcA 46
Universal calibration curve 227
Urea, in solvents 198, 201, 208
Uric acid 199
Uridylic acid 211
Urine 217
—, mucopolisaccharides of 217
—, proteins of 188

Van Dyke protein 166
Vasopressin 202
Viscosity 218
— correction 118
—, effect of 104
— effect on flow velocity 136
—, intrinsic 227, 278
Vitamin B_{12} 166
Volume
—, elution 56, 57
— fraction
Volume
—, inner 50
—, mixing 124
—, outer 50
—, retention 56
—, sample 57
—, swelling 59

Volume
—, total 56
Volumetric
—, determination of 145, 153
— distribution coefficient 58, 60, 61
— equations 60
— parameters 56
— theories 55

Wall effect 124
Water regain 59

Xerogels 187
Xylodextrines 221, 222

Zeolith-permutites 11
Zinc 257, 262, 263
Zone precipitation 162, 186